45

THE NEUTRON'S CHILDREN

The Neutron's Children
Nuclear Engineers and the Shaping of Identity

SEAN F. JOHNSTON
University of Glasgow

OXFORD
UNIVERSITY PRESS

OXFORD
UNIVERSITY PRESS

Great Clarendon Street, Oxford, OX2 6DP,
United Kingdom

Oxford University Press is a department of the University of Oxford.
It furthers the University's objective of excellence in research, scholarship,
and education by publishing worldwide. Oxford is a registered trade mark of
Oxford University Press in the UK and in certain other countries

© Sean F. Johnston 2012

The moral rights of the author have been asserted

First Edition published in 2012

Impression: 1

All rights reserved. No part of this publication may be reproduced, stored in
a retrieval system, or transmitted, in any form or by any means, without the
prior permission in writing of Oxford University Press, or as expressly permitted
by law, by licence or under terms agreed with the appropriate reprographics
rights organization. Enquiries concerning reproduction outside the scope of the
above should be sent to the Rights Department, Oxford University Press, at the
address above

You must not circulate this work in any other form
and you must impose this same condition on any acquirer

British Library Cataloguing in Publication Data

Data available

Library of Congress Cataloging in Publication Data

Data available

ISBN 978-0-19-969211-8

Printed and bound by
CPI Group (UK) Ltd, Croydon, CR0 4YY

Links to third party websites are provided by Oxford in good faith and
for information only. Oxford disclaims any responsibility for the materials
contained in any third party website referenced in this work.

```
621 .48 JOHNSTO    2012

Johnston, Sean, 1956-

The Neutron's children
```

PREFACE

During the mid-twentieth century, new technical specialists responsible for atomic energy materialized seemingly from nowhere. Their origins can be traced to the discovery of the neutron and the cascade of concepts and applications triggered by it. First defined by the project to develop an atomic bomb, nuclear experts were further shaped by their post-war explorations of nuclear engineering and eventually emerged to public view with the promise of electrical power generation.

But these children of the neutron grew unseen during their early years, nurtured by bountiful government funding and segregated by the concerns of the Cold War. Their isolation shaped them. In technical cloisters in the USA, Britain, and Canada, the scientists, engineers, and skilled workers who had first collaborated during the war grew into divergent nuclear breeds. Their skills and working cultures owed much to their distinct political, cultural, and occupational contexts, and their experiences provided templates for countries that had not had their head start. The result was a remarkable discipline distinguished by national experiences and goals but shaped enduringly by state intervention. Those origins have influenced public perceptions of these half-hidden specialists over the intervening decades.

For some subjects, the transition between non-existence and universal recognition is relatively clear-cut and uncontentious. They can be commemorated by anniversary banquets and founding fathers. But such history can equally become ossified, imprisoned in unanimity and overlooking the features of their professional identities that were once contentious. Nuclear engineering was, and is, different. Like nearly all technical expertise, there was a time at which it did not exist at all. Yet while this discipline gained a foothold in some institutional domains—being recognized at some technical sites, taught at some universities, and accorded status by some other professional groups—its attributes, products, and experts were disputed in others.

This book explores the gestation, incubation, and emergence of the new nuclear experts, and how they established unstable public identities in three countries. The story focuses on how their special knowledge was recognized, challenged, categorized, and spread. The building of nuclear expertise—of concern to professionals, politicians, and public alike—has been a decisive factor in the practice and popular understandings of late twentieth-century science, industry, and culture.

Shaped by governments and special working environments from the Manhattan Project to Fukushima, this interstitial subject and its specialists have followed a tortuous trajectory over three generations. Still unsettled and challenged, their experiences and border disputes reveal the cultural factors that shape knowledge into particular packages.

Sean Johnston
Dumfries, Scotland
October, 2011

ACKNOWLEDGEMENTS

I would like to thank the archivists at American, British, and Canadian institutions who provided access to a wide range of materials, and the nuclear specialists who related their experiences, including Trevor Barrett, Mike Brey, Morgan Brown, Dr Bill Buyers, Neil Craik, Chris O'Dwyer, Alistair Fraser, Professor Tony Goddard, Colin Gregory, Peter Higginson, Ernie Lillyman, Dan Meneley, Ian Pearson, P. N. Rowe, David Rowse, Simon Rutherford, Gilles Sabourin, Colin Tucker, Dr Lynn Weaver, and Professor Michael Williams. I am grateful to colleagues, reviewers, and editors who have commented on material presented at conferences and in journal articles and to Sönke Adlung (Senior Commissioning Editor, Physics) and Jessica White (Assistant Commissioning Editor, Physical Sciences) and their colleagues at Oxford University Press who guided the publication. This book is dedicated to the members of my family, who always provide much valued motivation and support. The funding of this work by an Economic and Social Research Council grant RES-000-22-2171 is also gratefully acknowledged.

CONTENTS

1 Introduction: The neutron and its progeny 1
    1.1 Core knowledge 2
    1.2 Accounts of the nuclear age 7
    1.3 Nuclear specialists and the shaping of expertise 10

PART A: GESTATION

2 New knowledge for new purposes 19
    2.1 A twentieth-century field 19
        2.1.1 Players, problems, and research products 20
        2.1.2 Fission as an engineering field 23
    2.2 The trigger of war 24
        2.2.1 Chain reactors and atomic bombs in the USA 26
        2.2.2 Britain and the nuclear energy machine 29
        2.2.3 Canada and the heavy-water boiler 31

3 Implanting industrial cultures 33
    3.1 The Anglo-Canadian project and the challenges of technical collaboration 34
        3.1.1 The National Research Council and an engineering perspective 35
        3.1.2 The Montreal laboratory and allied cultures 37
        3.1.3 Cooperation across borders 40
    3.2 Du Pont and the gestation of nuclear specialists 44
        3.2.1 Engineering and science: Negotiating the atomic pile 47
        3.2.2 Wilmington engineers versus Chicago scientists 53
        3.2.3 Training the first experts 60
        3.2.4 Multiplying nuclear expertise 64
        3.2.5 The industrialization of plutonium: Challenges at Hanford 68

PART B: INCUBATION

4 The atomic nursery 75
    4.1 Hesitant steps 75
    4.2 National Laboratories in post-war America 78
        4.2.1 Defining Oak Ridge specialists 79
        4.2.2 Ascendancy at Argonne 82
    4.3 Chalk River for Canadians 90
    4.4 The British atomic bomb and beyond 92
    4.5 Shaping secret programmes 95

	4.5.1	Compartmentalization	98
	4.5.2	Scrutiny	103
	4.5.3	Filtering	105

5 'Like children in a toy factory'   109
    5.1 The core of a new discipline   110
        5.1.1 Early explorations   110
        5.1.2 Chain-reactor potential   115
        5.1.3 Engineering mysteries   122
        5.1.4 Engineering values   124
    5.2 Walter Zinn and nucleonics at Argonne   126
    5.3 Anti-discipline: Christopher Hinton and the British nuclear worker   128
        5.3.1 Post-war legacies   128
        5.3.2 A British workforce   131
        5.3.3 A British specialism   133
    5.4 Controlling information flow   135
        5.4.1 American training under the cloak of secrecy   136
        5.4.2 Spreading nuclear knowledge in the UK   143
    5.5 Atomic industry for defence: The Savannah River Plant   144
        5.5.1 Industrial paradigm: Du Pont's Atomic Energy Division   146
        5.5.2 Exploring American pile networks   149
        5.5.3 Continuing friction in reactor design   154

PART C: EMERGENCE

6 A state-managed profession   161
    6.1 Triggered release: Declassifying nuclear knowledge   161
    6.2 Freedom to publish   166
    6.3 Professional fallout   170
    6.4 Altering cross-sections: The first university-educated generation   176
        6.4.1 Controlling colleges in the UK   177
        6.4.2 Shaping British curricula   180
        6.4.3 Finding a niche at American universities   185
        6.4.4 Contesting curricula in the USA   188
        6.4.5 Developing a demand in Canada   193

7 Nuclear specialists at work   197
    7.1 Nuclear unions: Segregation and identity   201
        7.1.1 Patriotism and dissuasion in the USA   202
        7.1.2 British organizations and nuclear subversion   207
        7.1.3 Canada: Occupational labels for all   208
    7.2 Risk and radioactivity   210
        7.2.1 Hazards of operation   211
        7.2.2 Radiation rites   216
        7.2.3 Risk and identity   222

PART D: REPRESENTATIONS

8 Unstable impressions — **227**
  8.1 Popular atomics and public recognition — 227
    8.1.1 Hidden heroes: The atomic scientists — 228
    8.1.2 A mistrusted elite — 230
    8.1.3 Ingenious engineers — 234
    8.1.4 Role models — 240
  8.2 The neutron's grandchildren — 242
    8.2.1 The troubled generation: Nuclear engineers and commercial power — 242
    8.2.2 Unwelcome voices — 247
    8.2.3 Mistrust and responsibility — 250
    8.2.4 Identity by accident: Three Mile Island and Chernobyl — 252
    8.2.5 Fukushima and its fallout — 256
    8.2.6 Nuclear generations — 259

9 Conclusions: Careers from the Manhattan Project to Fukushima — **264**
  9.1 Fertile environments — 264
  9.2 Fragile ecosystems — 268
  9.3 Critical conditions — 269
  9.4 Between anonymity and rhetoric — 272

Appendix I: Acronyms, organizations, and year of origin — 274
Appendix II: Nuclear engineering periodicals since 1945 — 276
Appendix III: Archival sources — 278

Abbreviations used for archive citations — **279**

Bibliography — **280**

Index — **301**

1

INTRODUCTION: THE NEUTRON AND ITS PROGENY

How does technical expertise grow? For nuclear specialists, one source was seminal: the neutron, first identified by physicist James Chadwick during experiments at Cambridge University in 1932. Within a decade that discovery was being channelled into an unprecedented effort to create new weapons, new schemes for power generation, and new biological effects. And with them came new technical experts. As summarized in the first public account of the atomic bomb project,

> The one characteristic of neutrons which differentiates them from other subatomic particles is the fact that they are uncharged. This property of neutrons delayed their discovery, makes them very penetrating, makes it impossible to observe them directly, and makes them very important as agents of nuclear change.[1]

Spawned by the neutron, nuclear workers inherited some of its properties. As new types of technical specialist within secret government programmes, they were unusually powerless. They remained hidden for a decade and penetrated to the core of post-war security, benefiting from unprecedented funding to explore a new scientific landscape. And, within a generation of the neutron's discovery, they provided their governments with technologies that promised not just international esteem, but also the means to effect intellectual and social change.

The analogy of child development is apt. The growth and maturation of this progeny in dissimilar environments produced distinct offspring. In some locales, ample resources and privileged activities marked a cosseted upbringing; in others, the troubled relationships and identity crises of adolescence were played out. Nuclear engineers developed as an amalgam of influences through their relationships with other technical specialists from the 1940s, shaped as much by the environments of post-war politics, military secrecy, and commercial cultures as by disciplinary refinement. Their expertise, gestated at nuclear reactors or chemical separation facilities, was incubated in isolated environments and corporate subdivisions to emerge belatedly to public visibility a full decade after the war. Over the years that followed, their products were simultaneously experimental and industrial in scale, and discussed within government committees as much as at scientific meetings. Containment shaped and constrained their enigmatic identity. During a brief window of time, 'atomic scientists' achieved celebrity and were vocal in offering advice to governments and citizens. But their siblings, the engineers and technologists, achieved tardy public recognition, praise, and, ultimately, notoriety.

---

[1] Smyth, Henry D., 'A general account of the development of methods of using atomic energy for military purposes under the auspices of the United States Government 1940–1945', 1 July 1945, LAC MG30-E533 Vol. 1. For an early summary, see Feather, Norman, 'The production and properties of neutrons', *Science Progress* 33 (1938): 240–56. On the range of archival sources and their abbreviations, see Appendix III.

Nuclear specialists established themselves in a range of rapidly maturing occupations, disciplines, and professions during the early 1960s. These three distinct aspects of working life deserve careful tracing. Job roles are the most obvious characteristic defining a skilled person, allowing the expert a measure of autonomy and responsibility in the workplace that is recognized by co-workers. The associated *discipline* combines intellectual and social features: it comprises the body of knowledge and skills that the expert applies to job tasks, and usually is defined by formal training and education. Disciplines are broadly represented by academic departments that collect together peers sharing and disseminating an intellectual tradition, and granting certification that is recognized beyond their own institution. And as Stephen Turner argues, disciplines are usually linked closely with a market for their know-how.[2] By contrast, *professions* refer to the wider social world in which experts are granted a social position and status in relation to other types of expert and to lay society. Their presence is defined neither by the workplace nor the teaching institution alone, but instead by learned societies and professional bodies.[3]

But such neat divisions become blurred when we investigate a new job or nascent subject. Occupational labels for these 'workers', 'specialists', or 'experts' of the nuclear field were often locally defined, contentious, and malleable, and so I use these terms as general descriptors and refine their meaning in particular contexts.[4] Some technical niches developed from prior expertise, fissioning when their new knowledge became too unwieldy for existing fields. Others were created as exotic new products. And yet others emerged as meta-stable states, only to decay back to more familiar species over time. The problems of applying atomic energy required the skills of nuclear engineers, technologists, and technicians, 'atomic scientists', radiochemists, and health physicists, before they bore those names.

Over some seventy years, then, a family of nuclear specialists was created, scintillated, briefly stabilized, and transmuted. That ephemeral process, in the first three countries that collaborated on the subject—the UK, USA, and Canada—is the subject of this book.

1.1 Core knowledge

While the heart of this emerging subject is easy to locate, its boundaries were not. The novelty of nuclear engineering was subtle to some but obvious to others. Employed at a

---

[2] See, for example, Turner, Stephen, 'What are disciplines? And how is interdisciplinarity different?' in: P. Weingart and N. Stehr (eds.), *Practising Interdisciplinarity* (Toronto: University of Toronto Press, 2000), pp. 46–65.

[3] Abbott, Andrew D., *The System of Professions: An Essay on the Division of Expert Labor* (Chicago: University of Chicago Press, 1988). See also Jarausch, Konrad, *The Unfree Professions* (Oxford: Oxford University Press, 1990).

[4] This ambiguity is needed in order to track how expertise developed, and contrasts with, for example, the more formulaic labels of 'engineer', 'technician', and 'worker' that Gabrielle Hecht was able to adopt in order to contrast the special jobs of established French nuclear specialists of the 1960s [Hecht, Gabrielle, 'Rebels and pioneers: technocratic ideologies and social identities in the French nuclear workplace 1955–1969', *Social Studies of Science* 26 (1996): 483–529]. The distinct organizational circumstances of the French state, its education system, institutions, and labour representation also limit the amount of generalization possible, but allows for intercomparison [Abbott, Andrew D., *The System of Professions: An Essay on the Division of Expert Labor* (Chicago: University of Chicago Press, 1988), pp. 201–3].

scale never imagined before the war, radiation would be applied to fundamental research, destructive force, biological applications, and power generation. All these possibilities were enabled by the nuclear reactor. Its new specialists became adept at designing, operating, and applying this new creation to seemingly unlimited purposes.

But what is a nuclear reactor? Depending on the age of the reader, the term will be cloaked in meanings and laden with connotations of progress, secrecy, complexity, danger, or failure. For many, the term will also conjure up an image not of an engineering design but of a social and political system.[5] Understanding the origins of these jarring overtones requires them to be stripped off to examine the history not just of the reactor, but of the specialists who grew to design and tend it.[6]

On the face of it, the first reactors were mundane-looking devices. Their original name—*piles*—mirrored their construction, originally a compact assembly or lattice-work mountain of materials. Their perceptible characteristics were also unexciting: such piles generated heat, which was removed either to keep them cool or to generate useful electrical power. For power generation, some well-established engineering principles were applied—principles developed over the previous two centuries to collect and transfer the heat via an exchange medium such as water, steam, or gas, and to convert it to mechanical motion with turbines, which in turn generated electrical power. Other design principles were even more traditionally established: how to package the materials in mechanical structures that were mechanically, structurally, chemically, and thermally stable; and how to ensure reliable operation of these factory-sized environments of interlinked mechanical, electrical, and thermal systems.

But the first reactors, developed during the Second World War, had a distinctly unfamiliar purpose and invisible characteristics, too. The trigger for their development was the experimental confirmation of *fission*, the division of an atomic nucleus into two or more parts following the absorption of a neutron. Their original goal was to be a radioactivity amplifier—generating a sustained *chain reaction* of radioactive materials—and to use this controlled fission to transmute those materials into new elements to be used in bombs. This deeper function, revealed publicly after the war, was the source of their new name: *chain reactors* or, soon after, *reactors*.[7] As a lyrical contemporary newspaper account put it,

---

[5] On the claim that the adoption of nuclear power engenders or requires a particular political environment, see Winner, Langdon, 'Do artifacts have politics?' in: L. Winner (ed.), *The Whale and the Reactor: A Search for Limits in an Age of High Technology* (Chicago: University of Chicago Press, 1986), pp. 19–39.

[6] This notion, that various creators and users may define their technologies in dissimilar ways, is dubbed *interpretive flexibility* and is a central theme of the social construction of technology [Pinch, Trevor J. and Wiebe E. Bijker, 'The social construction of facts and artefacts: or how the sociology of science and the sociology of technology might benefit each other', *Social Studies of Science* 14 (1984): 399–441].

[7] This understanding is supported by the patent deposition filed on 19 December 1944 by Leo Szilárd and Enrico Fermi for a 'neutronic reactor' [US Patent 2,708,656, granted 17 May 1955] and by early titles such as Wigner, E. and A. M. Weinberg, *Physical Theory of Neutron Chain Reactors* (Chicago: University of Chicago Press, 1958). An alternate etymology credits wartime chemical engineers, familiar with chemical reactors, with making an analogy between chemical and nuclear technologies. The competing origins are noteworthy for our story, as they attribute authority over the new domain to different technical specialists.

> Call it a pile if you wish to think of it in terms of a pile of uranium rods, plus aluminum, plus helium, plus cadmium, and the whole thing cooled off with Ottawa River water and steadied down with heavy water... Call it a reactor if you wish to think of it in terms of being a great furnace 35 feet wide and 35 feet high wherein these things react to one another so that mass disappears and becomes heat and radiation.[8]

A reporter's description of Britain's first reactor evoked similar sentiments:

> First sight of Gleep may not be impressive; it looks a rough affair, rather like a giant packing case. But after a few moments one cannot help thinking that here is the first device ever built in Britain to produce energy by the annihilation of matter. Behind those concrete shields seethe neutrons by the myriad; the whole contrivance is as silent as the sky and yet here is a source of power potentially more concentrated than anything this side of the sun. The pile itself, huge and rough, is in striking contrast to the cool control room, with its great display of dials which show what's happening inside... from here the vertical safety rods can be made to drop into position to extinguish the atomic furnace.[9]

Underlying the wonder and potential of the chain reaction lay rapidly growing knowledge about the physics of atomic nuclei, but an even greater wave of novel engineering experience. That proficiency developed swiftly from 1940 to 1945 in isolated environments and high secrecy to understand how radioactive materials could be selected and combined with others to optimize the chain reaction.

As the varieties of reactor system proliferated, their characteristics and often unexpected problems were focused on the engineering practicalities introduced by this high science. The first plutonium production reactor at Hanford, Washington, for example, proved to be choked by the production of an unexpected fission product which strongly absorbed neutrons and so diminished the chain-reaction rate (Figure 1.1). How did irradiation alter the structural properties of the materials used in a reactor? Uranium rods swelled, buckled, and ruptured their containers when heated and irradiated, requiring compensatory design changes and operational procedures. Water-cooling pipes became brittle and cracked after irradiation. Nuclear power reactors, despite functioning as mere heat generators, were qualitatively different from conventional coal- or oil-fired power plants: the heat could be generated in a more compact volume, and much more rapidly increased; the materials and dimensions employed had to be compatible with the nuclear chain reaction; and the materials themselves were or could become radioactive, posing a biological danger and requiring increasingly elaborate safety systems. These hidden factors introduced unknown and often unanticipated variables into the arts of metallurgy, chemical engineering, thermal design, and mechanical engineering. Existing specialists were transformed.

The nature of this hybrid expertise, then, was defined by the nuclear reactor and its multiple design parameters and products. The technical challenges introduced professional strains. Metallurgists and materials engineers were faced with unknown processes of distortion, embrittlement, and transformation, each of which had to be tested experimentally on materials immersed in a reactor environment. Civil and mechanical engineers saw the

---

[8] Smith, I. Norman, 'The magic and reality of nuclear energy: a layman takes a look', *Ottawa Journal*, 1953, in LAC MG40 B59.

[9] Dick, William E., 'The hangars hide uranium piles', *Discovery* 9 (9), 1948: 281–5; quotation p. 284. Perhaps the most accurate and informative popular account of nuclear engineering of the period was the pragmatic and unromantic Woodbury, David O., *Atoms for Peace* (New York: Dodd, Mead & Co., 1955), which exemplified the Eisenhower administration's political agenda for nuclear energy.

Fig. 1.1 Workers preparing the face of the first Hanford reactor, 1944.[10] (Courtesy of US Department of Energy, Declassified Document Retrieval System.)

reactor as a peculiarly precise large-scale construction, in which mechanical tolerances were extremely tight to ensure operating efficiency, on the one hand, and to guarantee that radioactive fuel elements and other components did not jam, on the other. Thermal engineers were crucial for devising means of extracting heat from the energy-dense core, and to ensure that these heat-transfer systems operated reliably in extreme environments barred to direct human manipulation. Chemical engineers faced the new problem of developing processes to chemically separate nearly identical elements. The efficient extraction of plutonium and untransformed uranium from fuel rods was crucial in accumulating fissile materials for bombs and for reuse in planned reactors.

Control engineers had the novel task of designing fail-safe systems to ensure stability of the finely balanced chain reaction—a responsibility more difficult than initially anticipated. Practical experience with these new creations limited the speed of technical innovation. Damage, and even destruction, of early reactors by overheating and contamination incidents was not rare, although the later Three Mile Island (1979), Chernobyl (1986), and Fukushima (2011) events gained much greater attention. The Canadian NRX reactor was closed down for fourteen months to decontaminate and rebuild it following a 1952 accident; future American

[10] 'B reactor front face', 1944, DOE DDRS N1D0029053.

President Jimmy Carter participated in its clean-up operation as a nuclear technician in the American Navy, an experience that shaped his attitudes regarding the technology a quarter-century later. In 1955, the American EBR-1 reactor suffered a partial meltdown. The British Windscale reactor was breached by a fire in 1957, leading to the contamination of farmland and produce. At the Santa Susana Field Laboratory near Los Angeles in 1959, an experimental sodium-cooled reactor was damaged but continued to operate for two weeks with a significant radiation leak, and in January 1961 three men died in a thermal explosion of the SL-1 reactor in Idaho. The Fermi 1 Breeder Reactor near Detroit suffered a partial meltdown in 1966 and was not returned to service for nearly four years, and a more modest event at the Chapelcross facility in Scotland in 1967 closed one of its four reactors for two years.[11]

Such engineering uncertainties dominated the early explorations of nuclear reactor concepts, and strongly influenced the geographical distribution of experimental reactors and their experts. The possibility of such events—accompanied by melting or combustion, explosion, rupturing of containment systems, and release of radiation—was feared by the engineers and scientists developing the first reactors, leading them to site the reactors at remote locations and, over a period of decades, to progressively ramp up safety measures.[12] Early experimental reactors were situated far from population centres to obviate the risk of explosion or radioactive release—in the hills of Oak Ridge, Tennessee; at Hanford, Washington, surrounded by the northern desert of the Pacific Northwest; in the forests at Chalk River on an isolated stretch of the Ontario/Quebec border; and at Dounreay at the northern tip of mainland Scotland, some 60 km from John O'Groats.

Nuclear engineering thus developed in a context that was both seductively familiar and thoroughly mysterious, an uncomfortable situation for most technologists. Some professions—pipefitters and civil engineers, for example—felt confident to perform their traditional roles with only slightly extended knowledge. Others, such as metallurgists and chemical engineers, were required to absorb new technologies and unfamiliar constraints, transforming themselves in the process. Still other domains, such as reactor engineering and its experts, were wholly new. And familiar or not, each of these types of specialist worker operated in the unfamiliar context of invisible and lethal irradiation.

They explored this technical ground in a peculiar political terrain: wartime and Cold War secrecy. Expertise in nuclear reactors was closely affiliated with knowledge useful for bomb production. Nevertheless, there was an early division of labour between reactor designers and bomb designers. While both operated initially in secrecy, their intellectual terrain rapidly diverged as they pursued the distinct goals of their government clients.[13]

---

[11] Other countries had similar experiences: In 1969, coolant-system failure in an experimental reactor in Lucens, Switzerland, contaminated its underground cavern; a 1977 fueling accident at Jaslovské, Czechoslovakia, led to significant radioactive release; in France, the Saint Laurent power plant suffered a partial meltdown in 1980; in Buenos Aires, Argentina, in 1983, a criticality accident irradiated and killed the reactor operator and contaminated 17 others. Accidents followed by radioactive release also have occurred at chemical separation and reprocessing plants, e.g. in Tomsk, Russia, in 1993, and in Tokai-Mura, Japan, 1999.

[12] For an insider's account see Weinberg, Alvin M., *The First Nuclear Era: the Life and Times of a Technological Fixer* (New York: AIP Press, 1994), p. 191.

[13] The analogy of Darwinian evolution is apt: like the Galapagos species isolated by widely separated islands and favoured by local conditions, nuclear expertise diverged and found stable environmental niches.

Bomb design was a particularly unfertile endpoint. It led to increasingly sophisticated weapons designs, but a dearth of other applications under its self-limiting shroud of concealment. On the other hand, nuclear chain reactors could generate a spectrum of new isotopes of interest to scientists, engineers, and doctors. Reactors encouraged the development of commercially valuable chemical and physical separation processes and materials handling techniques. They also hinted at open civilian benefits, from new forms of medical diagnosis and treatment to power generation. The broader promise of atomic energy, then, offered a convergence of interests to unite physicists, engineers, chemists, and biologists in a new and dynamic field.

1.2 Accounts of the nuclear age

A vast library has accumulated since the Second World War to document and explain the impact of the atomic bomb and nuclear power. Reflecting the broad interest and profound cultural dimensions of the wartime project, the literature is wide-ranging and falls into several distinct categories.

The earliest of these are administrative histories and scientific overviews. The closing weeks of the Second World War produced the first historical summary by American physicist Henry deWolf Smyth, followed by official accounts of the American and British projects.[14] They were succeeded by organizational explanations documenting, and usually lauding, the rise of national nuclear power programmes.[15]

Nevertheless, histories of the wartime work in the UK, USA, and Canada to build the first atomic bomb have underplayed the growth of engineering knowledge that accompanied it. The first volume of the official American history, for example, noted that its policy-centred approach left voids preventing 'a comprehensive history of any single organization, project, discipline or period'.[16] This was to be expected, given the popular interest in the momentous decisions so important for post-war politics, and the relatively early availability of public archives documenting the actions of government departments. From the standpoint of science and technology studies, however, the early accounts imposed naïve explanations for the creation of the new subjects of atomic energy, often describing an unproblematic sequence of discovery, application, and consequent effects.

---

[14] Smyth, Henry D., *Atomic Energy for Military Purposes: The Official Report on the Development of the Atomic Bomb under the Auspices of the United States Government*, 1940–1945 (Princeton, NJ: Princeton University Press, 1945); Hewlett, Richard G. and Franciscus Duncan, *A History of the United States Atomic Energy Commission* (3 vols) (University Park, Penn.: Pennsylvania State University Press, 1969); Gowing, Margaret, *Britain and Atomic Energy, 1939–1945* (New York: St Martin's Press, 1964); Arnold, Lorna, *Britain and the H-Bomb* (London: Palgrave MacMillan, 1979).

[15] E.g. Eggleston, Wilfred, *Canada's Nuclear Story* (Toronto: Clarke Irwin, 1965); Pocock, Rowland Francis, *Nuclear Power: Its Development in the United Kingdom* (London: Unwin Brothers, 1977); Hurst, D. G. and E. Critoph (eds.), *Canada Enters the Nuclear Age: A Technical History of Atomic Energy of Canada Limited as Seen From its Research Laboratories* (Montreal: Atomic Energy of Canada Ltd, 1997); Gingras, Yves, 'The institutionalization of scientific research in Canadian Universities: the case of physics', *Canadian Historical Review* 67 (1986): 181–94.

[16] Hewlett, Richard G. and Jack M. Holl, *The New World, 1939–1946* (Berkeley: University of California Press, 1969), p. xii.

This mischaracterization was more pronounced in discussions of the post-war atomic energy programmes and their specialist workers. Too often, they reiterated the late twentieth century theme that 'hard work and money yield progress'.[17]

Among the first historical accounts that provided a glimpse of nuclear workers themselves were the autobiographies and reminiscences of scientists who had led the secret wartime research. These vary significantly in relevance and candidness, and have focused on scientists identified immediately after the war as pivotal in technical or political decision-making. Examples include the autobiographies of Arthur Compton, Enrico Fermi, Otto Frisch, Rudolf Peierls, and Edward Teller.[18] A second wave of narratives focused on pen portraits of individuals, sometimes little-known, working in atomic energy.[19] These amounted to journalistic impact studies, documenting how research and development were progressing at the fringes of public awareness. And a third wave beginning in the late 1950s again focused on biographies of key historical actors, providing further analysis of their historical context, although often with only a gently critical dimension.[20]

The pendulum of historical and sociological analysis began to swing the other way during the 1970s. Accounts of nuclear energy focused increasingly on critical assessments of the political and economic dimensions of civilian power and nuclear weapons, leaving nuclear specialists—now identified as engineers rather than scientists—to be categorized as a shadowy and homogeneous contingent linked to questionable motives or outcomes.[21] These accounts, flavoured by their own contemporary perspectives, redirected the historical analysis to reinterpret earlier periods according to a simplistic—and generally negative—trajectory.

With the end of the Cold War and the opening of more archives, a distinct body of literature has sought to re-examine and narrate the multiple factors and events in a more holistic

---

[17] Historians of technology have long resisted this deterministic view of technology and its conflation with social progress. See, for example, Smith, Merritt Roe and Leo Marx, *Does Technology Drive History?: The Dilemma of Technological Determinism* (Cambridge, Mass.; MIT Press, 1994) and Staudenmaier, John M., *Technology's Storytellers: Reweaving the Human Fabric* (Cambridge, Mass.: MIT Press, 1985).

[18] E.g. Compton, Arthur Holly, *Atomic Quest: A Personal Narrative* (Oxford: Oxford University Press, 1956); Fermi, Laura, *Atoms in the Family: My Life with Enrico Fermi* (Chicago: University of Chicago Press, 1954); Frisch, Otto, *What Little I Remember* (Oxford: Oxford University Press, 1979); Peierls, Rudolf, *Bird of Passage: Recollections of a Physicist* (Princeton, NJ: Princeton University Press, 1985); Teller, Edward and Judith L. Shoolery, *Memoirs: A Twentieth-Century Journey in Science and Politics* (Oxford: Perseus Press, 2001).

[19] E.g. Lang, Daniel, *Early Tales of the Atomic Age* (New York: Doubleday, 1948); Hope, Nelson W., *Atomic Town* (New York: Comet, 1954). In the same vein but adopting a more critical tone were Caulfield, Catherine, *Multiple Exposures: Chronicles of the Radiation Age* (London: Secker & Warburg, 1989); Hall, Jeremy, *Real Lives, Half Lives* (London: Penguin, 1996).

[20] E.g. The first and best known of these was Jungk, Robert, *Brighter Than A Thousand Suns: A Personal History of the Atomic Scientists* (San Diego: Harcourt Brace, 1956). See also Hartcup, Guy and T. E. Allibone, *Cockcroft and the Atom* (Bristol: Adam Hilger, 1984); Fawcett, Ruth, *Nuclear Pursuits: The Scientific Biography of Wilfrid Bennett Lewis* (Montreal: McGill-Queen's University Press, 1994).

[21] E.g. Pringle, Peter and James Spigelman, *The Nuclear Barons* (London: Joseph, 1982).

fashion. Given the scale of activity, these accounts have focused on bomb development,[22] the consequences of the nuclear complex,[23] lower-tier workers,[24] public engagement[25] and on the post-war context and infrastructure that sustained them.[26] Insightful prior research from the top-down perspective of policy studies has seldom addressed questions of technical identity, i.e. from a bottom-up perspective.[27] And as noted by the previous generation of

[22] E.g. Rhodes, Richard, *The Making of the Atomic Bomb* (London: Simon & Schuster, 1986); Holloway, David, *Stalin and the Bomb: The Soviet Union and Atomic Energy, 1939–1956* (New Haven, Conn.: Yale University Press, 1994); and, Rhodes, Richard, *Dark Sun: The Making of the Hydrogen Bomb* (New York: Simon & Schuster, 1995). Anthropological accounts of weapons designers include Parfit, Michael, *The Boys Behind the Bombs* (New York: Little and Brown, 1983), Rosenthal, Debra, *At the Heart of the Bomb: The Dangerous Allure of Weapons Work* (Reading, Mass.: Addison-Wesley, 1990) and Masco, Joseph, *The Nuclear Borderlands: The Manhattan Project in Post-Cold War New Mexico* (Princeton, NJ: Princeton University Press, 2006); Furman, Necah Stewart, *Sandia National Laboratories: The Postwar Decade* (Albuquerque: University of New Mexico Press, 1990); Gusterson, Hugh, 'The death of the authors of death: prestige and creativity among nuclear weapons scientists', in: Biagioli, Mario and Peter Galison (eds.), *Scientific Authorship: Credit and Intellectual Property in Science* (New York: Routledge, 2003), pp 281–307; Hoddeson, Lillian, Paul W. Henriksen, Roger A. Meade and Catherine Westfall, *Critical Assembly: A Technical History of Los Alamos During The Oppenheimer Years, 1943–1945* (Cambridge: Cambridge University Press, 1993).

[23] E.g. Carlisle, Rodney P. and Joan M. Zenzen, *Supplying the Nuclear Arsenal: American Production Reactors 1942–1992* (Baltimore: Johns Hopkins Press, 1996); Gerber, Michele S., *On the Home Front: The Cold War Legacy of the Hanford Nuclear Site* (Lincoln: University of Nebraska Press, 1997); and Egorov, Nikolai N., *The Radiation Legacy of the Soviet Nuclear Complex: An Analytical Overview* (London: Earthscan, 2000).

[24] E.g. Johnson, Charles W., *City Behind A Fence: Oak Ridge, Tennessee 1942–1946* (Knoxville: University of Tennessee Press, 1981) and Olwell, Russell B., *At Work in the Atomic City: A Labor and Social History of Oak Ridge, Tennessee* (Knoxville: University of Tennessee Press, 2004).

[25] E.g. Weart, Spencer R., *Nuclear Fear: A History of Images* (Cambridge, Mass.: Harvard University Press, 1988); Balogh, Brian, *Chain Reaction: Expert Debate and Public Participation in American Commercial Nuclear Power, 1945–1975* (Cambridge: Cambridge University Press, 1991); Wellock, Thomas R., *Critical Masses: Opposition to Nuclear Power in California, 1958–1978* (Madison: University of Wisconsin Press, 1998).

[26] E.g. Wang, Jessica, *American Science in an Age of Anxiety: Scientists, Anticommunism, and the Cold War* (Chapel Hill, NC: University of North Carolina Press, 1999); Forman, Paul, 'Behind quantum electronics: national security as basis for physical research in the United States, 1940–1960', *Historical Studies in the Physical and Biological Sciences* 18 (1987): 149–229. Focusing on the built environment rather than institutions is a cultural geography of the wartime and post-war sites, Hales, Peter B., *Atomic Spaces: Living on the Manhattan Project* (Urbana: University of Illinois Press, 1997). For a survey of more recent work, see Hughes, Jeff, 'Essay review—Deconstructing the bomb: recent perspectives on nuclear history', *British Journal for the History of Science* 37 (2004): 455–64. Since then, research has begun to focus on civil nuclear policy and national programmes, particularly in Europe and after the export of American, British, and Canadian reactor expertise from the late 1950s.

[27] E.g. Krige, John, 'Atoms for Peace, scientific internationalism and scientific intelligence', *Osiris* 21 (2006): 161–81; Kevles, Daniel, 'Cold War and hot physics: science, security and the American State, 1945–1956', *Historical Studies in the Physical and Biological Sciences* 20 (1990): 239–64; Galison, Peter and Barton J. Bernstein, 'Physics between war and peace', in: M. R. S. Mendelsohn and P. Weingart (eds.), *Science, Technology, and the Military* (Dordrecht: Kluwer Academic, 1988), pp. 47–86. Two exceptions

researchers of science, technology, and society (STS), there is a 'bigness bias': 'a preference for investigating the top structures of science and technology'.[28]

This book, on the other hand, aims to shift focus directly back to the experts that grew and mutated to shape these disparate activities from the inside: the specialists behind the wartime reactors, post-war atomic energy, and late twentieth century nuclear power. The contexts examined by previous analyses are explored with a new goal: to reveal how they shaped the identities of nuclear workers and their new professions. A second theme is that the history of nuclear specialists must be understood in relation to changing contexts. The atomic bomb, atomic energy, and nuclear power were distinct goals amassing heterogeneous groupings of specialists with dissimilar configurations at different times in separate countries. They cannot be understood according to a narrow account of technological process, intellectual advance, economic history, or national policy. Professional identity, now recognized as a malleable component contributing to multiple expressions of individuality, is shaped not only by one's self-categorization, but also by evolving communities of practice (peer groups) and social structures such as the institutional context and career constraints.[29] In short, this book complements the top-down accounts and analyses of the nuclear era to reveal the contingencies that shaped nuclear experts themselves.

1.3 Nuclear specialists and the shaping of expertise

This book can be described as an interdisciplinary history. It deals with events, concepts, individuals, and institutions of interest to social, cultural, labour, and political historians, sociologists of the professions, and sociologists and philosophers of scientific knowledge. The study of nuclear specialists and their claims of expertise is particularly apt for illuminating these domains.[30] During the twentieth century, science itself entered a professional phase, and engineering was established and diversified as a series of academic fields. As a

---

are Gusterson, Hugh, *Nuclear Rites: A Weapons Laboratory at the End of the Cold War* (Berkeley: University of California Press, 1996), Chap. 4, which discusses the role of secrecy in shaping the identities of American weapons scientists in the Livermore National Laboratory, and Masco, Joseph, 'Lie detectors: on secrets and hypersecurity in Los Alamos', *Public Culture* 14 (2002): 441–60, a complementary account of secrecy at Los Alamos, also dedicated to weapons development experts.

[28] Spiegel-Rosing, I., 'The study of science, technology and society (SSTS): Recent trends and future challenges', in: Spiegel-Rosing, Ina and Derek de Solla Price (eds.), *Science, Technology and Society: A Cross-Disciplinary Perspective* (London: Sage, 1977), pp. 28–9.

[29] Wenger, Etienne, *Communities of Practice: Learning, Meaning and Identity* (Cambridge: Cambridge University Press, 1998); Stryker, Sheldon and Peter J. Burke, 'The past, present, and future of an identity theory', *Social Psychological Quarterly* 63 (2000): 284–97.

[30] These shared interests have not always been recognized or cited within separate disciplines. See, for example, Scranton, Philip, 'None-too-porous boundaries: labor history and the history of technology', *Technology and Culture* 29 (1988): 722–43. On practitioner-centred research, see Collins, H. M. and Robert Evans, 'The third wave of science studies: studies of expertise and experience', *Social Studies of Science* 32 (2002): 235–96. On a categorization of types of expertise, see Turner, Stephen, 'What is the problem with experts?' *Social Studies of Science* 31 (2001): 123–49. An earlier approach focusing on scientific disciplines is Geison, Gerald L., 'Scientific change, emerging specialties, and research schools', *History of Science* 29 (1981): 20–40.

result, their specialists became ever more important parts of existing networks of industry, government, and academe, and increasingly visible to the wider public. The expansion of technical proficiency raised questions of legitimacy: how were the new experts to be assessed and how could their competence be certified, when their fields were young and malleable? This deep embedding in social and cultural currents explains the growing interest shown by historians of science and technology, as well as by scholars from other disciplines, in such twentieth-century professionals.

Interest in the professionalization of technical specialists has focused on both external and internal aspects. Sociologist Andrew Abbott provided a robust framework for understanding this process of professional differentiation, characterizing it as a jockeying for intellectual space, jurisdiction, status, and identity among occupational groups in a shifting ecology of professions.[31] Both knowledge claims and occupational jurisdiction are perhaps most actively contested when new intellectual subjects are being defined, and most readily discerned from historical studies. For this reason, inter-professional power relations are a valuable perspective for understanding emergent groups such as chemical engineers during the early twentieth century.[32]

A more nuanced approach can allow investigation of intra-professional relations, i.e. the development of distinct identities within occupational groups. Peter Galison, for example, argued for the evolution of diverging methodologies and understandings within communities engaged in particle physics, based on the visual and computational approaches required for different forms of particle detector. The hardware that they adopted shaped the world-views of their specialist groups—so much so, that communicating their methodologies and concepts required intermediaries able to understand and translate their technical dialogue via the equivalent of pidgin dialects between human languages. The particle physics community has remained a heterogeneous one, forming not so much a stable discipline as a multi-hued and shifting collection of enmeshed technical experts, varying from site to site and problem to problem.[33]

To understand nuclear specialists, then, both the inter-professional scale championed by Abbott and the intra-professional scale highlighted by Galison are valuable. This study applies these insights, building upon previous work on other technical professions and employing extensions of their methods.[34]

---

[31] Abbott, Andrew D., *The System of Professions: An Essay on the Division of Expert Labor* (Chicago: University of Chicago Press, 1988).

[32] Divall, Colin and Sean F. Johnston, 'Scaling up: the evolution of intellectual apparatus associated with the manufacture of heavy chemicals in Britain, 1900–1939', in: A. S. Travis, H. G. Schröter and E. Homburg (eds.), *Determinants in the Evolution of the European Chemical Industry, 1900–1939: New Technologies, Political Frameworks, Markets and Companies* (Dordrecht: Kluwer Academic, 1998), pp. 199–214; Reynolds, Terry S., 'Defining professional boundaries: chemical engineering in the early 20th century', *Technology and Culture* 27 (1986): 694–716.

[33] Galison, Peter, *Image and Logic: A Material Culture of Microphysics* (Chicago: University of Chicago Press, 1997). On the parallel construction of theoretical concepts, see Pickering, Andrew, *Constructing Quarks: A Sociological History of Particle Physics* (Edinburgh: Edinburgh University Press, 1984).

[34] The emergence of new technical professions in different hinterlands between science, industry, government, and wider culture has been a recurring theme of my previous research. Along similar lines, see Jesiek, Brent K., *Between Discipline and Profession: A History of Persistent Instability in the Field of Computer Engineering, circa* 1951–2006, PhD thesis, Princeton University (2006).

More firmly than other technical specialists, nuclear workers were a product of, and had a strong influence on, particular contexts of the twentieth century. Only by comparing the evolving experience of nuclear specialists in different countries can exogenous factors be assessed. The study of their growth, mutation, and shifting authority reveals their key roles. Promoted by the wartime allies, nuclear workers emerged from secrecy in the decade following the Second World War to become increasingly self-aware cohorts by the early 1960s. This book charts and analyses their transition in the three countries that first devoted significant resources to nuclear technologies in an engineering context: the USA, UK, and Canada. By hosting the first intensive research and development of nuclear engineering, these countries created templates that influenced the creation of subsequent national programmes.

This book does not, however, seek to document all nuclear workers over all time. Two other countries significant in the early development of atomic energy, for example, have not been centrally included in this study: the USSR and France. This is not meant to slight the achievements in those countries, but instead to clarify the distinct objectives of this work. French scientists dominated the early exploration of radiation and nuclear physics as illustrated by five Nobel Prize winners. During the four-year wartime occupation of France, however, the bulk of French nuclear research—and virtually all of the country's contributions to the field of nuclear engineering—was in effect hosted by Canada. Post-war nuclear developments in France, informed by the experiences of French scientists who had worked in the Anglo-Canadian project, were nationally distinctive, as shown by the excellent studies of Gabrielle Hecht.[35] Even so, that wartime proficiency was not invested profitably during the 1950s, when reactors similar to those developed in the UK were pursued instead. Hecht's insights provide valuable tools for the present study, but this book does not tread directly in her footsteps: where she has emphasized co-construction of nuclear technology, politics, and French national identity, I focus instead on the factors that shaped expertise in disparate national contexts, seeking to tease out shared and unique factors.[36]

The Soviet experience, explored in depth by other authors, is similarly significant but beyond the central concerns of this book.[37] Unlike France, the USSR instigated an

---

[35] Hecht, Gabrielle, *The Radiance of France: Nuclear Power and National Identity after World War II* (Cambridge, Mass.: MIT Press, 1998). This, and some of her earlier work, deals with issues examined in this book for the Anglo-American countries, aspects of employment, labour categories, and their relationship to national policy and working conditions: Hecht, Gabrielle, 'Political designs: nuclear reactors and national policy in postwar France', *Technology and Culture* 35 (1994): 657–85; Hecht, Gabrielle, 'Rebels and pioneers: technocratic ideologies and social identities in the French nuclear workplace 1955–1969', *Social Studies of Science* 26 (1996): 483–529; Hecht, Gabrielle, 'Enacting cultural identity: risk and ritual in the French nuclear workplace', *Journal of Contemporary History* 32 (1997): 483–507. See also Weart, Spencer R., *Scientists in Power* (Cambridge, Mass.: Harvard University Press, 1979).

[36] Where Hecht has focused on French engineers, I highlight and problematize the scientist–engineer relationship in the Anglo-Saxon countries. A further distinction is in methodology. Unlike France, the UK, USA, and Canada have made available a broad range of archival sources. On the other hand, I have found oral interviews relatively unrevealing with practitioners in this field, in contrast to my previous subjects of study.

[37] E.g. Holloway, David, *Stalin and the Bomb: The Soviet Union and Atomic Energy, 1939–1956* (New Haven, Conn.: Yale University Press, 1994); Graham, Loren, *What Have We Learned About Science and Technology from the Russian Experience?* (Stanford, Calif.: Stanford University Press, 1998); Josephson, Paul R., *Red Atom: Russia's Nuclear Power Program from Stalin to Today* (New York: W. H. Freeman, 1999).

independent wartime research effort but, owing to time and resource constraints, augmented its substantial home-grown expertise with espionage. Its wartime and post-war workers remained virtually invisible to Western practitioners until the mid-1950s, and so played little part in constructing early nuclear know-how in the Anglo-Saxon countries. After 1955, however, the internationalization of nuclear knowledge played a role in shaping technical identities beyond borders, a shifting context explored in the later chapters. Nevertheless, there are important comparisons to be made in relation to the role of the state. Scholars have suggested that the Soviet context simultaneously vaunted and eviscerated engineering identities, allying them to socialist progress and centralized governance. My work—which finds some broad similarities with the English-speaking democracies—tempers such claims.[38]

Just as I focus on particular national experiences, I also concentrate on the early decades of growth, during which national differences were most marked. The growing internationalization a decade after the war resulted in a flowering of nuclear energy programmes in some forty countries, a few of them eventually overtaking Britain and Canada. By the early twenty-first century, for example, India and China had ambitious plans for expanding their investment in nuclear power. That second wave was different in kind from the early experiences of nuclear engineering in the USA, UK, and Canada, and beyond the scope of a book of reasonable length and depth. While their respective national experiences were distinctive, the technologies and, more importantly, the characteristics of their technical workers, were influenced by established templates.

By examining the Anglo-American experiences of nuclear engineering, which began earlier and with sustained government funding unlike that in other countries, their shared (and sometimes unshared) construction of fluid technical identities can be compared and contrasted. My principal analytical aim is to intercompare the experiences of these first three collaborating countries in shaping new technical expertise. By so doing, the tensions between shared intellectual constructions and distinct national perceptions of technical professions can be explored. But, as illustrated so well by prior studies of Soviet and French work, I seek to incorporate into this analysis the insights from prior studies of individual countries.[39]

---

[38] On the cultural role of engineers in the Soviet and East German states, see respectively Graham, Loren, *The Ghost of the Executed Engineer: Technology and the Fall of the Soviet Union* (Cambridge, Mass.: Harvard University Press, 1993) and Augustine, Dolores, *Red Prometheus: Engineering and Dictatorship in East Germany, 1945–1990* (Cambridge, Mass.: MIT Press, 2007).

[39] The development of nuclear expertise has begun to receive belated attention from historians and sociologists. Beyond the French occupational dimensions analysed by Hecht is a valuable study of the origins of nuclear engineering education in Spain from the late 1950s, and including contemporary European comparisons: Barca Salom, Francesc X., *Els Inicis de L'Enginyeria Nuclear a Barcelona: La Càtedra Ferran Tallada (1955–1962)*, PhD thesis, Departament de Matemàtica Aplicada 1, Universitat Politècnica de Catalunya (2002) and sketched in Barca Salom, Francesc X., 'Nuclear power for Catalonia: the role of the official Chamber of Industry of Barcelona, 1953–1962', *Minerva* 43 (2005): 163–81. Adopting a similar focus are studies of UK education: Herran, Néstor, 'Spreading nucleonics: the Isotope School at the Atomic Energy Research Establishment, 1951–67', *British Journal for the History of Science* 39 (2006): 569–86 and Yeo, Frances E. M., *Nuclear Engineering Education in Britain*, MSc thesis, University of Manchester (1997).

Nuclear specialists developed during and after the war in unique contexts that combined rapid growth with unprecedented intellectual isolation and monopolistic direction by their governments. A comparative study, then, has the rare opportunity to analyse these parallel historical experiments in distinct national contexts.[40] Tracking the dissimilar national trajectories for these specialists reveals the social, political, and cultural contingencies in the creation of new intellectual disciplines. The barriers created by classified knowledge also help to reveal how knowledge and expertise circulated. This was not simply diffusion from a single centre, but instead involved adapting and mutating skills to new circumstances at each post-war site.

The working identities of these now-hidden, now-revealed specialists altered repeatedly. Depicted as shadow creators of wondrous technologies by their governments, the public image of nuclear workers was constructed largely from a vacuum of fact. Wider cultural perceptions were flavoured enduringly by their secretive early years when they had been a voiceless collection of experts, their achievements trumpeted second-hand by their governments. And, some thirty years after their birth, public judgements of their products became increasingly critical. Over a few decades they transmuted from the heroic scientific geniuses of the war to unreliable eggheads having dubious allegiances, to even more distant engineering experts characterized as the pawns of corporate interests. The evolving identity was punctuated by high-profile accidents. And their own voices were unusually muted and, on occasion, actively stifled by their governments and employers. This creation of balkanized identities mirroring their social contexts is an important part of the story. At its heart were the shifting conceptions of the terms 'nuclear' and 'engineer', and how these understandings were constructed in distinct places and times.

National and regional contexts were significant in shaping different visions of the disciplines, professions, and occupations of nuclear expertise. In the following chapters I argue that four factors shaped technical identities more profoundly than in other fields. First, concerted government management was seminal in imposing a wartime merger of academic science and industrial cultures. Second, the context of secrecy was significant in gestating and incubating the nascent specialists. Third, trends in international politics during the mid-1950s fostered a rapid declassification of knowledge and created opportunities—and, indeed, considerable government demand—for a nuclear power industry and an openly taught discipline to support it. And fourth, public representations of the new specialists were shaped by the rhetoric of Cold War national science and technology. In short, the role of the state in creating technical alliances, controlling the circulation of knowledge, and mediating professional roles was of significant importance in creating and nurturing these new specialists.

The book correspondingly traces a chronological narrative, with its first three parts documenting the wartime gestation, post-war incubation, and professional emergence of the new specialists.

---

[40] A model for this approach is Hughes, Thomas Parke, *Networks of Power: Electrification in Western Society,* 1880–1930 (Baltimore: Johns Hopkins University Press, 1993). On the value of comparative studies of technical specialists, see Kranakis, Eda, 'Social determinants of engineering practice', *Social Studies of Science* 19 (1989): 5–70.

In opposition to this coherent weaving, however, specific events unravelled public perceptions of nuclear specialists in the following decades, the focus of the final section of the book. Three Mile Island, Chernobyl, and Fukushima challenged public confidence about the nuclear age and its experts in ways that nuclear weapons had not. These unplanned and spasmodic events, too, are an important part of the story of how technical identity is shaped, and a sobering example of the contingencies facing emerging professions of the twenty-first century.

PART A

Gestation

'Ships may be propelled by nuclear energy before the end of the war...the present ideas and research work should be developed by a firm in the United Kingdom for the British Empire, whatever may be done in other parts of the world.'
MAUD Committee, UK, March 1941[1]

'If it is successful it will be one of the most spectacular things of the war.'
C. J. MacKenzie, National Research Council of Canada, February 1942[2]

'"It sounds like what Buck Rogers reads about when he reads"...Lord help us'.
Crawford Greenewalt, Du Pont de Nemours Company, USA, December 1942[3]

---

[1] Imperial Chemical Industries Ltd, 'Report by M.A.U.D. Committee on the Use of Uranium for a Bomb, Appendix VII: Nuclear energy as a source of power', in: Gowing, M., *Britain and Atomic Energy 1939–1945* (London: Macmillan, 1964), pp. 435–6.
[2] 19 February 1942 diary, Mackenzie, C. J., LAC MG30-B122 Vol 1.
[3] Greenewalt, Crawford H., 'Manhattan Project Diary, Vol II', 1942–43, Hagley 1889, p. 5, 26 December 1942.

2

NEW KNOWLEDGE FOR NEW PURPOSES

This chapter surveys the intellectual and geographical terrain and introduces themes that will be at the core of the book. In setting the scene, it argues that international nuclear science became a fertile and attractive field from the turn of the twentieth century, and one that quickly encouraged a close intermingling with application and commerce. But the technical environments seeded by the Second World War brought scientists into closer proximity to engineers and gestated new expertise, allowing it to grow rapidly into exotic forms at a few locations. These wartime hot-houses fostered unique national visions and specialists collaborating in a domain variously dubbed 'atomic energy', 'nucleonics', and 'nuclear engineering'. Yet these labels—and the attributes of the experts themselves—were to remain fluid and contested; the descriptions adopted for their domain were to define *them* as much as *it*.[1]

2.1 A twentieth-century field

One of the best appreciated narratives of progress for mid-twentieth-century audiences was the blossoming of nuclear physics. In the space of fifty years, the subject had grown from hesitant experiments on radioactivity to represent the epitome of modern knowledge and the template of the future. The coming of the Atomic Age paralleled the growth of other fields that linked intellectual insights with societal improvement: the rising sophistication of medicine and industrial chemistry, burgeoning mass production, transport, and communications technologies. Part of that popular story of explosive advance involves the subject of this book—but only part of it. Rather little about nuclear experts can be fitted convincingly to those accounts of discovery, genius, and idealism. Much more focuses on cooperation, contingency, and constraints.

The fount of this new knowledge was the concept of the chain reaction, and its intimate relationship with engineering materials, processes, and products. Nuclear physics and chemistry spawned knowledge and potential that fed nuclear engineering; engineering, in turn, pulled science in new directions and provided it with new experimental environments. They became the most expensive and extravagantly funded domains in physical research, for which

---

[1] Because of this long-standing ambiguity, tracking 'nuclear engineers' also involves following scientists, technicians, and process workers, and assessing their changing representations. A suite of historical actors and social contexts—not an objective definition—determined how these experts distinguished themselves. On the related notion of the social construction of the 'nuclear', see Hecht, Gabrielle, 'The power of nuclear things', *Technology and Culture* 51 (2010): 1–30.

the moniker 'big science' was coined.[2] Both exemplified leading-edge research and development. But nuclear engineering became more tightly wrapped up with national defence, aspirations, and progress. Entwined for a time, the paths of the two subjects diverged increasingly thereafter. Their parting was not along directions of 'pure science' and 'application', as popular accounts might have it, but in different paths of industrialized *technoscience*.[3]

2.1.1 Players, problems, and research products

The scientific queries that motivated nuclear engineering were bracketed by two markers: the beginning of the twentieth century, and the beginning of the Second World War. Over those four decades, nuclear research grew to become the paramount field in physical science.

At the turn of the century, the practice of physics was crossing several boundaries. Its content was being revitalized by the study of profoundly new phenomena; a growing number of physicists were working not just at universities, but for industrial and other sponsors; and growing public awareness of this cadre of experts began to link them with the trappings of modernity. The style, scale, and pace of work were transformed.

During the nineteenth century physics had been remarkably successful in explaining the physical world. For many practicing scientists, the subject represented the culmination—and near completion—of an intellectual project begun nearly 300 years earlier: the experimental detailing and mathematical explanation of natural processes. Isaac Newton's seventeenth-century concept of gravitation had been extended successively to all of mechanics; the new concepts of energy and electromagnetism had been proposed, tested, and refined. By the end of the century Lord Kelvin (1824–1907) in Britain could argue that knowledge unsupported by measurement and unexpressed by numbers was both meagre and unsatisfactory. The prominent American experimentalist, Albert A. Michelson (1852–1931), suggested that physics in the twentieth century might involve mainly a mopping-up exercise, seeking more subtle phenomena beyond the next decimal place.[4]

Such confidence was being challenged by a younger generation investigating new phenomena, however. Electrical phenomena had been an enduring subject of enquiry through the nineteenth century and they generated a new class of physical effects. *Cathode rays* were identified by a blue glow appearing around a charged electrode of a new apparatus, the Crookes tube, and could be diverted by a magnetic field. *Roentgen rays* (or X-rays, identified in 1895), emitted by similar apparatus, could pass through flesh to leave exposure shadows

---

[2] Weinberg, Alvin M., *Reflections on Big Science* (Cambridge, Mass.: MIT Press, 1967); Seidel, Robert W., 'A home for Big Science: the Atomic Energy Commission's laboratory system', *Historical Studies in the Physical and Biological Sciences* 16 (1986): 135–75; Galison, Peter and Bruce Hevly, *Big Science: The Growth of Large-Scale Research* (Stanford, Calif.: Stanford University Press, 1992); Hughes, Jeff, *The Manhattan Project: Big Science and the Atom Bomb* (New York: Columbia University Press, 2003).

[3] On the enmeshed socio-technical aspects of the modern world, see Latour, Bruno, *Science in Action* (Cambridge, Mass.: Harvard University Press, 1987) and Latour, Bruno, *We Have Never Been Modern* (Hemel Hempstead: Harvester Wheatsheaf, 1993).

[4] For overviews including *fin de siècle* physics, see Brown, L., B. Pippard and A. Pais (eds.), *Twentieth Century Physics*, Vol. I (Bristol: Institute of Physics Publishing, 1995); Krige, John and Dominique Pestre (eds.), *Science in the Twentieth Century* (London: Routledge, 1997); Kragh, Helge, *Quantum Generations: A History of Physics in the Twentieth Century* (Princeton, NJ: Princeton University Press, 1999).

on photographic film. New rays with distinct properties, dubbed *alpha*, *beta*, and *gamma* by their investigators, appeared in quick succession over the following decade. They could variously travel centimetres, metres, or hundreds of metres, through air, water, or concrete. They were found to be emitted from certain rocks, notably uranium salts, and later by elements dubbed *radium* and *polonium* by their discoverers, French physicists Marie and Pierre Curie (1867–1934 and 1859–1906), who also coined the term *radioactivity.*

As subsequent generations of schoolchildren learned, these new rays were linked profoundly with the properties of matter. J. J. Thomson (1856–1940) in England proposed that cathode 'rays' were in fact 'corpuscles' later dubbed *electrons*; at the beginning of the new century, there was growing evidence that radioactivity was the emission of particles from an unstable atom. Imaginative experiments and novel theorization led to rapid developments in the field. By the early 1930s—a single scientific generation—the rays had been recast as a collection of fundamental particles relating to the concept of the atom and its nucleus, and the new subject of *quantum mechanics* was built to describe their unintuitive and radically unclassical behaviours.[5]

With new knowledge came experts. The rapidly growing subject of radioactivity, and later the field of nuclear physics, cultivated a team approach reliant on a dense international network of collaboration. The groups shared findings and proficiency via journal publications, international conferences, and research students, who commonly studied at a handful of European centres and moved to new sites as their careers advanced. The New Zealand physicist Ernest Rutherford (1871–1937) and English chemist Frederick Soddy (1877–1956), for example, explored radioactivity at McGill University in Montreal, Rutherford as a Professor (1898–1907) and Soddy as a Demonstrator (1900–1903). Later, at the University of Manchester and the Cavendish Laboratory, Cambridge (where he succeeded physicist J. J. Thomson as director), Rutherford mentored a generation of internationally linked investigators. Among this international generation was James Chadwick (1891–1974), who found conclusive evidence for the existence of the neutron in 1932. By the 1930s, similar research clusters had grown around Niels Bohr (1885–1962) in Copenhagen, Frédéric and Irène Joliot-Curie in Paris, Werner Heisenberg (1901–1976) in Leipzig, Enrico Fermi (1901–1954) in Rome, Ernest Lawrence (1901–1958) in Berkeley, and Igor Kurchatov (1903–1960) in Moscow.[6] Their collective product was a new conception of

---

[5] On early nuclear physics, see Mladjenovic, Milorad, *History of Early Nuclear Physics (1896–1931)* (Singapore: World Scientific, 1992); Keller, Alex, *The Infancy of Atomic Physics: Hercules in his Cradle* (Oxford: Clarendon Press, 1983); Jammer, Max, *The Philosophy of Quantum Mechanics: The Interpretations of Quantum Mechanics in Historical Perspective* (New York: John Wiley, 1974).

[6] On the evolving community of radioactivity specialists eventually overtaken by nuclear physicists, see Hughes, Jeff, 'Radioactivity and nuclear physics', in: M. J. Nye (ed.), *The Cambridge History of Science: The Modern Physical and Mathematical Sciences* (Cambridge: Cambridge University Press, 2003), pp. 350–74 and Hughes, Jeff, '"Modernists with a vengeance": changing cultures of theory in nuclear science, 1920–1930', *Studies In History and Philosophy of Science Part B: Studies In History and Philosophy of Modern Physics* 29 (1998): 339–67; Hughes, Jeff, 'The French connection: the Joliot-Curies and nuclear research in Paris, 1925–1933', *History and Technology* 13 (1997): 325–43. See also Sclove, Richard E., 'From alchemy to atomic war: Frederick Soddy's "technology assessment" of atomic energy, 1900–1915', *Science, Technology, & Human Values* 14 (1989): 163–94 and Sime, R. L., 'From radioactivity to nuclear physics: Marie Curie and Lise Meitner', *Journal of Radioanalytical and Nuclear Chemistry* 203 (1996): 247–57.

subatomic reality and, with it, a subject pregnant with new possibilities. The field as a whole became the centre of modern physical science, with over two dozen individuals awarded Nobel Prizes by 1939.[7]

Just as the practice of physics was being transformed, its experts, too, were changing. Most physicists (who only a couple of generations earlier had referred to themselves as *natural philosophers*) were members of university science departments, an organizational innovation of the late nineteenth century. But some were employed alongside other technical experts directly by their governments, notably in government laboratories in Germany, Britain, and the USA. And, in each of those countries, physicists and chemists were joining large firms and industrial laboratories.[8] This intermingling of turn-of-the-century specialists of rays and radiation also linked research with practice, blurring the faint demarcation that was just becoming established between university research, industrial application, and popular understandings. Seeking to measure radioactivity, for example, Marie Curie's laboratory fostered an industry based on radium and other radioactive materials.[9]

The new field extended beyond physical science and engineering, too. The emissions of X-ray machines, rapidly reproduced from 1896, were poorly understood but widely employed. Not only could X-rays reveal hidden bullets or broken bones, but remove unwanted facial and body hair—encouraging medics, therapists, and charlatans alike to make use of them. The rays of radium and polonium were also discovered to have biological effects, often interpreted as curative or rejuvenating. But alongside these benefits, dangers soon became apparent: new forms of occupational affliction resulted, with a disturbing number of X-ray operators developing untreatable injuries. Both positive and negative outcomes of radiation motivated scientific attention to the underlying physical and biological mechanisms, as well as to techniques of measurement and control. The products of radioactivity and nuclear physics offered a new and interdisciplinary 'scientific' industry to be

---

[7] 1901: W. Roentgen, for X-rays; 1903: A. H. Bequerel, P. Curie, and M. Curie, for the discovery of radioactivity; 1905: P. von Lenard, for cathode-ray investigations; 1906: J. J. Thomson, for conduction in gases; 1908: E. Rutherford, for the chemistry of radioactive elements; 1911: M. Curie, for the discovery of radium and polonium; 1918: M. Planck, for energy quanta; 1921: A. Einstein for the photoelectric effect, and F. Soddy for isotope investigations; 1922: N. Bohr for the structure of atoms, and F. W. Aston for isotopes in non-radioactive elements; 1923: R. A. Millikan, for the determination of the elementary charge; 1925: J. Franck and G. L. Hertz, for the laws of electron–atom impact; 1927, A. H. Compton, for X-ray scattering by electrons; 1929: L.-V. de Broglie, for the wave nature of electrons; 1932: W. Heisenberg for the creation of quantum mechanics; 1933: E. Schrödinger and P. A. M. Dirac for atomic theory; 1935: J. Chadwick for the discovery of the neutron, and Frédéric Joliot and Irene Joliot-Curie for the synthesis of new radioactive elements; 1936: C. D. Anderson for the discovery of the positron; 1938: E. Fermi for the discovery of nuclear reactions by slow neutrons; 1939: E. O. Lawrence for the cyclotron.

[8] See, for example, Kevles, Daniel, *The Physicists: The History of a Scientific Community in Modern America* (New York: Knopf, 1977).

[9] Boudia, Soraya, 'The Curie laboratory: radioactivity and metrology', *History and Technology* 13 (1997): 249–65 and Roqué, Xavier, 'Marie Curie and the radium industry: a preliminary sketch', *History and Technology* 13 (1997): 267–91. See also Hughes, Jeff, 'Plasticine and valves: industry, instrumentation and the emergence of nuclear physics', in: J.-P. Gaudillière and I. Löwy (eds.), *The Invisible Industrialist. Manufactures and the Production of Scientific Knowledge* (London: Macmillan, 1998), pp. 58–101.

simultaneously exploited and regulated, tying even arcane leading-edge science to commercial outcomes and social consequences.[10]

The novelty and scale of the new science encouraged public engagement. While men wore radium belts for invigoration of their libidos and women visited beauty salons for X-ray depilation treatments, they could read about scientific advances in atomic and nuclear physics. Popular magazines recounted the life story of physicist Robert A Millikan (1868–1953), who had won the Nobel Prize for measuring the charge of the electron; showed snapshots of the Curie family; popularized the baffling ideas of Albert Einstein (1879–1905), Nobelist for his explanation of the photoelectric effect and a celebrity from the 1920s; and revealed images of the first atom smashers, which infected popular culture as much through news magazines as science-fiction cinema.[11]

By the late 1930s, then, the subject of atomic physics and nuclear science was large-scale, team-oriented, goal-directed, and thoroughly international in scope. Its participants provided a template for an aspirational and increasingly recognized twentieth-century scientific community. The content and trajectory of new nuclear knowledge changed abruptly, however, at the beginning of the Second World War.

2.1.2 *Fission as an engineering field*

The neutron provided a way of unlocking atomic energy on an industrial scale. The discovery of nuclear fission and its corresponding release of energy, confirmed by the experiments and analysis of Otto Hahn (1879–1968), Lise Meitner (1878–1968), Fritz Strassman (1902–1980), and Otto Frisch (1904–1979) in Germany and published seven months before the outbreak of the European war, offered new opportunities. At the Collège de France in Paris, the team led by Frédéric Joliot (1900–1958) and including Hans Halban (1908–1964) and Lew Kowarski (1907–1979) demonstrated that this fission usually released two or more neutrons.[12] This significant detail gave the fission of uranium nuclei not only scientific but also potentially engineering interest: if absorbed by a neighbouring uranium atom, each extra neutron would cause its nucleus to fission, too, releasing more energy and yet more neutrons. The newly liberated particles might cause a sustainable series of chain of reactions; in the right circumstances, the reactions might multiply in an exponential expansion, leading to a proportional release of energy. Researchers began attempts to investigate the feasibility of a chain reaction in the laboratory.

Enrico Fermi, having emigrated from Mussolini's Italy and now working at Columbia University in New York, began a methodical series of experiments from early 1940 using uranium and graphite bars interspersed in structures soon dubbed *piles*. Also that year,

---

[10] Caulfield, Catherine, *Multiple Exposures: Chronicles of the Radiation Age* (London: Secker & Warburg, 1989).

[11] 'Life story of Robert A. Millikan told in pictures', *Startling Stories*, May 1939: 29–31; 'Eve Curie writes a biography of her mother', *Life*, 22 November 1937: 75–8; 'Einstein explains relativity: his new book explains science to the layman', *Life*, 11 April 1938: 48–9; 'Atom smasher aids research in many fields', *Popular Mechanics*, February 1939: 238–9.

[12] Meitner, Lise and Otto R. Frisch, 'Disintegration of uranium by neutrons: a new type of nuclear reaction', *Nature* 143 (1939): 239–40; Von Halban, Hans, Frederic Joliot and Lew Kowarski, 'Number of neutrons liberated in the nuclear fission of uranium', *Nature* 143 (1939): 680.

George C. Laurence (1905–1987) at the National Research Council in Ottawa, Canada, attempted to achieve a chain reaction by building an assembly of uranium and carbon. Held in paper bags, the packages of uranium oxide powder and coke (like graphite, a source of almost pure carbon) were stacked in a lattice arrangement. At Imperial College in London, Phillip B. Moon (1907–1994), directed by George P. Thomson (1892–1975), unsuccessfully tried a combination of uranium oxide and paraffin. Joliot's group acquired *heavy water* from Norway, which they planned to use instead of graphite.[13] No chain reaction resulted from any of these initial trials.

These experiments were motivated by similar insights. Graphite, coke, paraffin, and heavy water had each been chosen as promising *moderators* for the chain reaction. The fission of a uranium nucleus liberates high-speed (energetic) neutrons which were recognized to be very unlikely to be captured by another nearby uranium atom. Fermi and his colleagues in Italy had found, though, that if the neutrons passed through certain materials they could be slowed by collisions with other nuclei to the 'thermal' speeds comparable to the random jostlings of the uranium atoms themselves. Within moderators such as graphite this billiard-table process was efficient: the neutrons were slowed, but not 'pocketed', by carbon nuclei. The lower energy neutrons could then wander by multiple collisions towards uranium atoms, where they were much more readily captured to produce further fission. Most materials, however—including the numerous impurities in supplies of graphite, coke, paraffin, and heavy water—absorbed neutrons so readily that they would snuff out any potential chain reaction.

The key to a chain reaction, then, was to produce more than one—and preferably significantly more than one—fission per fission, even in the presence of myriad losses. This so-called multiplication factor, or k value, was generally below 0.90 in the first experiments. If k were slightly below 1.0, the reaction would fizzle out as the chain of nuclear breakups failed to reproduce itself sustainably; k equal to 1.0 would yield a steady-state chain-reaction signalled by the steady emission of neutrons and heat; but k even slightly greater than 1.0 would support a divergent chain reaction in the pile, producing ever-increasing energy. By choosing appropriate moderator and fissile materials having adequate purities and suitably interleaved dimensions, a well-designed pile could conceivably produce a k greater than unity. Then, by using movable neutron absorbers inserted into the pile, the 'compound interest rate' could be reduced controllably, tuning the rate of the chain reaction.[14] The result would be copious neutrons, large-scale nuclear transmutations of materials, and excess energy—each of which was expected to have scientific and engineering value.

2.2 The trigger of war

Despite initial difficulties in obtaining sufficiently pure moderators and adequate supplies of natural uranium, many of the investigators were convinced that a chain reaction should eventually be achievable. Confidence in the possibility was pertinent beyond the borders of

---

[13] Heavy water is water composed of oxygen and deuterium atoms, where deuterium or *heavy hydrogen* is an isotope of hydrogen in which the nucleus consists of a proton and neutron instead of a proton alone.

[14] The movable absorber was dubbed a *control rod* in the first chain reactor having a k greater than one in 1942.

physical science, though. Several appreciated even before experiments began that the chain reaction could yield a source of heat (and potentially electrical power derived from it) if the reaction rate could be controlled, and a powerful bomb if it became rapidly exponential.

Physicists Leo Szilárd (1898–1964) and Albert Einstein (1879–1955) consequently wrote to President F. D. Roosevelt in August 1939 to urge the US government to sponsor research on the potential for a powerful explosive device based on the chain reaction.[15] Roosevelt authorized an ad hoc Uranium Committee a month later to begin small-scale research. Scientists in Britain lobbied their own government at about the same time. At the University of Birmingham in early 1940, physicists Otto Frisch and Rudolf Peierls (1907–1995), émigrés, like Einstein, from Hitler's Germany, estimated that an intense explosion probably could be produced by a small quantity of one variety of uranium (the so-called U-235 isotope, having an atomic mass of 235). The fast neutrons emitted by the fissioning of atoms of this purified radioactive material would cause further fissions of neighbouring atoms. If a sufficient quantity were concentrated into a small volume to capture a sufficient fraction of neutrons, an uncontrolled chain reaction would occur, likely releasing orders of magnitude more energy than a conventional chemical explosive.[16]

Mark Oliphant (1901–2000), professor of physics at Birmingham, communicated the report of Frisch and Peierls to the Committee on the Scientific Survey of Air Defence chaired by chemist Henry Tizard (1885–1959). Unlike the tepid American reaction, an official response in the UK was rapid: a concerted British effort began in April 1940, when a special committee labelled with the code name MAUD was set up by the Ministry of Aircraft Production to investigate the feasibility of such a uranium-based explosive. The six-member committee was chaired by physicist George Thomson (1892–1975), and included his Imperial College colleague P. B. Moon, Oliphant, James Chadwick, discoverer of the neutron, and Cambridge physicist John Cockcroft (1897–1967), then the Assistant Director of Scientific Research in the Ministry of Supply. In September 1940, Tizard visited the USA to discuss the exchange of military technologies. He judged that all current fission research—that of Hans Halban and Lew Kowarski, now working at Cambridge on their concept for a heavy-water reactor, George Laurence in Ottawa, and Enrico Fermi in New York on uranium-graphite schemes—would not likely influence the outcome of the war.

---

[15] Soviet lobbying was effective only two years later: in April 1942 Georgii Flerov, as representative of the State Defence Committee for science, convinced Stalin to support similar research. A year further on, three laboratories were planned—later becoming the Ukrainian Physico-Technical Institute in Kharkov, the Kurchatov Institute of Atomic Energy, and Institute of Theoretical and Experimental Physics, both in Moscow; the uranium project began in 1944, when Kurchatov's laboratory was completed [Josephson, Paul R., *Red Atom: Russia's Nuclear Power Program from Stalin to Today* (New York: W. H. Freeman, 1999), pp. 16–17]. On the parallel German research see, for example, Goudsmit, Samuel A., *Alsos: The Failure of German Science* (New York: Sigma Books, 1947); Hentschel, Klaus (ed.) and Ann M. Hentschel (translator), *Physics and National Socialism: An Anthology of Primary Sources* (Berlin, 1996); Rose, Paul L., *Heisenberg and the Nazi Atomic Bomb Project: A Study in German Culture* (Berkeley: University of California Press, 1998).

[16] The crucial difference between such a bomb and a chain-reacting pile is that the bomb would use purified or enriched U-235 so that enough *fast* neutrons are produced to yield a chain reaction in most of that uranium, despite their low probability of being captured by nuclei. The necessarily compact geometry and lack of moderator made such a bomb much smaller than the planned piles.

But the first MAUD report, distributed in March 1941, was far-reaching in its conclusions and recommendations. Indeed, it set the trajectory for British involvement in nuclear energy over the following decades. The report judged a uranium-based fission explosive to be technically feasible, and an appendix discussed the possibilities of nuclear energy as a source of power that could have military and civilian purposes.[17] The MAUD report represented an important shift in direction: not only did it steer nuclear physics in Britain into an application of immediate wartime significance, but it was a detour reliant on the collaboration of scientists, industrial engineers, and government.

2.2.1 Chain reactors and atomic bombs in the USA

The MAUD report was crucial in galvanizing American scientists, too, who campaigned that autumn for an all-out development programme.[18] In October 1941, a committee of the American National Academy, chaired by University of Chicago physicist Arthur Compton (1892–1962), brought together a contingent of physical scientists and engineers. Compton himself had completed his PhD at Princeton University and had gained industrial experience working for two years in industrial research for Westinghouse, followed by a research fellowship under Rutherford at Cambridge.[19] His engineers—seconded from General Electric, Westinghouse, and Bell Telephone—suggested that some three-to-five years would be required to develop an atomic weapon.

By early February 1942, two months after American entry to the war following the bombing of Pearl Harbour by the Japanese, there was a concentration of expertise and administrative focus at the University of Chicago under the cover name 'The Metallurgical Laboratory'. There, physicists from Columbia, Princeton, Indiana, and other universities came together, including Compton (its Director), Fermi, Eugene Wigner (1901–1995), John Wheeler (1911–2008), and Edward Teller (1908–2003).[20]

---

[17] Imperial Chemical Industries Ltd, 'Report by M.A.U.D. Committee on the Use of Uranium for a Bomb, Appendix VI: Nuclear energy as a source of power', in: Gowing, M., *Britain and Atomic Energy 1939–1945* (London: Macmillan, 1964), pp. 433–4. ICI's own postwar plans for nuclear development and power generation are further detailed in 'ICI Progress Reports (General Chemistry Division) 1941–43', NA AB 1/331.

[18] By contrast, Soviet physicists appear not to have alerted their government about potential military uses, nor to have organized research on nuclear fission as early [Holloway, David, *Stalin and the Bomb: The Soviet Union and Atomic Energy, 1939–1956* (New Haven, Conn.: Yale University Press, 1994), p. 58]. See also Josephson, Paul R., 'Atomic-powered communism: nuclear culture in the postwar USSR', *Slavic Review* 55 (1996): 297–324.

[19] Compton's reputation as a physicist—and his 1927 Nobel Prize—was founded on his investigation of the scattering of X-rays by electrons (the *Compton effect*), providing evidence that X-rays themselves can be understood as particles. By the beginning of the war, Compton was also recognized as a competent administrator, having served as the head of the physics department of Washington University in St Louis and subsequently at the University of Chicago. Indeed, his father and two brothers also had careers as senior academic administrators.

[20] In a field that rapidly became imbued with cultural meanings, their biographies and autobiographies are clear products of their times: Compton, Arthur Holly, *Atomic Quest: A Personal Narrative* (Oxford: Oxford University Press, 1956); Fermi, Laura, *Atoms in the Family: My Life with Enrico Fermi* (Chicago: University of Chicago Press, 1954); Wigner, Eugene P. and Andrew Szanton, *The Recollections of Eugene P. Wigner* (London: Plenum, 1992); Teller, Edward and Judith L. Shoolery, *Memoirs: A Twentieth-Century Journey in Science and Politics* (Oxford: Perseus Press, 2001).

The government decision to fund the American bomb project was guided by two highly placed administrators who subsequently played an important role in defining its new specialists and their scope. Vannevar Bush (1890–1974) had been appointed Chair of the National Defense Research Committee (NDRC) in 1940 by Roosevelt. Bush had a career background that wedded high engineering to scientific utility. Obtaining a PhD in engineering from the Massachusetts Institute of Technology (MIT) in 1917, he had worked during the remainder of the First World War on methods of submarine detection. He became professor, then Dean of Engineering, at MIT between the wars, where he developed a mechanical analogue computer, or 'differential analyser', that was applied to scientific calculations. As the European war accelerated and America's entry began to seem more likely, Bush lobbied for the creation of the NDRC to improve cooperation between civilian science and military aims. He appointed James B. Conant (1893–1978) as one of the four scientists to the committee, with responsibility for overseeing explosives development. Conant was then President of Harvard (a mile further up the Charles River from MIT) and, like Bush, had earlier studied at, and become a professor of, his university. When a more powerful coordinating body, the Office of Scientific Research and Development (OSRD) was established in 1941, Bush was appointed its Director. The NDRC was subsumed within the new organization, and Conant became its Chair. The wide-ranging remit of the OSRD included the developing atomic bomb project. Bush, supported by Conant, reported directly to the President and had essentially unlimited access to resources for wartime research and development.

Over the next year, experimental groups across the country (notably the group under Robert Oppenheimer (1904–1967) at the University of California, Berkeley) collaborated to consider potential designs for an atomic bomb. The scientific participants, increasingly confident that an atomic bomb was feasible, recognized that available knowledge was nonetheless inadequate to ensure a successful outcome. They consequently divided effort into parallel research and development projects divided along disciplinary lines.

The most straightforward approach was the idea favoured by the MAUD committee. It relied on separating the fissile U-235 isotope from the much more prevalent U-238. A bomb could then be created by explosively combining two pieces of this purified metal into a *critical mass*. A sufficient quantity of the material would undergo a rapid chain reaction and release a quantity of heat much greater than any chemical explosive. The principal problems were broadly in the domain of chemistry, but their novelty led to distinct proposals from chemical engineers and physicists. Each focused on separating the isotopes of uranium, which had identical chemical properties and nearly identical mass.

Three approaches were conceived. The first—a diffusion process in which the lighter U-235 atoms in gaseous form percolated more quickly through a labyrinth of minute holes—was familiar in principle to chemical engineers, and favoured by the ICI industrial chemists on the MAUD committee. The second concept relied on electromagnetic separation, capturing the U-235 atoms that travelled through a strong magnetic field in a tighter arc than their heavier U-238 counterparts. The scheme was familiar to physicists with experience of mass spectrometers. A third method, attractive to chemists and physicists alike, was based on separation by centrifuge.

A fourth branch of parallel research was distinctly different. This bomb concept was based on a longer string of unknowns and relied on a radioactive material that was predicted theoretically but which had been scarcely detected. Certain isotopes of the heavier element plutonium were expected to undergo fission and energy release like uranium. It was predicted that plutonium would be a by-product of chain reactors, created from uranium atoms that absorbed one or two neutrons instead of splitting. Although a chain reactor had not yet been developed, the advantage of plutonium for a bomb was that it could be more readily generated from uranium supplies than could U-235. Thus the practical resolution of the problem of creating an atomic bomb traded off uncertainty in industrial chemistry with uncertainty in engineering physics. Together, these unknowns collectively defined a new intellectual terrain.[21]

As Arthur Compton later reflected, the goals of the plutonium project required new hybrid capabilities drawing on both scientific knowledge and engineering expertise:

> It is the tradition in American large-scale industry that research, development and production shall be carried on by separate departments...But here was an exceptional situation. There was no established industrial art into which would fit either the nuclear reaction or the chemical separation of highly radioactive materials. Nearly all that was known of these arts was in the research laboratories where the processes were being developed. To whom could we turn?[22]

The new domain was expanded by a forced marriage. In June 1942, project control was passed to an Army organization dubbed the Manhattan Engineer District and directed by General Leslie Groves (1896–1970)—see Figure 2.1. By the following April, control of the Manhattan Engineer District (later colloquially the 'Manhattan Project') was concentrated in the Army's hands. The government–science axis almost immediately incorporated an industrial dimension, however, when American engineering companies were folded in. Thus the constituency and practices of the project were shaped forcefully by a handful of administrators acting under wartime pressures. The expertise that they collected provided the nuclei of the new art of nuclear engineering.

New sites, which were to accommodate many of the specialists for the following three years, were acquired by the Army Corps of Engineers. Near the Clinton River in Oak Ridge, Tennessee, site X was chosen for its hydroelectric power, necessary to separate U-235 using electromagnetic equipment multiplied on an extravagant scale; in a neighbouring valley grew the Clinton Laboratories, tasked with building a small pilot-scale pile and developing the chemistry of plutonium separation.[23] Los Alamos, New Mexico, became

---

[21] There was in fact a fifth line of research, pursued by the Anglo-Canadian project in Montreal. Their concept of a heavy-water reactor was expected to be the most efficient type for generating plutonium from uranium. It had its own chemical engineering problem, though: the need for large quantities of heavy water. Its production was of similar difficulty to that of separating uranium isotopes, and demanded large-scale plant which had a discouragingly slow rate of production. The single North American plant, in Trail, British Columbia, was owned by an American company and allocated to the American project; the limited supply available to the Anglo-Canadian team delayed their completion of a heavy-water reactor until the end of the war. The Canadian project is discussed at length below.

[22] Compton, Arthur Holly, *Atomic Quest: A Personal Narrative* (Oxford: Oxford University Press, 1956), p. 108.

[23] Johnson, Charles W., *City Behind A Fence: Oak Ridge, Tennessee 1942–1946* (Knoxville: University of Tennessee Press, 1981); Johnson, Leland and Daniel Schaffer, *Oak Ridge National Laboratory: The First Fifty Years* (Knoxville: University of Tennessee Press, 1994).

Fig. 2.1 General Groves with workers at the Hanford plutonium production reactors, 1944.[24] (Courtesy of US Department of Energy, Declassified Document Retrieval System.)

site Y, the home for the scientists and engineers developing the bomb.[25] And Hanford, Washington, on the banks of the Columbia River, was chosen for large-scale atomic piles that would be used to generate plutonium as a by-product of uranium. The 'factories' created there, generating the relatively miniscule quantities of chemical product for the atomic bombs, fostered an equally explosive new mixture of science and engineering.

2.2.2 Britain and the nuclear energy machine

While the American effort was beginning in new secret facilities, work continued in British university and industrial laboratories. The MAUD committee had been superseded by a new 'Tube Alloys Directorate' in the autumn of 1941, a division of the Department of Scientific and Industrial Research (DSIR) attached to the Ministry of Supply. Physicists at universities in Birmingham, Cambridge, Liverpool, Oxford, and Imperial College in

---

[24] DOE DDRS N1D0029066.
[25] Badash, Lawrence, Joseph O. Hirschfelder and Herbert P. Broida (eds.), *Reminiscences of Los Alamos, 1943–1945* (Dordrecht: Reidel, 1980).

London were contributing to the British programme, their numbers augmented by émigré scientists from Germany and France.

A novel feature of wartime work, though, was collaboration between scientists and engineers at an unprecedented pace and scale. The result was the exploration of their social boundaries and, periodically, disputes about intellectual and occupational jurisdiction. The clashes and accommodations refined the established role of nuclear scientists and defined a new form of specialist, the nuclear engineer.[26]

Even before Groves' inclusion of American industry into the Manhattan Project, the MAUD committee had recognized the necessity for disciplinary collaborations. The committee relied on the two British companies that were large enough to support research staff, having related knowledge and industrial capacity for the planned work: Imperial Chemical Industries (ICI) and Metropolitan-Vickers.[27] Unlike the American work, however, ICI had made an early play for development of nuclear energy beyond a mere explosive. The 1941 MAUD report had included appendices written by ICI research staff that evaluated Hans Halban's scheme for a 'nuclear energy machine'—a distinct concept for the chain reactor that Fermi conceived as 'a pile'. The analysts judged the scheme to be feasible but requiring some 20 tons of heavy water and an equivalent amount of purified uranium, an exorbitant and then unavailable amount. The report had also emphasized the importance of 'power production in peace and war', with the likelihood that 'ships may be propelled by nuclear energy before the end of the war', and argued that 'the present ideas and research work should be developed by a firm in the United Kingdom for the British Empire, whatever may be done in other parts of the world'.[28]

ICI was crucial in providing early engineering validation of the scientists' concepts. The most feasible means of producing a bomb, they argued, was by purifying U-235 from natural uranium. ICI chemists selected gaseous diffusion as the most practicable means; determined the suitable chemistry (using uranium in the form of uranium hexafluoride gas); and investigated how suitable diffusion membranes could be fabricated. The separation process would require forcing the corrosive and toxic uranium hexafluoride gas through multiple cascades of such diffusion filters, an unparalleled industrial challenge. In effect, the company redefined the high-science atomic bomb concept almost entirely as an *engineering* project in which the principal problem was amassing sufficient U-235 for a bomb. The separate chain-reaction technique of generating plutonium was deemed potentially interesting but less certain in outcome and probably impracticable for wartime promotion.

It was equally obvious to administrators that the resources to achieve these goals would be difficult to obtain in wartime Britain, which did not have the resources to refine uranium on a scale adequate for the proposed bomb design. The MAUD report suggested that Canada, with known supplies of uranium ore, would be a logical haven for Halban's team

---

[26] On jurisdictional disputes, see Abbott, Andrew D., *The System of Professions: An Essay on the Division of Expert Labor* (Chicago: University of Chicago Press, 1988), Chapter 3.

[27] Rose, Hilary and Steven Rose, *Science and Society: The Chemists' War* (Harmondsworth: Penguin, 1970), pp. 63–5.

[28] Imperial Chemical Industries Ltd, 'Report by M.A.U.D. Committee on the Use of Uranium for a Bomb, Appendix VII: Nuclear energy as a source of power', in: Gowing, M., *Britain and Atomic Energy 1939–1945* (London: Macmillan, 1964), pp. 435–6.

empowered by ICI engineering skills. Although Tube Alloys initially disdained collaboration with the less-advanced and loosely affiliated American groups, the project turned gradually towards complementary and associated tasks with its neighbour. During that first year of independent research, though, the American groups had made substantial progress. Under the Army's management, the Manhattan Engineer District was by then reticent to accept foreign collaboration. Discussions in early 1942, followed by negotiations in Canada, therefore led the bulk of the British team of scientists and engineers being moved *en masse* to Montreal under Hans Halban in early 1943.

2.2.3 *Canada and the heavy-water boiler*[29]

As in other countries, nuclear physics in Canada had attracted researchers from the turn of the century; Rutherford's 1908 Nobel Prize was for his Montreal work, which had confirmed that radioactivity signalled the disintegration of atoms and was characterized by a *half-life*, or typical decay rate for each type of atom. Soddy's 1921 Nobel Prize recognized his formulating at Montreal of the concept of *isotopes* for variants of an element having different atomic weights, later attributed to differing numbers of neutrons in the atomic nucleus. Both subsequently engaged in research in Britain, with Rutherford working at Manchester and Cambridge, and Soddy at University College London, Glasgow, Aberdeen, and Oxford. Rutherford's labs fostered many of the key actors in nuclear fission and its application, including Neils Bohr, James Chadwick, John Cockcroft, Mark Oliphant, W. Bennett Lewis (1908–1987), and Canadians George Laurence, David Keys (1890–1977), and B. W. Sargent (1906–1993).

Serious Canadian involvement in atomic energy was nevertheless a wartime contingency. At the beginning of the war, the National Research Council of Canada in Ottawa was the site of relevant, but almost unfunded, work: there, as noted earlier, physicist George Laurence began studying the possibilities of the nuclear chain reaction in 1940. While unusually secret, this was nevertheless a low-priority and low-budget project: the acting president of the NRC, Chalmers Jack Mackenzie (1888–1984), initially focused attention and funding on research perceived to be of direct and immediate importance to the war. The situation changed when British interest in Laurence's work became evident.[30]

Tube Alloys represented just the kind of project that Mackenzie's NRC aspired to undertake. Mackenzie accepted the new project matter-of-factly as a wartime requirement with potential post-war benefits. It promised economic benefits along with a strong component of internationally worthy science, fitting well with the Council's remit of 'fostering the scientific development of Canadian industry for Canadian needs and for the extension and expansion of Canadian trade at home and abroad'.[31] C. J. Mackenzie was correspondingly impressed with the British plans for the project, 'talking in terms of very large plants—something in the vicinity of 40 or 50 million dollar plants and if it is successful will be one

---

[29] The term used by the Montreal group during the war. See, for example, Auger, P. V., H. H. Halban, R. E. Newell, F. A. Paneth and G Placszek, 'Research programmes for development of heavy water boiler', memo, 30 December 1943, LAC RG77 Vol. 283.

[30] John Cockcroft took an interest in Laurence's work while in North America during the summer of 1940 as a member of the Tizard mission; at the same time, Laurence was able to learn about parallel American work. When Cockcroft returned to the UK, his appeal to ICI led to a $5000 grant for Laurence.

[31] Commemorative plaque, entrance hall of NRC laboratories, 100 Sussex Drive, Ottawa.

of the most spectacular things of the war'.[32] Following further visits from a half-dozen members of Tube Alloys (Mackenzie referring to 'the very hush hush project' alternately as 'the uranium business', 'problem S-1', the 'U project', 'the radiological problem', and 'the corrosion project'),[33] Lesslie Thomson, appointed Comptroller and Secretary, began seeking a Canadian base for the operation in November 1942.

At that point, the British and Canadian projects merged. The Anglo-Canadian arm of the Tube Alloys Project hosted Halban's team in Montreal and, later in the war, at a new facility in Chalk River, Ontario, to develop and build the heavy-water boiler.

In Britain, the USA, and Canada, then, the beginning of the war submerged nuclear science from public view, and focused the minds of specialists on the most recent and potential-filled of their findings: the engineering of nuclear fission to produce chain reactions, which in turn could yield the energy for powerful bombs or power sources. In quick succession universities, governments, their militaries, and industrial firms absorbed the subject, wrapping it in layers of administration and secrecy. Swaddled and carefully tended within, a new breed of technical specialist was nurtured to grow the concept into an engineering reality.

---

[32] 19 February 1942 diary, Mackenzie, C. J., LAC MG30-B122 Vol. 1.

[33] Each label corresponded to an organizational initiative or euphemism. In October 1939 an Advisory Committee on Uranium (the U-project) had been initiated under Lyman Briggs of the American National Bureau of Standards. This became the S-1 Project of the NDRC in July 1941. Tube Alloys was the British codename for the follow-on to the MAUD Committee reports of July 1941, for which 'the corrosion project' was a convenient euphemism and the 'radiological problem' a genuine concern.

3

IMPLANTING INDUSTRIAL CULTURES

Wartime experiences seeded the new nuclear specialists, and these environments gestated their post-war opportunities. The laboratories and production sites merged industrial and scientific cultures. Hybrids—neither engineers nor scientists in the pre-war pattern—were bred and nurtured there. The working contexts fostered newly valued expertise and supported jurisdictional claims, as suggested by Andrew Abbott.[1] But the viability of the new breeds remained disputed through the war and beyond. Disputes between established disciplines—primarily physics and chemical engineering—initially ceded no territory to upstart experts.

Promotion of such claims was supported, though, by participants having a foot in both camps. The administrators of the wartime atomic energy projects were sympathetic to an engineering perspective and, significantly, were often senior academics, too. Electrical engineer Vannevar Bush's first employment, for example, had been as a test supervisor at General Electric. With James Conant, a chemist, he dominated the academic hierarchy of Massachusetts' principal engineering institutions by the beginning of the war. Similarly Arthur Compton, scientific director of the Met Lab, had worked in his early career as a research engineer at Westinghouse.[2] The members of the British MAUD Committee—all academic physicists—consulted closely with ICI in generating their report, and the company became closely integrated with the subsequent Tube Alloys directorate. Building on such backgrounds and contexts, science and engineering were melded to explore a new discipline.

Even so, difficulties of merging distinct intellectual traditions rapidly became apparent to all the participants. Within the Anglo-Canadian project, the role of ICI proved abrasive for some scientists, but was largely papered-over by a pragmatic working relationship inspired by the working culture of the National Research Council. In the USA, three cultures jostled: that of university scientists, of Army engineers, and of industrial managers. The University of Chicago Met Lab, organized some four months before Army involvement, had already divided the identified problems according to scientific discipline. According to their categorization, 'pure science' would direct 'applied science' and be rendered into operational form by 'engineering'—a pecking order of early-twentieth-century provenance. Unsurprisingly, then, Arthur Compton recalled that he was faced by 'a near

---

[1] Abbott, Andrew D., *The System of Professions: An Essay on the Division of Expert Labor* (Chicago: University of Chicago Press, 1988), pp. 92–3.

[2] On the company's adaptation to new fields, see Lassman, Thomas C., 'Industrial research transformed: Edward Condon at the Westinghouse Electric and Manufacturing Company, 1935–1942', *Technology and Culture* 44 (2003): 306–39.

rebellion' by the project scientists when it was first suggested that engineers would direct the project. Compton characterized those most urgently pressing the case for keeping the control in the hands of the scientists as merely unversed in the ways of American industry, 'chiefly the younger research men who had had little contact with industrial organizations' and foreign scientists such as Leo Szilárd and Enrico Fermi, who mistrusted large corporations to act in the national interest.[3]

The urgent development of new technical expertise introduced novel strains. The engineering companies were to participate not merely in production management, but in conceptual design, too, and even in contributing to investigation of new scientific phenomena.[4] This crossing of boundaries and challenging of existing hierarchies gave participants their first hints of a field-in-the-making.

The wariness was not one-sided, though. In the same vein, Glenn Seaborg, head of the Met Lab plutonium chemistry group, characterized General Groves as unfamiliar with 'scientists, their motivations, and their way of thinking'.[5] And the engineers of Du Pont—the company deemed to have the industrial capacity and the closest match to the required work—were equally unenthusiastic about collaboration with the Met Lab. Even a decade later, Compton could not downplay the difficulties of integrating these disparate groups. He recalled that although 'the degree of harmony in the co-operation between the scientists and the Army at this stage was rather surprising', for the scientists and engineers 'there remained a state of tension that caused continual concern'.[6] The transition from an academic pursuit to industrial product—and from scientists to redefined engineers tentatively creating a new discipline—consequently was contested and difficult for all parties.

3.1 The Anglo-Canadian project and the challenges of technical collaboration

The new environments were rocky and unpromising. The wartime atomic energy projects allocated technical workers to sites and production tasks in a way that fostered separate clusters of proficiency while impeding communications between them. In this terrain of compartmentalized and segregated intellectual cells, there was no natural centre; each site was on the periphery, according to one perspective or another. Each could, and did, contribute a unique version of know-how to the nascent subject of nuclear engineering. Let us begin, then, with the least explored contributors to the Manhattan Project: the specialists working in Canada.

While American piles and their specialists were to develop in near isolation (Section 3.2 below), the Anglo-Canadian Tube Alloys Project grew under distinct cultural influences. The National Research Council and ICI were central to the Anglo-Canadian exploration of

---

[3] Compton, Arthur Holly, *Atomic Quest: A Personal Narrative* (Oxford: Oxford University Press, 1956), pp. 109, 112.

[4] On comparison of the UK and USA, see Brown, John K., 'Design plans, working drawings, national styles: engineering practice in Great Britain and the United States, 1775–1945', *Technology and Culture* 41 (2000): 195–238.

[5] DOE, OpenNet, 'History of the activities of the Manhattan District Research Division, October 15, 1945—December 31, 1946', 31 December 1946, DOE ON NV0714682.

[6] Compton, Arthur Holly, *Atomic Quest: A Personal Narrative* (Oxford: Oxford University Press, 1956), pp. 105, 169.

the new field. As the first Canadian and British organizations involved with the new subject of atomic energy, they played a seminal role in determining the post-war British conceptualization of nuclear specialists and the orientation of Canadian research.

3.1.1 *The National Research Council and an engineering perspective*

The developing Canadian project was overseen and administered by a handful of individuals, most with engineering backgrounds. The national shaping, its personnel, and post-war ambitions were contingent on their past experience and interactions. Canada's Ministry of Munitions and Supply, headed by C. D. Howe (1886–1960), was formed in 1940 with a remit rather like that of Britain's Ministry of Supply, i.e. to oversee the production and delivery of armaments. Howe—formerly a professor of civil engineering—appointed Lesslie R. Thomson (1886–1958), a mechanical engineer and pre-war professor of civil and fuel engineering at two Canadian universities, as Comptroller and Secretary. The Ministry was approached by the British government in early 1942 about hosting aspects of the Tube Alloys Project. Howe consulted C. J. Mackenzie—also a senior academic and civil engineer before joining the National Research Council of Canada (NRC)—as the most senior scientific administrator to oversee the work. The Canadian government offered to pay all project costs except for salaries of its contributors from Britain, and in October 1942, Howe assigned Thomson to the project as administrator and liaison officer.

The personnel staffing Mackenzie's NRC—an organization with an established engineering affinity—were a collection of scientists, engineers, and technicians having a distinctly twentieth-century contour. This combination of specialists able to marry scientific research with economically valuable outcomes was becoming familiar in the national standards laboratories that appeared at the turn of the century—the Physikalisch Technische Reichsanstalt in Germany (1887), the National Physical Laboratory in Britain (1900), and the National Bureau of Standards in the USA (1901)—and in industrial laboratories such as those of General Electric, Bell, and Westinghouse in the USA, and GEC, British Thomson-Houston, and Metropolitan Vickers in Britain.[7]

In these environments, neither science nor engineering had a permanent ascendancy in the hierarchy of status and power: either could assume authority depending on the task at hand, a fluidity that was to prove amenable to the exploration of atomic energy. Mackenzie

---

[7] General Electric (1892), for example, opened the first corporate research laboratory in the USA in 1900; Du Pont established laboratories to develop new chemical products before the First World War. In the UK, British Thomson-Houston (1894), Metropolitan Vickers (1919), and Imperial Chemical Industries (1926)—each of them, like GE, founded by a series of mergers—became important exemplars of science-based industry. On industrial research, see Reich, L. K., *The Making of Industrial Research, Science and Business at GE and Bell* (New York: Cambridge University Press, 1985); Clayton, R. J. and J. Algar, *The GEC Research Laboratories 1919–1984* (London: Institution of Engineering and Technology, 1989); Reader, W. J., *Imperial Chemical Industries: A History* (Oxford: Oxford University Press, 1975). On national standards laboratories, see Cahan, David, *An Institute for an Empire: The Physikalisch-Technische Reichesanstalt 1871–1918* (Cambridge: Cambridge University Press, 1989); Cochrane, Rexmond C., *Measures for Progress: A History of the National Bureau of Standards* (Washington, DC: US Department of Commerce, 1966); Pyatt, Edward C., *The National Physical Laboratory: A History* (Bristol: Institute of Physics Publishing, 1982).

sought to avoid a superior–subordinate relationship between divisions and between technical workers as far as possible, noting that 'in classic organizational terms, the chart was kept flat and horizontal'. This intentional institutional culture was nurtured by Mackenzie and his successors. As summarized by the next NRC Director, chemist E. W. R. Steacie (1900–1962), 'the main thing to do is to develop a character and an atmosphere which distinguish the organization from all others'.[8]

The organization of the NRC advanced this professional profile. Neither of Mackenzie's predecessors—Henry M. Tory (1864–1947), an educator responsible for founding several Canadian universities, and General Andrew McNaughton (1887–1966), trained as an engineer and responsible for modernizing the Canadian Army—had engaged in research themselves, but had a record of promoting it in other contexts. Founded in 1916 to serve wartime ends, the National Research Council had played primarily an advisory role for the Canadian government through the 1920s. During that period, it informed policy by surveying Canadian research strengths, funding committees to investigate specific problems, and providing university fellowships. Although founded ostensibly to coordinate and promote scientific and industrial research, it had soon discovered that there was little to promote. Canada, it seemed, was a country focused on raw materials and application: when the NRC was founded there were an estimated fifty 'pure' scientists in the country engaged in research.[9]

But under Tory, who oversaw the completion of the Council's first research laboratories in Ottawa in 1932, the organization began to conduct more applied research of its own in the national interest. His 'Temple of Science' was populated with some fifty scientists and engineers during the Depression years, but working almost exclusively on industrial problems. As a Crown Corporation, the NRC was not a Department of Government, and its staff organization did not follow Civil Service norms. Indeed, some staff worked without pay during the Depression. Nor was the organization modelled closely on universities. The first reorganization in 1929 created a Division of Physics and Engineering, lumping together fields that in other institutions were held more firmly apart.[10] Under McNaughton's four-year direction to 1939, staff numbers doubled and working culture remained firmly oriented

---

[8] Doern, G. Bruce, *Science and Politics in Canada* (Montreal: McGill University Press, 1972), quotations pp. 42 and 43.

[9] On the history of the NRC, see Eggleston, Wilfred, *National Research in Canada: The NRC, 1916–1966* (Toronto: Clarke, Irwin, 1978); Middleton, W. E. Knowles, *Physics at the National Research Council of Canada, 1929–1952* (Waterloo, Ont.: Wilfrid Laurier University Press, 1979); and, Dandurand, Louise, *The Politicization of Basic Science in Canada: NRC's Role, 1945–1976*, PhD thesis, History, University of Toronto (1982).

[10] Redhead, Paul A., 'The National Research Council's impact on Canadian physics', *Physics in Canada* 56 (2000): 109–21. This union was enduring. The Division became Physics and Electrical Engineering in 1936 and Physics alone only in 1947. It split into separate divisions of Physics and Applied Physics in 1952, when all NRC activities relating to atomic energy were taken over by the new Atomic Energy of Canada Limited, but such disciplinary titles were abandoned in 1990 as too academic. See also Doern, G. Bruce, *Science and Politics in Canada* (Montreal: McGill University Press, 1972) and Gingras, Yves, 'The institutionalization of scientific research in Canadian universities: the case of physics', *Canadian Historical Review* 67 (1986): 181–94.

towards scientific–engineering collaborations. Through the wartime and immediate post-war years, Mackenzie was to expand the organization's staff to some two thousand, with a significant fraction eventually committed to the nuclear project. Throughout his tenure, he emphasized productivity, seeking to maintain 'the realistic view which all members of the staff here take. We all feel keenly that unless our endeavours produce equipment and findings... we will not be achieving our fundamental purpose'.[11]

Although he knew something of the British nuclear work from previous visits by John Cockcroft, Mackenzie first met Sir George Thomson, Chair of the MAUD Committee, and industrial engineer Wallace Akers (1888–1954), the Director of Tube Alloys, in February 1942.[12] He warmed to Akers immediately as an engineer like himself. His diary entries characterize Akers, the Research Director of ICI and now head of a British Government department, as 'an extraordinarily able and impressive man. He is sound scientifically and has a very pleasant personality and practical sense. He has extensive industrial experience also'.[13] The commonality of background was to play an important role in developing their joint perspective on the project and the nature of its workers.

3.1.2 *The Montreal laboratory and allied cultures*

During late 1942 working quarters for the first few members of the Anglo-Canadian group were accommodated in a Montreal house, but a more suitable location was found in an empty wing of the newly constructed hospital of the Université de Montreal, and occupied the following March. Over the following months the 'Montreal Laboratory' was populated with a growing number of workers. With them came a small stock of heavy water produced in Norway, obtained by the French group and transported with them to Cambridge in May 1940 after the fall of France.[14]

While often referred to as the 'Anglo-Canadian project', the Montreal Laboratory team was cosmopolitan, consisting of the French and Commonwealth scientists and ICI engineers, and equivalent Canadian personnel from the National Research Council of Canada and universities.[15] Some of the design engineers were recruited from the Central Register

---

[11] Mackenzie, C. J., *The Mackenzie–McNaughton Wartime Letters* (Toronto: University of Toronto Press, 1978), p. 78.

[12] Akers' obituaries, describing him as a physical chemist rather than engineer, scarcely mention his wartime MAUD Committee and Tube Alloys work [Lord Waverley and Alexander Fleck, 'Wallace Alan Akers. 1888–1954', *Biographical Memoirs of Fellows of the Royal Society* 2 (1956): 1–4; 'Obituary: Sir Wallace Alan Akers', *Atomic Scientists' Journal* 4 (1955): 257–8].

[13] 19 February 1942 diary, Mackenzie, C. J., LAC MG30-B122 Vol. 1.

[14] That original stock appears never to have been used even in Canada; although a heavy-water reactor was eventually constructed at the end of the war, the supply had been sent for reprocessing to the only heavy-water plant then existing in North America, in Trail, British Columbia, and was eventually repatriated to France in 1948. See Smith, Pat, 'On the trail of Drum T-7', *AECL Inter-Comm*, 2 June 1989.

[15] By October 1944, when John Cockcroft took control of the Anglo-Canadian project, there were some 140 graduate scientists and engineers, half being Canadian, with 22 British, seven New Zealanders, and four French [Cockcroft, John, 'Montreal staff', NA AB 1/278. See also Galbreath, Ross, 'The Rutherford connection: New Zealand scientists and the Manhattan and Montreal projects', *War in History* 2 (1995): 306–21].

in the UK, but Wallace Akers noted that 'the Canadians, who have been found for us, are of a very high standard indeed'.[16]

The Montreal-based workers had frustratingly little contact with their American counterparts until late 1943, during which time American expertise had developed rapidly. Without adequate supplies of uranium or heavy water, and isolated from experimental findings, workers at the Montreal Lab consequently developed a local Canadian variant of nuclear knowledge, devoting most of their effort to relatively sophisticated theoretical studies of heavy-water based reactor design.[17] But, as in the USA, initial tendencies were to adopt conventional clusterings of knowledge. As one member recalled, 'hierarchy prevailed, and the atmosphere was in some ways more military than academic'.[18]

And the project became steadily less academic. All of the administrative leaders of the project—C. D. Howe, C. J. Mackenzie, Lesslie Thomson, Wallace Akers, and the scientific leader of the team, Hans Halban—remained embroiled in the scientific, administrative, and political details of the Canadian project through the war. For each of them—all except Halban, an engineer by training—the tension between scientific research and engineering accomplishments was evident. They learned that the same realization was becoming apparent to the American administrators, too. Following a meeting with Akers, Mackenzie commiserated,

> I gather that things are not going so well in certain parts of the American programme as the groups of physicists, particularly at Chicago, do not realize the wisdom of calling in engineers when it comes to plant design etc. These high grade physicists sit around for hours discussing problems which are to be solved in the first chapter of elementary engineering texts. They are beginning to realize that in the States now and are beginning to correct it. In other of the projects engineers from the Kellogg Co. and the Du Pont Co. have been called in early in the game and these projects are going very well.[19]

Meeting James Conant of the American NDRC a few weeks later, Mackenzie picked up the same perspective more directly:

> Conant was quite concerned about the whole work and said that it was very difficult to get a sound opinion as to the merits of the various projects. He said that his difficulty was to get the opinion of a detached nuclear physicist...He agrees with Akers' contention that it is largely an engineering development or at least the major difficulties will be engineering...They have now a special committee investigating all the projects from an engineering standpoint. The subcommittee is really a group of Du Pont engineers.'[20]

In some respects, this profile of experts was better assimilated in Canada and the UK. Tube Alloys included the collection of engineers and scientists seconded to Canada and the USA,

---

[16] '1943 Canadian organization: personnel', NA AB 1/380.

[17] Williams, Michael M. R., 'The development of nuclear reactor theory in the Montreal Laboratory of the National Research Council of Canada (Division of Atomic Energy) 1943–1946', *Progress in Nuclear Energy* 36 (2000): 239–322.

[18] Wallace, Philip R., 'Atomic energy in Canada: personal recollections of the wartime years', *Physics in Canada* 56 (2000): 123–31; quotation p.126.

[19] 19 November 1942 diary, Mackenzie, C. J., LAC MG30-B122 Vol. 1. Kellogg had created a new company, Kellex, for the job; see 'K-25 Production Team', in: Hewlett, Richard G. and Jack M. Holl, *The New World, 1939–1946* (Berkeley: University of California Press, 1969), pp.120–3.

[20] 29 November 1942 diary, Mackenzie, C. J., LAC MG30-B122 Vol. 1.

but also research staff at British universities and ICI facilities, Canadian scientists and engineers in other provinces, and the National Research Council in Ottawa.

Like the NRC involvement, ICI's participation also provided a positive model of engineering organization. By late 1943 Tube Alloys managed 276 research workers: 30 at Birmingham University, 23 at Cambridge, 22 at Oxford, 10 at Liverpool, and 67 at the new Montreal Laboratory.[21] ICI staff, accounting for 93 of the total, remained intimately involved through its assistance to individual scientists associated with the project, and by September of that year ICI representatives had joined the Technical Sub-Committee.[22] As the largest and widest-ranging chemical manufacturer in the UK, ICI was involved in every aspect of the early developmental work. Via its fertilizer and synthetic products division at Billingham in the north-east of England, the General Chemicals and Alkali Divisions in the north-west of England and the Metals Division in the Midlands, the company was studying heavy-water production processes, and would later produce the chemicals and supervise the production of special membranes for the pilot American diffusion plant and uranium metal for the first test reactors.[23]

In the Montreal group, scientists gained the most prominent, but not necessarily the most influential, roles. From March 1943, forty UK-based professionals were based at the Montreal Lab, notably Head of Engineering, Ronald E. Newell (1905–); Head of Physics, Pierre Auger (1899–1903); Head of Theoretical Physics, George Placzek (1905–1955); and Head of Chemistry, F. A. Paneth (1887–1958). As Newell saw it it, the assembled group was constructing a new field of expertise:

> Owing to the unusual nature of the work the great majority of the additional staff did not have the full specialized knowledge necessary and a period of training was required. In the case of engineering, for example, this meant development of a completely new branch of engineering. Generally speaking, the engineering work is similar to that of chemical engineering and it was from this branch of engineering that most of the design staff have been obtained. However, a great deal of new knowledge had to be acquired by the engineers and to some extent this has also applied to the theoretical and experimental physics sections and the chemistry section.[24]

It is no coincidence that Newell identified the engineering as similar to his own background. This perceptual pigeon-holing was to be repeated by other administrators responsible for defining the nature of nuclear specialists through the 1950s. Newell's words paralleled the views of Crawford Greenewalt, his Du Pont counterpart, and echoed in the war-end summary of the American project:

> Evidently the operation of a full-scale plant of the type planned would require a large and highly skilled group of operators. Although Du Pont had a tremendous background of experience in the operation of various kinds of chemical plant, this was something new and it was evident that operating personnel would need special training.[25]

---

[21] '1942–1945 Staff, general', NA AB 1/246.

[22] Clark, R. W., *The Birth of the Bomb* (London: Phoenix, 1961), pp. 127–39, 155–8.

[23] Reader, W. J., *Imperial Chemical Industries: A History*, Vol. 2 (Oxford: Oxford University Press, 1975), pp. 287–96.

[24] Newell, R. E. to L. R. Thomson, letter, 11 November 1943, LAC MG30 E533 Vol. 1.

[25] Smyth, Henry D., 'A general account of the development of methods of using atomic energy for military purposes under the auspices of the United States Government 1940–1945', 1 July 1945, LAC MG30-E533 Vol. 1.

To Newell and Greenewalt, the new skills seemed closest to industrial chemistry because they concerned the manufacture of chemical products by industrial, albeit unfamiliar, processes. The importance of chemical separation, thermodynamics, and mathematical analysis to the project made the expertise of university trained chemical engineers particularly relevant to the solution of its development problems.[26] Yet as Newell noted, this engineering field generated *new knowledge* that would be *applied* to fundamental physics and chemistry. The new subject of atomic energy, he argued, was reversing conventional understandings of the linkage between science and engineering.

There was a hesitancy of scientists to merge in this interdisciplinary work, though. Working cultures and hierarchies were threatened. As Peter Hales has noted, compartmentalization 'was a means to redesignate scientists and engineers as workers'.[27] Absorbed in particular tasks, the members of the Montreal theoretical group tended to collaborate little, even with each other. And, interested in 'discussions of broader aspects of physics', they socialized separately from those focusing on 'the technical problems with which the project was preoccupied'.[28] Indeed, as one of the Canadian members recalled,

> We did not have seminars, nor general discussions. Self-confidence is hard to sustain in such circumstances. Secrecy is the main fetish in the wartime military culture.
>
> It was science in a closet. So we worked more or less in our separate corners. Under stress, we could not afford the luxury of seeing the broad picture, and became technicians in our separate cells.[29]

3.1.3 *Cooperation across borders*

Those cells developed impermeable walls. Security isolated the would-be specialists. Early in the project, there was relatively little for engineers to do at the Montreal Laboratory. Detailed design work, construction, and testing were in abeyance because, as Mackenzie had suspected, the Americans were reluctant to provide the necessary heavy water and uranium for Hans Halban's group to build a chain-reacting 'boiler' or pile. American (and Canadian) administrators continued to distrust the heterogeneous European participants at the Montreal Lab; as Mackenzie summarized it,

> the Montreal group...is really not an Anglo-Saxon group, and...they felt there was no guarantee that the various nationals – French, Austrian, Russian, Czecho-Slovakian, German, Italian, etc. could be guaranteed for any length of time. I think there is a great deal to be said for their point of view.[30]

---

[26] The technical requirements are discussed by Kuhn, James W., *Scientific and Managerial Manpower in Nuclear Industry* (New York: Columbia University Press, 1966), p. v. By contrast, the post-war Argonne National Laboratory found that merely 7% of its technical staff were chemical engineers, distributed between the Chemical Engineering, Reactor Engineering, and Metallurgy Divisions [Roberson, John H. to M. Carbon, letter, 3 November 1959, UI box 86]. This again reflects differing institutional goals and cultures.

[27] Hales, Peter B., *Atomic Spaces: Living on the Manhattan Project* (Urbana: University of Illinois Press, 1997), p. 118.

[28] Wallace, Philip R., 'Atomic energy in Canada: personal recollections of the wartime years', *Physics in Canada* 56 (2000): 123–31; quotation p. 127.

[29] Wallace, Philip R. to M. M. R. Williams, letter, 21 July 2000, SFJ collection.

[30] C. J. Mackenzie, 18 January 1943 diary, LAC MG30-B122 Vol. 1.

Groves, Conant, and Bush were equally uneasy about UK involvement in what they saw as an American development project. The commercial risks particularly exercised them: Wallace Akers, the Director of Tube Alloys, was a senior member of ICI staff; his deputy M. W. Perrin, senior engineers R. E. Newell and D. W. Ginns were also seconded ICI employees, and their company had promoted a potential post-war British nuclear industry in the MAUD report. As a result, the Montreal group found itself increasingly excluded from American information, with its members pleading for action by the Canadian administrators and pursuing increasingly arcane, but as yet unverifiable, theoretical studies of heavy-water reactor design.

Collaboration with the American programme improved but remained difficult. The information flow to the American teams was aided by a military liaison officer, except for a brief period in late 1943 when Mackenzie and Akers, frustrated at the stonewalling by Groves, Conant, and Bush, restricted the Montreal group from further scientific contact with their American cousins; the theoretical studies by the Montreal Laboratory were sometimes more sophisticated than those produced by the Met Lab in Chicago.[31] A direct discussion between Churchill and Roosevelt at Quebec in August 1943 led eventually to some relaxation of American restrictions. Nevertheless, as Mackenzie fumed a month later, General Groves was 'the dominant personality in the US group' and 'in effect a dictator' whose 'idea of collaboration seems to be to incorporate into the US project such sections of the British team as seem likely to promote a speedier and more certain realization of project'.[32]

Groves' decision-making was largely rubber-stamped by his industrial counterparts, though. When he consulted with Du Pont's Crawford Greenewalt to ask his advice 'as to the set up between the Canadians and Chicago', Greenewalt suggested to Groves that 'this was a matter of high policy but as a citizen I would think separate groups in both countries should carry on with visits between'.[33] In any event, James Chadwick, Mark Oliphant, Otto Frisch, and Rudolf Peierls soon were to work directly in the USA, and the parallel research in both countries led to the reassignment of a few members of the Montreal Lab—notably George Placzek, who assumed leadership of a Los Alamos theoretical group—to American labs in early 1945. The Anglo-Canadian participants were thus kept at arm's length but served as a labour pool for the American project, as Mackenzie had forecast.

The Anglo-Canadian decision in early 1944 to build a pile and chemical plants for heavy-water production and plutonium separation in Canada therefore was an effort to shift the centre of mass from the USA, and to transfer the project gradually from British to Canadian governance.[34] It also nurtured the ICI/NRC perspective on an appropriate focus for the engineering field.

---

[31] The liaison office was responsible for regulating the flow of information from the American project and acquiring information from the Anglo-Canadian project. Permitted information included Met Lab data on the physics and technology of their first heavy-water pile, but excluded all information on plutonium separation and the Hanford reactors. See Hewlett, Richard G. and Francis Duncan, *Nuclear Navy, 1946–1962* (Chicago: University of Chicago Press, 1974), pp. 283–4.

[32] 'Letters Re: Personnel—Organization at Chalk River', LAC RG77 Vol. 283, 24 September 1943.

[33] Greenewalt, Crawford H., 'Manhattan Project Diary, Vol. II', 1942–43, Hagley 1889, p. 333, 16 September 1943.

[34] '1944 Removal of Montreal Laboratory to UK', NA AB 1/149; '1946 Suggested re-organization of the Engineering Branch at Chalk River in the light of present and future responsibilities and the formation of a crown company to administer the Atomic Energy Project', NA AB 2/128.

At the new Chalk River site in southern Ontario, physicist John Cockcroft took over from Hans Halban the direction of the British/Canadian team to design the first reactor outside the USA. Halban's replacement had a reputation with both the Americans and British as a quietly efficient and tenacious administrator, having been Assistant Director of the Scientific Research for the Ministry of Supply and directing the Air Defence Research and Development Establishment (ADRDE) over the previous three years.

Cockcroft was well-suited to the emerging field. He had broad and relevant experience in nuclear physics with a flair for engineering. Like other senior administrators in the project, he had had earlier industrial experience, having worked as a student at British Thomson-Houston in Rugby. Starting university at Manchester as the First World War began and serving as a soldier during the war, he graduated as an electrical engineer and completed a College Apprenticeship at the Metropolitan-Vickers Company. Through its engineering apprenticeship scheme, Cockcroft completed MSc research on three problems in electrical power generation and analysis. Hoping to advance in the industry, he read mathematics at Cambridge and, after studying for a PhD at the Cavendish Laboratory there under Rutherford, demonstrated artificial disintegration of elements—atom smashing—with Ernest Walton.[35]

The independence of Cockcroft's Chalk River team from American work is noteworthy. The Canadian pile became operational in September 1945, four weeks after the Hiroshima uranium and Nagasaki plutonium bombs were dropped.[36] For the Anglo-Canadian group, focused on reactor theory, building the first Canadian reactor and proposing British developments in the months ahead, the use of the bombs themselves was marked by relatively little reaction. Cockcroft had spent the weeks preceding the Japanese events writing a scientific account of chain-reacting systems for Lord Cherwell, and the month before that had been devoted to a memo outlining post-war possibilities for generating power.[37] C. J. Mackenzie, on holiday during the bombings in early August, returned to his desk to pen congratulatory letters to Groves, Bush, Conant, and Chadwick.[38] And with Mackenzie's drafting, the Canadian government publicized this intense wartime effort within days of the news of the bombs on Japan.[39]

Despite such surface goodwill, however, this was clearly an unequal partnership. The American dominance of the Canadian project is illustrated by actions to shape the

---

[35] See Hartcup, Guy and T. E. Allibone, *Cockcroft and the Atom* (Bristol: Adam Hilger, 1984).

[36] For a detailed narrative of the American Manhattan project, see Rhodes, Richard, *The Making of the Atomic Bomb* (London: Simon & Schuster, 1986).

[37] Bauer, S. G. and J. Diamond, 'Note on piles for the production of useful power', memo, 4 June 1945, LAC RG77 Vol. 283; Cockcroft, John, 'The development of chain reacting systems', memo, 30 July 1945, LAC RG77 Vol. 283. Lord Cherwell, physicist F. A. Lindemann, was the chief wartime scientific advisor to Winston Churchill.

[38] E.g. 'the greatest scientific and technological achievement of all time' (Mackenzie to Conant, 14 August 1945) and 'it must have been a great thrill to see the experiment in the desert' (Mackenzie to Chadwick, 23 August 1945), in 'Letters Re: Personnel—Organization at Chalk River', LAC RG77 Vol. 283.

[39] Department of Reconstruction (Canada), 'Canada's role in atomic bomb drama', http://www.cns-snc.ca/history/history.html/1945Aug13PressReleasePart1, accessed 23 April 2009; Department of Reconstruction (Canada), 'Scientists who probed atomic secrets', http://www.cns-snc.ca/history/history.html/1945Aug13PressReleasePart2, accessed 23 April 2009.

Anglo-Canadian team during the last year of the war. Some eight months after Hans Halban's replacement by Cockcroft as project leader, General Groves complained of a trip he had made to liberated France in December 1944, during which he met with his colleague Frédéric Joliot. Head of their pre-war group and designated director of the new French atomic energy commission, Joliot was a devout communist. Halban's imperious and secretive manner had exacerbated his relations with his Anglo-Canadian co-workers as well as with Groves. The trip seemed to vindicate the decision to replace him as project leader, and also to warrant exclusion of the other French participants.

Groves was suspicious of later requests by Bertrand Goldschmidt (1912–2002), Lew Kowarski, and Jules Guéron (1907–1990) to visit France in April 1945 to discuss their eventual redeployment with Joliot in a post-war French nuclear institution.[40] In September 1945, Groves demanded that the remaining French members of the Montreal Lab/Chalk River team be immediately excluded from what was now a *post-war* project. With only token protest, C. J. Mackenzie complied, and the French departed by the end of the year.[41] James Chadwick, Head of the British Mission to the Manhattan Project, argued to Mackenzie that 'this consequence is inevitable' and that he was 'not prepared so seriously to prejudice our agreement on collaboration with the U.S. in order to relieve a temporary embarrassment'.[42] An unsigned memo from one of the senior Chalk River staff, however, complained bitterly about the loss of Goldschmidt, who had led their chemistry research:

> The morale of our chemists has always been adversely affected by the lack of exchange of information. We have been forced to do work which we know has been done already. The 'purging' of the leader of the group can hardly be expected to make for improved morale.
>
> In spite of the fact that the chemistry of 49 [i.e. plutonium] was well worked out, we were given little information [by American counterparts] other than a few vague hints from time to time. The result has been that we have been obliged with a group of about 40 men to do what the Americans did with several thousand…The position therefore is that the 'high command' refuse to give us help on 49 Chemistry, and as soon as we are well on the way to doing the job for ourselves insist on firing the man who has directed the work…
>
> Finally, I should like to point out that we have always been treated as the poor relation in this project, and I anticipate great difficulty in attracting good men to the project unless we can reach an international position which enables us to have some self respect.[43]

Jules Guéron, for his part, asked reasonably, if meekly, for 'clear cut indications as to the nature and extent of the secrecy regulations to which I am still committed', and what he would be able to take to the new French Commissariat à l'Énergie Atomique.[44] Given these

---

[40] Goldschmidt, as the last assistant of Marie Curie in 1933, was perhaps the only direct link in Canada at the time to the origins of nuclear science. Groves, mistrustful of joint activities with the Anglo-Canadian team, later boasted of having dragged his feet in requests for closer collaboration [Hewlett, Richard G. and Jack M. Holl, *The New World, 1939–1946* (Berkeley: University of California Press, 1969), p. 282].

[41] Mackenzie, C. J. to G. L. Groves, letter, 26 December 1945, LAC RG77 Vol. 283.

[42] Chadwick, James to C. J. Mackenzie, letter, 8 January 1946, LAC RG77 Vol. 283.

[43] Anonymous to C. J. Mackenzie, memo, 24 December 1945, LAC RG77 Vol. 283.

[44] Guéron, Jules to C. J. Mackenzie, memo, 13 December 1945, LAC RG77 Vol. 283. See also Hewlett, Richard G. and Jack M. Holl, *The New World, 1939–1946* (Berkeley: University of California Press, 1969), pp. 335–7.

divisions, the French programme was to remain isolated from work in America, Canada, and Britain after the war. But such security concerns were to remain problematic for Canadian workers, too, inhibiting both collaboration and independence through the 1950s.

This administratively decreed segregation helped to create a proto-profession along clearer national lines in Canada and the UK. In the USA, foreign nuclear workers were quick to adopt American citizenship, particularly in light of suspicions of their allegiances; Eugene Wigner (1937), Edward Teller (1941), Leo Szilárd (1943), and Enrico Fermi (1944) were prominent examples. Post-war contributions to nuclear research and development in Germany, on the other hand, were actively stifled by the prohibition of such activities by the allied powers.[45]

The episode surrounding the French scientists raises the issue of respect and comparative judgements of national contributions to the project, which were only hinted at publicly. For example, Arthur Compton, a decade later, proffered dismissively that 'Canada's principal contributions to the atomic project during the war were the mining of uranium ore in the Great Bear Lake region and the supplying of needed uranium materials', an echo of the nineteenth-century description of Canadians as 'hewers of wood and drawers of water'.[46] The on-again, off-again collaboration, soured by security concerns and mistrust about post-war commercialization, played its own part in creating distinct national identities for nuclear engineers. But operating at the fringes of Manhattan Project activities, the Montreal Laboratory and Chalk River had a relatively subdued influence on American conceptions of the new art of atomic energy.

3.2 Du Pont and the gestation of nuclear specialists

The USA provided a more fertile environment. Arguably the first engineering workers to emerge as self-recognized specialists were Du Pont and Met Lab employees who, from the late summer of 1942, began to investigate the production of plutonium on an industrial scale for eventual use in an atomic weapon.

Historian Richard Hewlett has provided an overview of how engineering followed closely on the heels of science during the Manhattan Project, but skirts the issue of disciplinary development. He argues that 'the atomic energy experience suggests...that direction of a technical project should at some point pass from the scientist to the engineer, even though both professions continue to be deeply involved in the work'.[47] Contesting conventional accounts—most originating with the Met Lab scientists catapulted to fame after the war—I would argue that the Du Pont Company was an equal player in defining the new

---

[45] In early 1946, Allied Control Law 25 decreed that research in nuclear physics and reactor construction could not be undertaken in Germany or Japan. Although the law was relaxed four years later, experimental research in the fields remained off-limits until 1955.

[46] Compton, Arthur Holly, *Atomic Quest: A Personal Narrative* (Oxford: Oxford University Press, 1956), p. 194; the quotation, from Joshua 9:21 and first attributed to Anthony Trollope [*North America*, Vol. I (1862)], referred alternately to Canadians in general and to French Canadians in particular.

[47] Hewlett, Richard G., 'Beginnings of development in nuclear technology', *Technology and Culture* 17 (1976): 465–78; quotation p. 477. See also Hewlett, Richard G. and Franciscus Duncan, *A History of the United States Atomic Energy Commission* (3 vols) (University Park, Penn.: Pennsylvania State University Press, 1969).

subject from its beginnings.[48] The role of Du Pont engineers in usurping the dominance of scientists in the young field is also a prime example of a jurisdictional skirmish that led, eventually, to a rebalancing of disciplinary hierarchies and a settlement in favour of engineering.[49]

Formed in 1802 as a gunpowder manufacturer, Du Pont had evolved to become an unusual American industrial company from the turn of the twentieth century, when it was sold to three grandsons of its founder. Establishing two of the first industrial laboratories in the USA and increasingly reliant on research, the company introduced successful products such as neoprene, nylon, Lucite, and Teflon, and the insecticide phenothiazine during the 1930s. It was also atypical in the American context in being relatively self-sufficient: it was normal practice for its plants to be designed, constructed, and operated in-house, rather than being subcontracted.[50] This combination of attributes, hinting at research competence, industrial accountability, and efficient security, made the firm attractive to General Groves.

Du Pont responsibility for plutonium production began tentatively and rather unwillingly. The company dipped its toes during the autumn of 1942, when the American government asked it to serve as a subcontractor to Stone & Webster, a Massachusetts-based consulting engineering company specializing in electrical engineering. Earlier that year, Stone & Webster had accepted responsibility for all aspects of uranium and plutonium fabrication. Since March, it had been subcontractor to the Standard Oil Development Company to work out preliminary designs and estimates for a U-235 separation plant using the centrifugal process. And that summer, the company's engineers under August C. Kline had begun collaborating with the scientists of Ernest Lawrence's Radiation Laboratory at the University of California to design efficient *calutrons* (electromagnetic mass spectrographs to separate uranium isotopes), a project closer to its industrial competences and involving the design of magnet and vacuum systems. At the same time, other Stone & Webster engineers were liaising with the Met Lab in the preliminary design of the pilot-plant and production plutonium piles. As later summarized by the company's promotional literature, 'two of our principal engineers were sent on a tour of the country's laboratories to be

---

[48] Other American companies joined the field of reactor design after the war, notably via the project to build nuclear submarines and nuclear-powered aircraft for the US Navy and Air Force, respectively. Firms included Westinghouse and General Electric (discussed below); Fairchild, and Electric Boat. See Hewlett, Richard G. and Francis Duncan, *Nuclear Navy, 1946–1962* (Chicago: University of Chicago Press, 1974); Schwartz, Stephen I. (ed.), *Atomic Audit: the Costs and Consequences of U.S. Nuclear Weapons since 1940* (Washington, DC, 1998). Du Pont was unique, however, in overseeing development of an integrated workforce ranging from designers and managers to reactor operators.

[49] Abbott, Andrew D., *The System of Professions: An Essay on the Division of Expert Labor* (Chicago: University of Chicago Press, 1988), pp. 69–79, 90–8.

[50] Thayer, Harry, *Management of the Hanford Engineer Works in World War II: How the Corps, Du Pont and the Metallurgical Laboratory fast tracked the original plutonium works* (Reston, VA: American Society of Civil Engineers Press, 1996). See also Chandler, Alfred D. and Stephen Salsbury, *Pierre S. Du Pont and the Making of the Modern Corporation* (New York: Harper & Row, 1971); Hounshell, David A. and John Kenly Smith, *Science and Corporate Strategy: Du Pont R & D, 1902–1980* (Cambridge: Cambridge University Press, 1988); Hughes, Thomas Parke, *American Genesis: A Century of Invention and Technological Enthusiasm, 1870–1970* (New York: Viking, 1989).

instructed by leading scientists in the principles of nuclear physics and the processes selected for producing fissionable materials,' but 'in the autumn of 1942, it was clear that the work would have to be subdivided'.[51]

With General Groves' organization of the Manhattan Engineer District, it had become evident that the various tasks were too physically separated and technically complicated for Stone & Webster alone to complete the various phases at the required pace. Du Pont was requested to play a more central role, liaising directly with the University of Chicago's Metallurgical Laboratory to design, construct, and operate the necessary plutonium plants. Crawford H. Greenewalt (1902–1993), a chemical engineer and Du Pont manager, was assigned to act as liaison with the Met Lab scientists. The result was important for gestating a new discipline: a single vertically integrated company assumed responsibility for collaborating with university scientists, and was soon building a burgeoning team of engineers to develop atomic piles as industrial factories.

The company proficiency was, however, firmly rooted in chemistry. As industrial chemists, Du Pont managers understood their task to be the development of the 'Chicago process' for manufacturing a scarce chemical product, plutonium.[52] They were briefed on the two other competing processes, which sought to manufacture uranium rather than plutonium: the 'California process' (electromagnetic separation of U-235 from U-238, pursued by Lawrence's Radiation Laboratory) and the 'New York process' (electromagnetic separation of uranium by thermal diffusion, pursued by Harold Urey's team at Columbia University). This collection of industrial processes, they reasoned, would be evaluated, refined, and optimized according to conventional management techniques.

Yet Arthur Compton's Metallurgical Laboratory had already conceived the problem in different terms. Compton accumulated and consolidated scientific staff with an eye to physics and engineering support. His team had already assumed responsibility for pursuing most of the problems associated with producing plutonium. In effect, they had defined the scope and required expertise as a hierarchy of subdisciplines. Compton's adopted remit included the subjects to be known after the war as reactor engineering and radioisotope separation, but initially categorizing them as extensions of the distinct fields of physics and chemistry.

The Met Lab staff consequently had been assigned to discipline-oriented solutions. Three groups of physicists were attacking the problem of designing a chain reactor; a separate chemistry division studied plutonium chemistry and uranium purification on a laboratory scale; and a production division, led by physicists Eugene Wigner and John Wheeler, was working on design studies for scaling up this knowledge for production plants. Compton included some engineers among his staff: the most senior

---

[51] Stone & Webster Inc., *Engineering for Atomic Power: The Role of Stone and Webster in Nuclear Development* (New York: Stone & Webster, 1957), p. 6. The M. W. Kellogg Co., another large American chemical engineering firm, was given responsibility for developing and operating the K-25 facilities at Oak Ridge that separated uranium isotopes by gaseous diffusion, a task for which it spawned a new wholly owned firm, Kellex. It, too, was reliant on close liaison with scientists, this time with Harold Urey's team at Columbia University.

[52] Greenewalt, Crawford H., 'Stine's memorandum', 27 November 1942, Hagley 1957 Series I Box 1.

was a chemical engineer from industry, Thomas V. Moore of Humble Oil and Refining Co., whom he appointed chief engineer of the Lab and Head of an Engineering Council (renamed the Technical Council in July 1942) to advise the Lab on plutonium production processes. The Council worked in collaboration with the pre-existing Nuclear Physics, Chemistry and Theory Divisions. At the end of June, Charles C. Cooper, the Assistant Director of Du Pont's Technical Division, was seconded for chemical engineering, and John P. Howe, from General Electric, for electrical engineering. A month later, Moore added Miles C. Leverett from Humble, for general engineering.[53] That autumn, Compton also appointed Martin D. Whitaker, former Chair of the Physics Department of New York University who had worked with Enrico Fermi, to head pile design and construction. And by October, the University of Chicago group began to explain their preliminary plans to nine senior Du Pont staff. Thus the Met Lab was unusually proactive in incorporating engineering skills into its academic group, while positioning them as subordinates in the hierarchy.

From the standpoint of the engineers, however, the lab was still decidedly academic in orientation. The problems and potential solutions of plutonium production were at that point entirely theoretical: no chain reaction had yet been demonstrated, and plutonium itself was available only in microgram quantities too small to reveal the chemical properties of the element or appropriate methods of separating it from other fission products following its creation in the chain reactor. As to the chain reactor itself, suitable materials, designs, and properties were then far too uncertain to make a production unit, or even 'pilot plant' (the term conventionally employed by chemists for a small-scale unit to test the feasibility of the industrial process). Expectations, then, were that the Met Lab scientists would direct the project with industrial assistance.

3.2.1 *Engineering and science: Negotiating the atomic pile*

The forced marriage between the Met Lab and Du Pont was crucial in generating the hybrid expertise of American nuclear engineering. It eventually allowed engineers to adopt an equal, rather than subordinate, role in defining the intellectual terrain of the new subject and the direction of applied research. This power shift, played out over the last three years of the war, began with a clash of technical cultures and ended with well-embedded occupational sites, entrenched intellectual clusters, and increasingly clear-cut disciplinary expectations.

Yet this production project was unlike Du Pont's typical industrial problems and required new knowledge. There were no specialists in large-scale nuclear processes akin to industrial chemists or chemical engineers to offer a suitable background for the peculiar new

---

[53] Miles C. Leverett (1911–2001), originally trained and practising as a chemical engineer, continued a career in atomic energy after the war as Technical Director of General Electric's Nuclear Energy for Propulsion of Aircraft (NEPA) project from the late 1940s. At Oak Ridge, he later recruited metallurgists for research on reactor materials, and led the team designing the Materials Testing Reactor. Described as a 'nuclear engineer' in newspapers as early as 1951, he also was active in organizing the profession, chairing the Nuclear Engineering Committee of the American Institute of Chemical Engineers (AIChE) after the war, and serving as sixth President of the American Nuclear Society, 1960–61.

problems to be faced. And the science itself lacked adequate detail on which to found a factory-scale process.[54]

Key Du Pont staff had no experience in nuclear physics but were quick to absorb it. Greenewalt devoted the first several pages of his notebook to a terse crib-sheet of nuclear facts relating to chain reactions, e.g.:

> ...Neutrons are of atomic number 0 and atomic weight 1 – they have no electrons associated with them and are of neutral charge...The liberated neutrons are caught by other U atoms and a chain reaction starts...the present project stems from that finding...Liberated energy interesting both for explosive and power depending on ability to control. Has been found that control can be exercised to give either explosion or smooth release of power. Group at Columbia investigating military use for atomic power. Looked expensive, and investigating committee appointed to determine whether or not this procedure looked practical....Are using C (graphite) in place of D_2O – C not as good but is available...could probably get power by using C as slow downer [i.e. moderator]. Investigating committee decided that military use of power was not sufficiently important for highest priority.[55]

The notes, an outcome of the first meeting between a nine-man Du Pont delegation and Met Lab physicists Compton and Norman Hilberry in early November 1942, communicated the ongoing activities and planning by the University of Chicago. Over the previous months, the Met Lab workers had been building roughly cubical piles ten feet on a side, and had so far found that the assembly multiplied the radioactivity of an artificial source of neutrons; this could be damped by control rods made of a steel alloy rich in boron, or by plated cadmium tubes. These experiments were being scaled up, with plans to demonstrate the feasibility of a self-sustaining reaction by the end of the year.

Anticipating a successful outcome, the Met Lab scientists were already planning a series of chain-reactors. Greenewalt learned that the demonstration reactor was to be built in the Argonne Forest near Chicago, and would comprise a spherical latticework some 25–28 feet in diameter of sixty tons of uranium oxide and 6 tons of uranium metal dispersed in 550 tons of purified graphite. A second pile, the pilot plant or 'semi-works', would be constructed at another site ('X'). Identical in form to the first, the pile would, however, allow the chain reaction to run more quickly for weeks at a time at a sustained power of some 1000 kW. To enable this, the scientists envisaged the pile being encased in a copper or steel jacket and filled with helium gas to conduct heat away and to contain radioactive gases and airborne material; the exterior would be sprayed with water and air-cooled by fans.[56]

---

[54] The industrialization of radioactivity did have pre-war exemplars, but they incorporated few relevant technoscientific aspects. Radium, for example, had been isolated from crude ores, and employed in medical treatments and commercial applications such as luminous paints; see Landa, Edward, *Buried Treasure To Buried Waste: The Rise And Fall Of The Radium Industry* (Golden, Colo.: Colorado School of Mines, 1987); Roqué, Xavier, 'Marie Curie and the radium industry: a preliminary sketch', *History and Technology* 13 (1997): 267–91; Rentetzi, Maria, 'The U.S. radium industry: industrial in-house research and the commercialization of science', *Minerva* 46 (2008): 437–62.

[55] Greenewalt, Crawford H., 'Manhattan Project Diary, Vol. I', 1942, Hagley 1889, pp. 2–3 (4–6 November 1942).

[56] For a complementary overview, see Hewlett, Richard G. and Jack M. Holl, *The New World, 1939–1946* (Berkeley: University of California Press, 1969), pp. 174–80.

The handful of Du Pont managers were disconcerted by signs of engineering naïveté. The Met Lab physicists conceived plutonium production with the planned Pile 2 as an impractical batch process rather than as a chemist's preferred continuous-flow pilot plant. After a month of operation, the pile was to be shut down for a week or two to allow radioactivity to decay sufficiently before being dismantled to recover the uranium fuel for chemical separation to yield a few grams of transmuted plutonium.[57]

Difficult chemical production processes were familiar to the company, but this batch process was unlike chemical plants involving toxic intermediate products and involved novel problems never faced by industrial chemists or, indeed, physicists. The most disturbing of these were radioactive dangers. The pile would generate some 100,000 curies of radioactivity—the equivalent of 100 kilograms of radium, far in excess of the world's current supply.[58] Compton consequently had set up a 'health group' alongside his pile designers to investigate the potential hazards and handling methods. Greenewalt's notes summarized the Met Lab's current understandings: that 'humans can take 0.1 "R" unit [roentgen] per day'; that operators would be rotated 'so as not to build up too much without recuperation'; and that a 'balance has been made between uranium danger and rapidity of construction'. Thus the project's health issues were settled by a combination of empirical estimates and compromise.[59] In the process, the seeds of a new specialism—health physics—were sown.

As Greenewalt also recorded, the Met Lab scheme for obtaining plutonium stretched these tentative safety limits. The plan was to tear down the accurately machined and meticulously positioned graphite 'bricks' by first remotely operating an elevator carrying a fine circular saw to cut away the jacket of copper. Next, a boom crane, also on elevator platforms, would be employed to latch onto each graphite block using vacuum-operated suction cups, depositing them on a moving belt to carry them away. Similarly, the uranium blocks would be picked up and dropped into a hopper and a tank, where they would be chemically processed. The process, still undetailed, would involve remote agitation, ventilation to disperse radioactive gases, and subsequent jacketing to isolate the results.

---

[57] Instead, the X-10 pilot pile, eventually built at Clinton, Tennessee, was operated until 1963 and later designated a National Historic Landmark.

[58] Comparing the world stock of pre-war radium with the mass of U^{235}, plutonium, and other fissile materials by the early 1950s [Parker, M. M., 'Radiation protection in the atomic energy industry: a ten-year review', draft conference presentation for meeting of the Radiological Society of North America, Los Angeles, 7 November 1954, DOE DDRS D198195963].

[59] Greenewalt, Crawford H., 'Manhattan Project Diary, Vol. I', 1942, Hagley 1889, p. 28. The *curie* (Ci) is a unit of radioactivity defined as 3.7×10^{10} decays per second, which is approximately the decay rate of one gram of the isotope radium-226. The *roentgen* is a measure of ionizing radiation, i.e. the ability of a radioactive particle or energetic ray to liberate electrons from atoms. One roentgen is defined as the radiation that will liberate one electrostatic unit (esu) of charge from a one cubic centimetre volume of dry air, which amounts to about 2.1×10^9 pairs of ions. To complicate matters further, the biological effect of radiation depends on its type (X- or gamma-ray, alpha or beta particle, or neutron, for example); the 'roentgen equivalent (in) man' (rem) unit was later introduced to account for this. Outside the USA, the most common unit of radiation quantity is the *sievert* (Sv), equal to 100 rem. The current US regulatory limit for occupational exposure is 5 rem/yr, equivalent to 5 roentgen of X-, gamma- or beta-radiation or 0.5 R of neutron radiation.

The oxide, carbon dust, and resulting solution of uranium and other fission products were radioactive and a hazard. The one predicted fission product that was volatile—radioactive xenon gas—was particularly dangerous: a Du Pont report later summarized that 'each cubic foot of radioactive gas must be diluted with as many as 100 trillion cubic feet of atmospheric air to assure safe conditions'.[60] Met Lab plans called for each separation plant to be four miles from other facilities, with the 'plant village for workers to be ten miles to windward of the nearest stack', a main highway or railroad ten miles away, and a town at least 20 miles from the nearest stack.[61]

The pile itself was to be surrounded by six feet of concrete; the agitation tanks for the liquid-based separation processes would be surrounded by seven feet of concrete. Radioactive solutions would be steam-siphoned into subsequent tanks for addition of hydrofluoric acid, precipitation, and centrifuging. The left-over soup of products, still highly radioactive, would be stored safely in shielded tanks. All these operations and chemical processes would be performed remotely, with vision restricted to a long water-filled observation hole and periscope, with a remotely adjustable mirror at its far end. The nuclear physicists, accustomed to laboratory research on radioactivity since the turn of the century, were relatively sanguine about such scaled-up activities.

But the reliance on such novel and untested schemes was unsettling for the Du Pont delegation. And, as they learned to their consternation, the Met Lab, with Thomas Moore and Miles C. Leverett of Humble Oil leading the design team, was already well advanced in planning a third, production pile. On the positive side, and in common with conventional chemical engineering practice, this third pile would enable a cyclical, rather than a one-time batch, production process—so avoiding some of the wastage and complexity of laboriously building and then remotely disassembling the pile. On the negative side, this pile, too, had to be unloaded periodically of its uranium fuel using a procedure that appeared complicated and dangerous.

The proposed Pile 3 would generate some one hundred times more heat (100,000 kW) than Pile 2 and a comparable increase in plutonium and radiation production. To do so, it was to be actively cooled by recirculating helium gas filtered to remove radioactive particles (Figure 3.1). The reactor proper would be cylindrical, allowing fuel rods to be withdrawn vertically by cranes and handled remotely for chemical processing.

Emphasizing the fluidity of design parameters, a Met Lab group surrounding Eugene Wigner, Gale Young, and Leo Szilárd also described plans for a distinctly different fourth pile design. Under consideration for scarcely a few weeks when Du Pont came on the scene, Pile 4 was even more speculative than the plans for Pile 3. It would be cylindrical and use hollow uranium tubes having an aluminium lining. The individual uranium tubes would be cooled by a flowing liquid—possibly water, liquid bismuth or a diphenylamine compound. The Du Pont engineers recognized that this design freedom also multiplied potential problems. Water had relatively high neutron absorption and would lower the reactivity of the pile, and thus its

---

[60] '100 Area facilities and operations', September 1945, Hagley 1957 Series III Box 58 folder 5. Subsequent experimental tests using oil smoke confirmed the potential dangers from inadequately dispersed stack emissions [Hales, Peter B., *Atomic Spaces: Living on the Manhattan Project* (Urbana: University of Illinois Press, 1997), pp. 144–50]. On the long-term concerns at the Hanford production reactor site, see Gerber, Michele S., *On the Home Front: The Cold War Legacy of the Hanford Nuclear Site* (Lincoln: University of Nebraska Press, 1997), Chapter 4.

[61] Greenewalt Diary, Vol. I, p. 101.

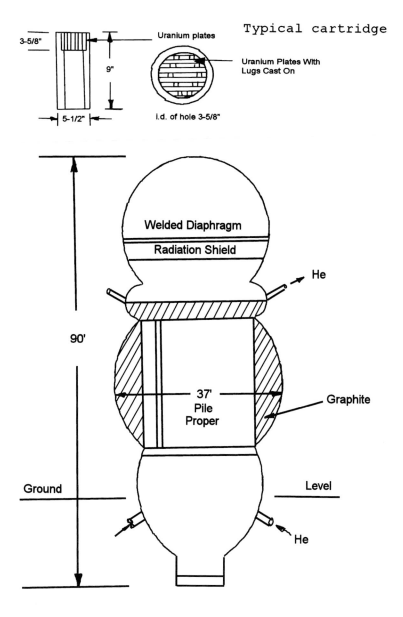

Fig. 3.1 Proposed Met Lab Pile 3, summer 1942.[62] (Courtesy of the Hagley Museum and Library; redrawn by S. F. Johnston.)

[62] Based on figure in Greenewalt, Crawford H., 'Manhattan Project Diary Vol. I', 1942, Hagley 1889, p. 32.

ability to transmute plutonium; it also reacted corrosively with uranium, requiring the fuel to be encased in a protective sheath of aluminium. On the other hand, bismuth, a greyish-white liquid metal at the reactor operating temperature, might be affected by high irradiation in unpredictable ways. And the diphenyl compound, a white crystalline hydrocarbon melting at 160 F and having good heat conduction, could polymerize or carbonize if overheated—clogging the tubes, reducing coolant flow, and potentially causing a thermal explosion.

Faced with the ambitious goals, misplaced confidence, and lack of engineering rigour in these plans, the Du Pont engineers were both unimpressed and increasingly concerned. In a private meeting the next day, they were divided on the feasibility of the pile designs, both for construction and operation. Tom C. Gary, Manager of Du Pont's Design Division, and F. W. Pardee Jr, the Supervising Engineer of its Engineering Department, felt the Met Lab's planned production reactor had neglected design flaws: the spherical top shield, for example, was impractical, and the scheme for loading and discharging uranium that would preserve a radioactive seal 'had practically no chance of working…We question in our own minds whether it is even possible to build this unit—much less operate it—as designed at present'. Vice President Charles Stine doubted that it could work at all, and would require at least two years to engineer.

The group was more impressed by the Met Lab's ideas about a chain reactor that would use heavy water as a moderator—at least in terms of its mechanical design and operating feasibility—and felt it had been prematurely rejected merely because no adequate supply of heavy water was currently available.[63] They returned to Arthur Compton with a series of points based on their engineering experience:

1. Step up U metal supply to the point that ore is bottleneck.
2. Study fabrication of U.
3. (a) commit for D_2O plant or plants
   (b) Physicists and engineers to design heavy water pile
4. Greater emphasis on engineering in existing pile program.[64]

In short, the Du Pont team placed their short-term confidence in U-235 separation rather than plutonium production and argued that a heavy-water pile *designed collaboratively between physicists and engineers* was the option most likely of success.

After further company discussions over the following days, Greenewalt summarized the group consensus: the Met Lab schedule, planning for full-scale production by mid-1943, was optimistic by probably a year, and that Du Pont should undertake the job. Even so, they were pessimistic about the engineering problems and did not consider assuming accountability for the science: Gary's view was that 'from engineering viewpoint if Du Pont takes on whole job both development and engineering but not theory (which is a physicist's job), the chances of mechanical success is perhaps 2/1'.[65] In the process of assigning

---

[63] The Du Pont contingent had learned a little from Compton of Hans Halban's interest in a heavy-water chain reactor, and heard that he was setting up a laboratory in Montreal where, by February 1943, Halban, with engineers, would be performing physical experiments and studying the problem of heat removal. Greenewalt Diary 2 February 1942, Vol. I, p. 104. See also Section 3.2.

[64] Greenewalt Diary Vol. I, pp. 45, 67–8.

[65] Greenewalt Diary Vol. I, p. 65.

responsibility, they shifted the balance of power in the new field, moving it away from precariously scaled-up science to a nascent engineering discipline constructed from scratch.

3.2.2 Wilmington engineers versus Chicago scientists

Nevertheless, given the sober engineering views, the expectation was that this unlikely new field would be nipped in the bud. The Du Pont delegation met with General Groves on 10 November 1942 and, a week later, Groves set up a committee of three Du Pont men to carry out a final review of Manhattan Project plans under Professor Warren K. Lewis (1882–1975) of the Massachusetts Institute of Technology.[66] With Lewis, a seminal American chemical engineer, the Du Pont committee members visited the university scientists at Columbia, Berkeley, and Chicago in late 1942 to assess ideas for gaseous diffusion, electromagnetic separation, and atomic piles, respectively. In effect, the disciplines of industrial chemistry and chemical engineering were applied to judge the feasibility of physicists' ideas, a significant placement for the later trajectory of nuclear engineering. The Lewis Committee judged gaseous diffusion—the original MAUD Committee recommendation, pursued separately at Columbia—to be the most likely of success, but both electromagnetic separation and the Met Lab's atomic piles generated strong reservations.[67]

Stine and Greenewalt, who served on the Lewis Committee, subsequently met with James Conant, Lawrence, and Groves, stating their judgement even more strongly than they had made it to Compton. Greenewalt summarized their strongly discouraging message as follows:

> ...the chances of ultimate success, leaving time out of the picture, were "fair" from a technical viewpoint only, and that in the view of the Du Pont Committee, the chances of ultimate success were much poorer from the commercial viewpoint [i.e. in producing the specified quantity of plutonium: '1 chance in 2 of technical success and 1 in 3 or 4 of commercial success', according to Greenewalt's private notes]. Dr Stine stated that in view of his thirty-five years of experience in chemical development work, he felt that the chances for successful reduction to practice of the Chicago process were somewhere between one in ten and one in a hundred [and]...viewing the Chicago project merely as the proposed reduction to practice of a piece of fundamental research, he felt that because of the complete lack of engineering data and because no pilot plant experiments had been made up to this time, the Du Pont Company would, in ordinary circumstances, not touch a project of this nature.
>
> The grave physiological hazards involved were emphasized and it was stated that, except for patriotic reasons, the Du Pont Company was opposed to having any connection with the attempt to construct and operate plants intended to produce X-10 [i.e. plutonium] according to the Chicago process.

Stine further emphasized that such development was likely to prove exceedingly expensive and the effort might very well turn out to be abortive. He stressed the desirability of a second line of defence in pile design and reiterated the merits of a heavy water pile, the project

---

[66] On Lewis, see Furter, William F. (ed.), *History of Chemical Engineering* (Pittsburgh: American Chemical Society, 1980) and Reynolds, Terry S., *Seventy-Five Years of Progress: A History of The American Institute of Chemical Engineers 1908–1983* (New York: AIChE, 1983).

[67] On physicists' colonization of electrical engineering, see Kevles, Daniel, *The Physicists: The History of a Scientific Community in Modern America* (New York: Knopf, 1977).

just then being started by the new Anglo-Canadian group. Greenewalt records that General Groves, Conant, and Stine withdrew from the Committee at that point for a private discussion, in which 'Conant stated to Dr Stine and General Groves that what Dr Stine had said about his lack of faith in the feasibility of reduction to practice of the Chicago process accorded exactly with Dr Conant's own opinion and with Dr Conant's doubts as to the feasibility of developing a manufacturing process along the lines of the Chicago project'.[68]

Nevertheless, as the varying estimates of success suggest, the Du Pont members continued to reassess their engineering conservatism in light of persistent scientific enthusiasm. When Arthur Compton took Crawford Greenewalt to the University of Chicago's West Stands on 2 December, where Fermi's prototype pile was approaching its first 'criticality', or operating point at which a self-sustaining nuclear reaction occurs, Greenewalt was swept along by the moment. He later described Fermi as 'cool as a cucumber—much more than his associates, who were excited or a little bit scared'. As an engineer, he noted a particular amalgam of observations: how the neutron count 'kept increasing by leaps and bounds'; how the pile 'responds beautifully to control devices'; and how the neutron concentration in the room during the final 'power flash' 'rose well above the tolerable limit and the gamma radiation to just above the tolerable limit' for a daily human exposure. Overall, he recorded, it was 'much better than expected. It was for me a thrilling experience'.[69]

Buoyed up by the demonstration, the Du Pont Executive Committee decided two weeks later to take on the Chicago project 'lock stock and barrel, or in other words design, construction and operation'.[70] They set up a new Division of the Explosives Department, to be known as 'TNX'. Edward B. Yancey (1888–1948), the General Manager of the Explosives Department, had the greatest responsibility for the project, but delegated much of it to Roger Williams (1890–1978), a chemical engineer who, with Greenewalt, had been a member of the Lewis Committee a few months earlier. Crawford Greenewalt's role as liaison between the Chicago Met Lab and Du Pont headquarters in Wilmington, Delaware, was formalized by his appointment as Technical Director of the new TNX Division. He met and explained the project to a small group of senior Du Pont engineers on Boxing Day 1942, recording their surprise and shock: 'one of them said, "It sounds like what Buck Rogers reads about when he reads". That's a fair characterization of this particular bit of fantasy. Lord help us'.[71]

Within days of Du Pont's signing of the contract, Greenewalt expressed his worry about the relationship between the Chicago scientists and Wilmington engineers. General Groves had wanted him to serve as Compton's executive assistant; Greenewalt opted instead to 'stay in the Du Pont setup' but agreed to spend most of his time in Chicago 'to "watch" Compton and see to it that the research went in a way that would provide the right technical information at the right time'. He realized, however, that he had 'no authority over the

[68] Greenewalt, Crawford H., 'Stine's memorandum', 27 November 1942, Hagley 1957 Series I Box 1; Greenewalt, Crawford H., 'Manhattan Project Diary, Vol. I', 1942, Hagley 1889, pp. 88–91.

[69] Greenewalt Diary Vol. I, pp. 111–14.

[70] Greenewalt, Crawford H., 'Manhattan Project Diary, Vol. II', 1942–43, Hagley 1889, p. 2.

[71] Greenewalt Diary Vol. II, p. 5, 26 December 1942. See also Thayer, Harry, *Management of the Hanford Engineer Works in World War II: How the Corps, DuPont and the Metallurgical Laboratory Fast Tracked the Original Plutonium Works* (Reston, VA: American Society of Civil Engineers Press, 1996).

Chicago crowd—but am to see to it by diplomacy and pleading that they do the right things at the right time and don't chase too many butterflies'. Fortunately, Compton seemed relieved with the liaison relationship, because Greenewalt knew he 'couldn't successfully "boss" the physicists; this can only be done by Compton, for whom they all have the greatest respect'.[72]

The disciplinary hierarchy understood in Chicago would need to be renegotiated. Just after Christmas, Greenewalt met with Compton and Norman Hilberry, Compton's most senior administrative colleague, to discuss general organization of tasks. He noted with concern that Compton 'has particular ideas as to the difference between "scientific" and "industrial" research. I started some missionary work to convince him that the difference was more of terminology than actuality but will have to do more'. Compton presented an organization chart identifying a head of 'developmental engineering' over which Greenewalt hesitated, arguing that the Chicago engineering group should be small and consulting rather than experimental, and that such a large project required more supervision.

Just as worryingly, that afternoon he discussed the design of the second, pilot-plant pile with Met Lab and Du Pont personnel. He was exasperated to find their ideas 'not at all worked out. Much confusion as to who is to head the work—no dove tailing of [plutonium] separation with pile operation—in fact no organization at all'.

Greenewalt also spoke privately to Compton's senior engineers, Thomas Moore and Miles Leverett, and learned that Moore was unhappy with his position in the Chicago structure, feeling that his engineering personnel were not properly used and too much under the domination of the physicists. Greenewalt suggested that he draw up a programme of work for the production pile and that, with Du Pont, they set up his staff to do the necessary engineering and development work. A couple of days later he floated the idea to his Wilmington superiors of transferring the Met Lab engineers to Du Pont administration. Although they gave no commitment, they agreed that Du Pont must lead the engineering. This was to be a colonization of engineering perspectives into a scientific stronghold: 'I believe', he recorded privately, 'we must infiltrate pile design in spite of the fact we aren't very welcome'.[73]

Within weeks, the Du Pont staff consequently began to make firm decisions to manage what they considered an undisciplined and poorly founded industrial project. Du Pont policy was for its own staff to prepare detailed engineering designs once the basic design data had been transferred for an industrial process, and by January 1943, things were looking clearer to the Du Pont team. The completion of Fermi's demonstration experiment in December had yielded an unexpectedly high value for the pile reactivity k, showing that even a large air- or water-cooled graphite pile would likely work as a chain reactor. Water was unappealing because of anticipated corrosion problems, which would be difficult to correct in a factory-sized radioactive assembly. And helium cooling seemed worst of all, with its unsolved problems in loading and unloading the uranium fuel, and the need for an air-tight enclosure and efficient blowers to circulate helium, which was in short supply during wartime. Roger Williams correspondingly ordered his Wilmington pile designers to

---

[72] Greenewalt Diary Vol. II, pp. 3–4.
[73] Greenewalt Diary Vol. II', pp. 5–12; quotations on pp. 5, 6, and 7.

plan for an air-cooled graphite and uranium pile for the next stage of development, the pilot plant to generate a few grams of plutonium.

Even more peremptorily, Greenewalt made choices affecting not just pile design, but locations, too. Supported by General Groves, Arthur Compton, and the Chicago Area Engineer, Captain Peterson, he argued that Fermi's pile must be moved from the university grounds to a new location in the nearby Argonne Forest, a safe distance from a population centre. He also directed that the second, pilot-plant pile could not be built at Argonne because of its much larger operating power—a thousand kilowatts rather than a handful—and possibility of catastrophic radiation release too near to Chicago. Greenewalt worried, too, about how a pile might fail inadvertently or by sabotage. Physicist John Wheeler, one of Compton's team, had calculated that if the control rods were pulled fully out, uranium metal could vaporize and deposit a lethal concentration of radioactivity over an area of five-mile radius, far too large for any feasible Army evacuation plan. Similar concerns were to determine the choice of Canada's first reactors.[74] Instead, the pile would be built at site 'X'—in Clinton, Tennessee, a few miles from the uranium separation facilities at Oak Ridge being planned for the Manhattan Project. Greenewalt argued that Argonne be devoted primarily to physics and Clinton to production.[75] This single siting decision had profound consequences for the fissioning of know-how: from that moment on, American nuclear engineering expertise would be divided between Chicago/Argonne and Oak Ridge, Tennessee—a decision having effects lasting through the century.

Such bold choices were made cautiously, though. Greenewalt was concerned that he had publicly usurped Compton's authority, making overt the new egalitarian partnership envisaged by the engineers. He mused that 'it will strain relations at Chicago since decision to move to site X was made in face of Compton's statement that Argonne was safe and on the basis solely of Wheeler's calculation. Milk is spilt and can only hope for the best'.[76]

As he forecast, the new arrangements aggrieved many of the Met Lab physicists. Norman Hilberry was gloomy about 'the Army's decision to move Fermi to Argonne' and the group working on the Argonne pilot plant threatened to resign if relocated to the backwaters of Tennessee. And Compton himself was upset that the Argonne site could not be used for both physics research and pilot-plant production.[77] In early February 1943, General Groves and Greenewalt met with the Met Lab group leaders to assuage the growing engineering–science rivalry. Greenewalt was able to advance his perspective even further,

---

[74] 'It must be pointed out that the Pilot Plant we propose to construct will have an output of approximately 10 times that of the American plant at X [i.e. Clinton, Oak Ridge]. Furthermore, our proposed plant will be water-cooled and its stability is uncertain under some conditions. For these reasons, the scientists and engineers of the National Research Council are of the definite opinion that, with the present state of knowledge, the plant should not be located closer than 4 miles to the village' [Bolton, B. K. to C. J. Mackenzie, letter, 18 August 1944, Ottawa, Ontario, LAC RG77 Vol. 283]. See also Bothwell, Robert, *Nucleus: The History of Atomic Energy of Canada Limited* (Toronto: University of Toronto Press, 1988); Kinsey, Freda, 'Life at Chalk River', *Atomic Scientists' Journal* 3 (1953): 18–23; Krenz, Kim, *Deep Waters: The Ottawa River and Canada's Nuclear Adventure* (Montreal: McGill-Queen's University Press, 2004).

[75] Greenewalt Diary Vol. II, pp. 20, 73–5.

[76] Greenewalt Diary Vol. II, p. 23, 8 January 1943.

[77] Greenewalt Diary Vol. II, p. 85.

casting the scientists as educators for the new field and as junior members of the production team. Compton argued that if the pilot plant was theirs the Met Lab would need to take a greater share in its design and operation. Greenewalt responded that this was exactly what was expected, but that 'operator training and process confirmation would have to go along with whatever research they decided to do'. In effect, the roles of scientists and engineers would have to be more closely—perhaps indistinguishably—combined. He reported the atmosphere as 'a bit tense'.[78]

The situation required active brokering. Among those who promoted collaboration was John Wheeler. Having already attempted design studies for plant, he worked with Du Pont, acting as a go-between for some of the more recalcitrant scientists such as Eugene Wigner (Figure 3.2). Yet the situation was not reducible simply to engineering versus science: more subtle hierarchical aspects of the Du Pont involvement rankled, too. Its engineering conservatism, combined with near-arrogance about project management, jarred with the more pragmatic experimental culture operating at the Met Lab. Wigner, leader of the Met Lab's Theory group and a physicist with chemical engineering training, felt that his team's expertise was essential to every stage to guide the process of designing, constructing, testing, and operating a chain reactor to generate plutonium. As discussed by Abbott, such contestations over methodology, competence, and authority are more likely in such heterogeneous environments.[79] Showing not a little arrogance of his own, Wigner later recalled:

> I was educated as a chemical engineer, which came in very well, because I learned a lot of chemistry and that was very useful... There are many, many things they were in error about because they were not at all familiar with nuclear reactions... Du Pont had to learn so many little things which we knew already... We objected, at least I strongly objected, to the Du Pont Company, which had no knowledge at all of nuclear physics, and very little knowledge of the other engineering problems... Well, eventually [it] worked, and Du Pont learned a lot. They did not close their eyes to facts of physics or engineering. [But] I thought it could have been done much cheaper, and much faster, much faster.[80]

Wigner made his initial misgivings clear to Greenewalt, suggesting over a fraught lunch that Du Pont was too 'diversified' to do a good job on design and construction. Stating his 'keen sense of responsibility' for the proposed water-cooled production piles (and their likely failure under Du Pont's management), Wigner wanted to resign 'rather than carry this burden'. Positions were entrenched. Over the following weeks, subsequent meetings with Wigner led Greenewalt to record privately that he was 'not too hopeful that he will ever see things our way'.[81]

Leo Szilárd, acting as a peripatetic consultant cum critic in the Met Lab, proved to be a chronic irritation to Du Pont managers. Greenewalt recorded arguments on the transfer of information, and doubts about his sincerity. By January 1943, Szilárd was threatening to

---

[78] Greenewalt Diary Vol. II, p. 88, 11 February 1943. See also Hewlett, Richard G. and Jack M. Holl, *The New World, 1939–1946* (Berkeley: University of California Press, 1969), pp.186–8.

[79] Abbott, Andrew D., *The System of Professions: An Essay on the Division of Expert Labor* (Chicago: University of Chicago Press, 1988).

[80] Sanger, S. L. and Robert W. Mull, *Hanford and the Bomb: An Oral History of World War II* (Seattle: Living History Press, 1989), p.16.

[81] Greenewalt Diary Vol. II, pp. 154 and 168, 20, and 31 March 1943.

Fig. 3.2 Eugene Wigner. (Courtesy of Argonne National Laboratory.)

resign from the project to 'file patent applications', but offering to stay 'if his salary is raised to somewhere between Wigner and Fermi'. A shrewd bargainer, General Groves recommended paying him his raise with the proviso that the Office of Scientific Research and Development gained all rights to his inventions before and after he joined the project.[82]

Szilárd and Enrico Fermi were equally keen about the importance of close contact between scientists and engineers even in design details, since some small point might violate physical principles. Fermi appeared satisfied with Greenewalt's assurances that he and colleagues would liaise closely, but Wigner and Szilárd remained doubtful. To Compton, Greenewalt recommended segregation: Du Pont engineers would stay in Wilmington, Chicago researchers would stay in Chicago, and his Technical Division would be the 'leg men' responsible for liaison and transfer of information.[83]

Greenewalt asserted industrial influence over the Manhattan Project in other respects, too. As Du Pont had been consulted in late 1942 about its assessment of the various schemes for generating fissile materials, Greenewalt had been granted full access to information and research resources. He consulted with not only Arthur Compton, but also Robert Oppenheimer at Los Alamos, Harold Urey at Columbia, Ernest Lawrence at the University of California's Radiation Laboratory, and General Groves himself. As early as January 1943, he discussed personnel for Los Alamos with Oppenheimer, urging him not to take Fermi from the Met Lab project, but recommending others there who could go if absolutely necessary. And Greenewalt's advice convinced Groves to continue to segregate the Anglo-Canadian workers from American reactor developments, and to poach them as needed for the

---

[82] Greenewalt Diary Vol. II, p. 73.
[83] Greenewalt Diary Vol. II, pp. 21–2.

American work. Such actions shaped not only wartime research careers, but also the profile of post-war clusters of expertise.

On one point of early planning, Du Pont broke with tradition. Besides controlling the design process, Du Pont corporate culture was also distinctive in that the company normally designed, constructed, and often operated its own industrial plants. Most other American chemical companies of the period subcontracted extensively. As a result, Du Pont engineers were not just a transitory nuisance for the Met Lab scientists; they would be in for the long haul. Greenewalt, however, was sensitive to the hazards of operating the Clinton pile, and also the engineers' lack of knowledge about the process itself. Instead, as described above, he proposed that the Clinton pile be managed and operated by Met Lab personnel, who would also be able to train Du Pont employees for the eventual large plutonium production pile to be built later. On the other hand, the large production piles, to be built at Hanford, in Eastern Washington State, would be operated directly by Du Pont personnel. As he observed, 'we get what we want and duck liability.... The plant gives a wonderful opportunity for pilot plant testing and later for operator training and instruction. This must be pushed'.[84] Like the siting choice itself, this was another crucial decision: it seeded the remit of the Clinton site. Not only would it be populated with a mixture of personnel having scientific provenance and Du Pont operations experience, but they would become responsible for training a wave of others in the new art. It also offset the growing dominance of Du Pont in the plutonium project, offering a genuine collaboration with the Chicago scientists in a new and expanding domain.

Through the first half of 1943, Greenewalt held numerous meetings with senior Manhattan Project staff and was beginning to wax optimistic on piles and their possibilities. He had discussed pile design for the Clinton site with Wigner and a junior colleague, Alvin Weinberg; he met with Fermi, Hans Halban, and others about heavy-water piles, and even mooted to Yancey, the head of the Explosives Department, the post-war possibilities for the '49' (plutonium) project for Du Pont. Greenewalt further mused about heavy-water piles as primary power sources at Du Pont chemical plants, and envisaged the design of small mobile units for resale (in turn, Yancey promised to look into the possibilities for a non-exclusive license to manufacture such power units as a government reward for their wartime efforts on the project). And in June, Greenewalt attended a celebratory dinner with Arthur Compton and others to celebrate the six-month anniversary of the first fission, with both talking expansively of peaceful applications such as power generation and radioactive tracers for organic chemistry, the use of the bomb for post-war peace-keeping, and the inevitability of government sponsorship. Months after reluctantly entering the project, then, engineers were developing a distinct vision for the future of atomic energy.[85]

Nevertheless, current realities kept Du Pont engineers busy. Relationships with Eugene Wigner, representing the scientific staff at the Met Lab, remained difficult. In August, Greenewalt defended Du Pont's activities to the 'P-9' Committee (yet another of the aliases used for plutonium), which had been exploring causes of dissatisfaction among the younger

---

[84] Greenewalt Diary Vol. II, p. 13.

[85] Greenewalt Diary Vol. II, pp. 88, 134, 239, 11 February 1942, 6 March and 6 June 1943; Smith, Alice Kimball, *A Peril and a Hope: The Scientists' Movement in America, 1945–47* (Chicago: University of Chicago Press, 1965), p. 19.

scientists. Wigner had written to Compton complaining of Du Pont's 'reluctance to cooperate'. Greenewalt used the opportunity to again argue for engineering parity in the new field. He responded with accounts of the 'ridiculous situation' of being 'turned down' in their efforts to get theoretical physicists to collaborate with them, but that '*all* our pile designs had been reviewed with Wigner's men from the very beginning and cooperation had been satisfactory'.[86] Some two weeks later a policy meeting settled the Met Lab's precise responsibilities in reviewing prints for the Hanford production piles, and involving all the pile designers: Wigner's men, Walter Zinn (working with Fermi), Miles Leverett, and Charles Cooper.

The uneasy collaboration began to generate more cooperative perspectives. The Clinton pile, designed principally by physicist Alvin Weinberg, another of Wigner's assistants, impressed the Du Pont engineers as much as the Met Lab scientists. Responding to the early Du Pont critiques, the design was the first engineered pile in existence. Fermi's 'CP-1' at the University of Chicago, the original pile to achieve a self-sustaining reaction, had been detail-designed and procured by Stone & Webster, the engineering company associated with the Met Lab before Du Pont's intervention, but it had been largely planned and fully assembled by Met Lab staff. The Clinton site was different. The pile itself, as Crawford Greenewalt observed, was meticulously planned and assembled. Touring the plant's construction, he recorded his satisfaction at its impressive attention to detail: 'Saw pile lay-up—going smoothly and with great precision. At 54th layer (out of 73) the surface was about 10 mils off of a perfectly flat surface, graphite machining giving a tolerance of $+0.002$'.[87]

Begun in late April 1943, the massive pile—incorporating 700 tons of accurately machined graphite—was started up in early November. Its four-storey building incorporated novel features that would serve as the template for future sites. It was mated with the Hot Laboratory Canal, a concrete L-shaped structure connecting it to the 'canyon', where chemical processing of the uranium rods would be carried out with remote manipulator apparatus. The cells in this long, thick-walled building were of standardized size and facilities to allow for the still unsettled processes of uranium slug dissolving, plutonium extraction, and decontamination—a design that was to be reproduced in all chemical separation facilities thereafter in the USA and beyond.

3.2.3 *Training the first experts*

As the Clinton pile came into operation and the first Hanford pile neared completion, there was a growing need to train operating staff in these new domains. The new skills demanded for this environment had to be refined, codified, and disseminated—an essential feature for grounding a new discipline. The original training in nuclear engineering began (and was later to be revived and long remain) at Clinton, the locus of the first genuine scientific and engineering collaboration. In addition to the training that supervisory and technical personnel received as a result of a succession of assignments, the Clinton Laboratories set up a training school for supervisory operating staff. The instructional staff, built around a

---

[86] Greenewalt Diary Vol. II, pp. 310–11. 13 August 1943.
[87] Greenewalt Diary Vol. II, p. 342. 21 September 1943.

nucleus of scientists and technicians brought from the Met Lab, provided them with several months' training before assignment to the Hanford Plant. The school operated between January and September 1944, reaching a peak enrolment of 116 in late spring. The first trainees were University of Chicago employees hired to operate the Clinton pilot plant. They were followed by hundreds of Du Pont employees destined to work in the Hanford plants, including 183 senior personnel scheduled to become supervisors of major Hanford operations. Another typical trainee group was Met Lab employees, who were given instruction in developing instruments for measuring radioactivity, before going to Hanford to build and operate the system installed in the production plants. Formal training of other supervisory personnel and of operators, maintenance, and power personnel took place at Hanford.[88]

For the operating staff, training was more specialized and restricted. The first employees on site received on-the-job training for the start up of the Test Pile, with later training on newly available radiation detection instruments. Like the other activities at Hanford, the training was intense: organized in groups of eight or ten, a full eight hours per day, six days each week, were spent in lectures and demonstrations. The original training course was laid out to cover a period of about four weeks, since the supervisors did not expect that a high degree of interest could be sustained for a longer period.

Defining the syllabus meant codifying the knowledge needed by newly identified specialists. As the supervisor's record noted,

> The general approach of the program was to break down each subject sufficiently to accelerate the learning process, with each step pointing toward the desired result of thorough understanding. Each group first was given lectures which illustrated the basic principles, and some of the possible uses. By demonstration and further lectures, the operating principles, method of calibration, and probable maintenance troubles were expanded upon.

But the large influx of operating personnel during that summer proceeded at such a high rate that the facilities of the training school were overtaxed. The available instructors at no time exceeded five men—mainly ex-Clinton personnel—and 'the individual load on each instructor was enormous'. Inevitably, training was further constrained. To reduce the training time for radiological instrument helpers, for example, it was decided to teach each one only a portion of the procedure, and to have enough devices in process so that one batch could follow another, going to each of the women in turn at successive stages. The first two of these operations were performed by the men who had been transferred from Chicago and Clinton; each of the others was taught to one of the four instrument helpers. With this division and cascade of labour, production rates increased slightly from week to week, mirroring the chain reaction itself.[89]

Such training imparted new skills to staff from Du Pont, the Met Lab, and other contractors. And, even more than other wartime projects, it became embroiled with security concerns. Du Pont managers, long familiar with military security on their many contracts, nevertheless disparaged the effects for this new field:

---

[88] 'Hanford Story, chap 17', Hagley 1957 Series V Box 50.

[89] Bowman, H. J., 'Instrument Department—Procurement and Training of Non-Exempt Personnel', memo to file, 4 August 1945, Hagley 1957 Series V Box 50 folder 16.

> Stringent security measures have governed all research, design, procurement, construction and operation procedure in order that the utmost secrecy may be maintained...Personnel has had to be divided into groups each possessing only such information as is necessary for its own use in connection with its own part of the work. Engineers and supervisors have had to be told what must be done but not why, stifling to a major extent the independence of thought and critical analysis of problems ordinarily considered essential to efficient prosecution of the work.[90]

Routine security measures had largely segregated design and construction employees, restricted the information given to each group. This was aided by the separate Du Pont divisions involved in each part of the project. But opposing this compartmentalization was the necessity of making quite complete information available to a large group about the pile operation processes so that they could 'contribute satisfactorily to the success of the operation and could provide adequately for the safety of the operating employees and the avoidance of catastrophe'. There were similar security concerns in the need to give the necessary legal notice to personnel about the existence of an Employees Benefit Fund which permitted special payments of $10,000 to employees who died or became permanently disabled as a result of their exposure to the special hazards of radioactivity. In an entirely untested engineering domain in which the dangers of intense radiation and potential explosion could not be estimated, the management militated against secrecy. As an anodyne Du Pont summary put it, 'all of this had to be accomplished without arousing undue alarm in the minds of employees which would result in lowered efficiency or resignations'.[91]

The conjunction of training with security inevitably constrained the understandings of the pile operations staff and shaped their occupational identity. This was evident in roles that mixed accountability with controlled knowledge; complex skills were distributed within a steep hierarchy of responsibilities. The new job of pile operator, for example, placed him in front of the main control panel where he would observe instruments which kept him constantly informed of power level and its minute deviations, control rod positions, and the identification of any of 28 conditions which might either automatically insert the control or safety rods into the pile or require investigation. The operator also had immediately at hand the switches for adjustment of control rod positions and for emergency manual insertion of the control and safety rods.[92] Bill Cease, a 'non-exempt' Du Pont employee who moved from a 'D' class operator responsible for monitoring the reactor's helium supply to an 'A' class 'pile operator', described the duties of managing the first production pile:

> At the control console, you keep the reactor at a certain level. It wasn't difficult but you couldn't go to sleep. You had a galvanometer in front of you, any minor movement of a control rod would move it. That measured the reactivity. You would look at it, and if the meter went to the left, you were losing power, so you would pull a rod. And if went to the right, you would poke a little back in...Somebody watched a panel with the couple of thousand process tubes. Each tube had a light and if a light came on something was wrong. That didn't happen very often. If a light came on, it could be a malfunction of the gauge, maybe an indication of a fuel element rupture and the tube was blocking up. If the tube blocked up, the pressure would go up. If a fuel element ruptured, you would shut down the reactor and try to push that tube. If you couldn't, you would call in maintenance. The training for reactor operation was on-the-job,

---

[90] 'History of TNX', Hagley 1957 Box 1 folder 1, p. 55.
[91] 'History of operations—Administrative', Hagley 1957 Box 1 folder 1, p. 3.
[92] '100 Area facilities and operations', September 1945, Hagley 1957 Series III Box 58 folder 5.

Fig. 3.3 Hanford control room operators.[93] (Courtesy of US Department of Energy, Declassified Document Retrieval System.)

there was no other way. I didn't have any idea what we were doing. It didn't bother me. I had a job, it was a war effort.[94]

One of his colleagues assigned to the third of the Hanford piles explained the duties of the first pile operators (Figure 3.3):

[93] DOE DDRS N1D0003425, dated 1954.
[94] Sanger, S. L. and Robert W. Mull, *Hanford and the Bomb: An Oral History of World War II* (Seattle: Living History Press, 1990), p. 139.

We didn't have the same type of classification you have now. Then, we did anything you were capable of doing. We had a lot of guidance, including from the world's most knowledgeable physicists. You didn't just go in there and run the darn thing up to full power. We followed procedures...The control supervisor was there to insure that you stayed awake and that he stayed awake...No less than three of us in a control room.[95]

3.2.4 Multiplying nuclear expertise

This exploration of the enmeshed histories of the Met Lab, Clinton, and Hanford is important to the narrative in two respects. First, the new technical problems identified, and the solutions employed, impressed the participants with the unexpected complexity and novelty of the new domain of pile technology, planting the seed for post-war identification of a new discipline but tempered by considerable pessimism concerning its practical future. And second, these participants became the first experts in atomic piles who, in American eyes at least, were recognized gradually as a distinct group, and quite unlike the industrial and scientific personnel attached to the Oak Ridge uranium separation factories and to the Los Alamos bomb-design group.

The first aspect—the design and construction challenges of the Clinton and Hanford piles—was problem enough, but the second—identifying technically competent staff—was another. The Clinton operation, under the responsibility of the Met Lab, was relatively straightforward: Greenewalt's battles in early 1943 had ensured that a considerable number of Met Lab personnel were relocated to Tennessee to test and operate the Clinton plant, and to train a series of operators for later transfer to Hanford. But the Clinton pilot plant was markedly different from the planned Hanford facilities, requiring much larger teams and refined experience. The Clinton pile was air-cooled; Hanford piles were water-cooled by the Columbia River. And Hanford was mammoth in scale of operation compared to the Clinton 'semi-works'. Three large piles were to operate in unison, each one hundred times the power of the Clinton pile. In this new industrial environment, Du Pont specialists argued for the senior role.

Greenewalt discussed personnel needs with Roger Williams, the Assistant Manager of the Explosives Division, and Arthur Compton. They agreed that, during the start-up of the first pile, at least four senior Chicago men would be needed, one for each shift of operation, and that other Chicago personnel should visit Hanford frequently, with Fermi 'on tap' to consult.[96]

Du Pont senior managers were even more involved. Norman Hilberry, the Met Lab's second-in-command, recalled the impact on Met Lab scientists:

the shift supervisors, when we got to operations, most of them were at least assistant plant managers, they hadn't done shift work for years. Du Pont said everything hangs on this. When we play, we play for keeps. They did, there wasn't any question. They put the best people they had on at every point.[97]

---

[95] Ibid., p. 141.

[96] Greenewalt, Crawford H., 'Manhattan Project Diary Vol. III', 1944–45, Hagley 1889, p. 18, 15 January 1944.

[97] Sanger, S. L. and Robert W. Mull, *Hanford and the Bomb: An Oral History of World War II* (Seattle: Living History Press, 1990), p. 26.

In spite of this example from the engineering counterparts, the Met Lab scientists were not eager to become industrial employees. The explanation of their Director, Samuel Allison, to Roger Williams two months later once again revealed the prickly status of relations between the two organizations:

> As you know, we are having considerable difficulty in finding physicists on our staff who are willing to enter the employ of the Du Pont Company and thus assist in the operation of the Hanford plant. In my opinion, the following attitudes are contributory factors in the situation:
> (1) The physicists consider their discovery of our process to be the greatest achievement of the science of physics, the culmination of three hundred years of research. They feel that an organization run by physicists should always have a prominent place in directing its development. It seems to them that in entering the Du Pont Company as individuals in subordinate positions, the prestige of physicists in our program is greatly diminished.
> (2) In the Metallurgical Laboratory, the efforts of the younger men were directed by physicists of world-wide reputations, whose great abilities had long been known to every member of the profession. They do not see any such men in prominent positions in the Hanford organization.
> (3) Most physicists, apparently in contrast to the chemists, greatly prefer an academic life to one on service to an industrial company. They feel that if peace should come while they are in the employ of the Du Pont Company at Hanford, their chances of obtaining an academic position from that location would be distinctly poorer than they would be at the Metallurgical Laboratory which has direct University connections.
> I am well aware that in the war emergency there are strong arguments contrary to the above which must be considered, but I am trying to report the feeling as it exists.[98]

Physicists, chemists, and engineers, he seemed to argue, had a clear sense of the intellectual hierarchy, and a natural preference for disciplinary segregation. Instead of a transfer of allegiances to an industrial organization, Allison consequently proposed a loan of the Chicago physicists. As recalled in 1959 by Nathan Sugarman, one of the young Met Lab scientists, such resistance had the further aims of eliminating the salary differential between Du Pont engineers and others 'at the same level' and, simultaneously, prevention of a post-war monopoly in atomic energy by private interests such as Du Pont.[99]

Hanford operating groups were defined rapidly. The highest proportions of scientists and professional engineers were put into its Technical Department. Most of its exempt salary personnel (those 'exempted' or 'deferred' from military service specifically because of their valuable job skills) had been transferred from other Du Pont departments or from Company-affiliated plants, with many having participated in development work at the Met Lab, Clinton Laboratories, or the Experimental Laboratory of the Grasselli Chemicals Department at Cleveland Ohio. As a rule, those who had not been assigned to one of these related laboratories received basic orientation at the Wilmington TNX office, involved with the more traditional industrial tasks of design and procurement liaison, organization planning, and preparation of technical manuals.

---

[98] Allison, Samuel B. to R. Williams, 13 March 1944, Chicago, Hagley 1889—Greenewalt Manhattan Project Diary, Vol. III.

[99] Smith, Alice Kimball, *A Peril and a Hope: The Scientists' Movement in America, 1945–47* (Chicago: University of Chicago Press, 1965), pp. 16–17. Indeed, Greenewalt had been instructed by General Groves 'not to discuss salaries with Chicago people since this brings about kicks as to disparity on salary matters' [Greenewalt, Crawford H., 'Manhattan Project Diary Vol. II', 1942–43, Hagley 1889, p. 298].

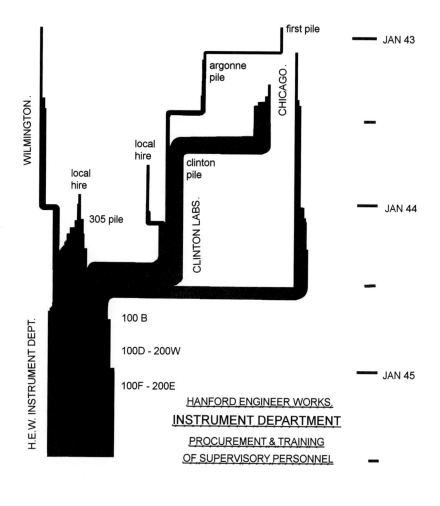

Fig. 3.4 Sources of technical personnel for the Hanford Instrument Department.
In this diagram reminiscent of a chemical engineer's flowsheet, the '305 Pile' was a small test pile at Hanford, and 100B, 100D, and 100F were the three wartime production piles there.[100] (Courtesy of the Hagley Museum and Library; redrawn by S. F. Johnston.)

The Instrument Department was the second locus for scientific–engineering personnel, and their responsibilities were unusually complex: designing, calibrating, and tending the sensors maintaining safety. The pile was instrumented to monitor some 5000 individual conditions at inaccessible locations, including the functioning of the helium gas system,

---

[100] Figure based on illustration in Overbeck, W. P., 'Instrument Department Functions and Organization to July 1, 1945', memo to file, 14 August 1945, Hagley 1957 Series V Box 50 folder 16, Exhibit A.

radioactivity in various parts of the building, and water pressure at the inlets of each of 2004 tubes, all linked to automatic control of the control rods.

Like the Technical Department, its exempt personnel came from senior staff at Wilmington's headquarters, a trickle directly from Argonne and the Met Lab, and a large contingent from the Clinton Lab, which also had sprung from Chicago. Indeed, half the new Department traced back to Arthur Compton's Met Lab. Their origins were carefully tracked by the Du Pont supervisor: the company was eager to ensure that experts were involved directly in identifying and solving start-up problems. His summary uniquely documented the otherwise hidden genealogy of scientists, engineers, and technicians that came to man the Hanford piles (Figure 3.4).

The allocation of staff illustrated the mind-sets of the wartime participants. The expediency of producing a viable pile to generate plutonium gave little time to reconsider labour categories or technical specialities. So, following Du Pont industrial practice, the organizational chart was headed by superintendents and supervisors—although they often occupied more junior roles than habitual. The variety of technical staff was considerably broader than typical Du Pont plants, but filled functional roles in conventional groupings of expertise. Hanford's Technical Department, for example, started in January 1944 with a physicist, a metallurgist, two analytical grade chemists, and a statistician-supervisor assembled to assist in starting a small Test Pile, which was to be used for evaluating uranium metal and graphite quality for the first large production pile. This 'Pile Engineering' Division provided technical assistance with problems in physics, engineering, and water/corrosion, and was intended to 'include a sizeable Physics group' so that 'a stable working unit of capable physicists might be an integral part of the plant organization'.[101] The Physics Section grew gradually in the early months of 1944, half of whom had been employees off the Metallurgical Laboratory. Seven had been engaged in the construction and operation of the experimental piles in Chicago, and the air-cooled unit at Clinton.[102] The addition of analytical chemists and chemical engineers expanded the group to 40 members by April, and by November 1944—when the first large pile was to be started—the Technical Department personnel had swelled to 1009, employing 32 Area and Assistant Area engineers, 13 physicists, 51 chemists, 12 physicians, 23 chemical engineers, 117 'assignment' engineers, 27 shift engineers, and 85 junior technologists, most of whom were men.[103] Lower-grade workers were rounded up from a wider variety of wartime sites.[104] This Hanford population, then, mapped onto existing disciplines and job histories.

[101] Marshall, John, 'Plant assistance (physics) to Jul 1 1945', 17 September 1945, Hagley 1957 Series III Box 58 folder 4.

[102] Bugbee, S. J., 'Technical Department Functions and Organization to 1 July 1945', memo to file, Hagley 1957 Series III Box 58 folder 4.

[103] 'Hanford organization', Hagley 1957 Series I Box 2 folder 8. A handful of women—sometimes physicists joining the organization with their physicist husbands—also came to Hanford for senior technical posts. Notable among them were Leona Marshall Libby, an associate of Fermi and the only woman present at the first sustained chain reaction [Libby, Leona Marshall, *The Uranium People* (New York: Crane, 1979)], and later an academic at the University of Chicago and New York University, and Jane Hamilton, later Assistant Director at Los Alamos. Some groups, such as Instrument Helpers assembling or repairing instruments, were entirely female.

[104] The 'non-exempt' or 'wage-rate' staff, responsible for maintenance and other operations, came from wider sources. Around 10% came directly from other Du Pont plants, and another 10% from local hire, but a quarter of them were transferred from Du Pont's Construction Division, and half came from the Denver Ordnance Plant.

The sole exception to these conventional staffing categories was health physics. The production of small quantities of plutonium at the Clinton Labs in early 1944 revealed its extreme biological toxicity. As a result, the Hanford staff was complemented by a health physicist and roentgenologist from Clinton.[105]

Despite this melting-pot environment, some divisions between Met Lab scientists and Du Pont engineers nonetheless remained. John Marshall recalled, 'we didn't have much time for a social life, except some with the other physicists'; David Hall observed that it was 'kind of cliquey, with not much intermingling with Du Pont. My friends were mostly academics, and the Du Pont engineers also tended to stick together socially.' But for Herbert L. Anderson, the close working association was a revelation:

> Friction between the Du Pont people and the Met Lab scientists was always a problem. It was handled beautifully by Greenewalt. His key engineers learned a lot, worked hard, asked lots of questions... Of course, there were always academic types who were very suspicious of big industries. It was a great eye-opener for me, an academic type with all the suspicions academics have about industry, to discover how competent these guys were. How important it was to have not only the competence but also the number of people who got involved in this planning, that really was necessary... So many details have to be followed, and only by having a huge engineering organization can you attend to that. I think the idea that Wigner had that he could manage that is just unrealistic.[106]

Given this need to allocate crucial responsibilities to a wide selection of employees, the administrators noted that Hanford's vulnerability to sabotage or enemy action increased as its operating period approached. In July 1944, they organized a separate Protection Department responsible for guarding, patrols, a site police force, and investigations. Within eight months, there were 832 employees in the Protection Department as compared with 662 employees in the Operating Department.[107] Arguably nuclear security itself, along with the unique threats and solutions it identified, became a new specialist occupation in the Manhattan Project. Its success is suggested by the surprise that most of the Hanford and Clinton scientists, engineers, and production workers felt when they eventually learned of the ultimate purpose of their new expertise: the atomic bombing of Nagasaki.[108]

3.2.5 The industrialization of plutonium: Challenges at Hanford

Months before that military revelation, though, came technical revelations aplenty. Hanford presented new problems, and on a scale never experienced. It is sobering to realize that Du Pont operated the Hanford piles for only nine wartime months.[109] During those months, fundamentally new physical phenomena—all presenting adverse effects for the operation—were discovered, explained, and counteracted. Science and engineering coexisted at the face of the big reactors. And new phenomena were being produced in mammoth factories,

---

[105] 'History of TNX', Hagley 1957 Box 1 folder 1, p. 54.

[106] Sanger, S. L. and Robert W. Mull, *Hanford and the Bomb: An Oral History of World War II* (Seattle: Living History Press, 1990), quotations pp. 129, 132, 63.

[107] 'History of operations—Administrative', Hagley 1957 Box 1 folder 1, p. 5.

[108] On the effects of this revelation on the experts and the wider public, see Chapter 8.

[109] Thereafter, its managers tried to extricate the company from further responsibilities, but passed it on to General Electric only in late 1946.

not laboratories: Hanford was the test-bed for a new engineering art that the Metallurgical and Clinton laboratories could not provide.

One of the best-known examples of the dangers latent in this unexplored terrain was the unexpected drop in power from the first Hanford reactor after its first few hours of operation in late September 1944 (Figure 3.5). After some days of bafflement, physicists John Wheeler and Enrico Fermi ascribed the problem to a product of fission—the isotope xenon-135—which proved to be a strong absorber of neutrons and so smothered the chain reaction. The reactor had been designed conservatively by Du Pont engineers, with an excess of channels through the graphite for uranium fuel rods. When fully used, these compensated for the 'xenon poisoning', driving the chain reaction on. Nevertheless, further experience with the reactor revealed that this poisoning was a temporary process: the xenon eventually is consumed by neutron absorption, and the chain reaction is correspondingly multiplied, leading to an increased power output.[110]

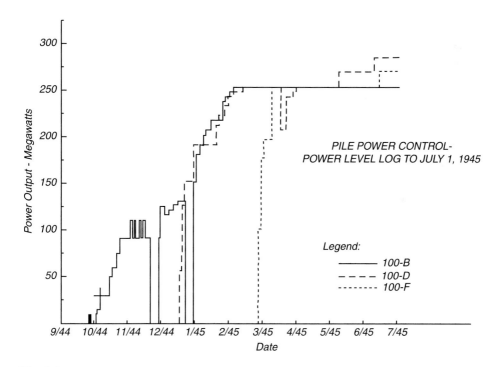

Fig. 3.5 Output power of the three Hanford piles.
The abortive start of Pile B, choked by xenon poisoning, is visible at the extreme left.[111] (Courtesy of the Hagley Museum and Library; redrawn by S. F. Johnston.)

---

[110] Gast, P. F. and C. W. J. Wende, 'Reactivity experience and control to July 1, 1945', memo to file, 20 September 1945, Hagley 1957 Series III Box 58 folder 4.

[111] Figure based on graph in Bugbee, S. J., 'Pile Operations Dept, Part II, 194446', memo to file, Hagley 1957 Series III Box 58 folder 2a; redrawn for clarity.

But this episode was merely the most public of a series of disturbing findings—and an oft-told tale that vaunted scientific insight and problem-solving. Most operational discoveries were less neatly incorporated into the rapidly growing skill-set of nuclear engineering. For example, the second worrying scientific discovery to be made at the Hanford Engineer Works, dubbed the 'Wigner disease', showed that the graphite bars comprising the reactor became distorted after intense irradiation by neutrons. After the first few months of operation, there was already perceptible bowing of the piles from expansion of the graphite structure. Fuel channels became bent, leading to stuck fuel elements and the potential for a ruptured element that could not be extracted.

But further engineering problems continued to surface. By August 1945, just days before the dropping of the atomic bombs on Japan, the supervisor of the piles, Du Pont engineer Hood Worthington, was warning of a series of other disturbing 'Wigner effects' on the graphite: a drop of 30% in thermal conductivity, and an increase of 30% in breaking strength. Both, he feared, could prove catastrophic. One of the primary design criteria of the massive piles was adequate removal of the heat generated by the absorption of energetic neutrons. A loss of conductivity—characterized by the Du Pont engineer as 'the most severe heat transfer problem ever encountered'—could generate overwhelming overheating; indeed, the recognized problem of corrosion or film deposition on the aluminium-coated uranium slugs had demanded extraordinarily careful design of the treatment system for the cooling water supplied by the Columbia River.[112] Worthington could see no evidence that the fall in conductivity was abating, and warned that a drop to less than 5% of its original value could occur, necessitating a reduction in power level and a proportional drop in plutonium production. Even more disturbingly, the increase in breaking strength for the graphite suggested that brittleness might ensue, and 'raised the possibility of future crumbling or disintegration of the graphite'. For this, no curative measure seemed likely apart from rebuilding the piles from scratch.[113]

Yet another major concern with the graphite was the 'Szilárd complication'. In August 1944, before the piles had been started up, Leo Szilárd had suggested that energy would be stored in the graphite by neutron collision in a fashion analogous to the cold working of metals. Local overheating might then release this energy suddenly with catastrophic effect. First indications from the Du Pont managers were that Szilárd's complication appeared a little more hopeful than Wigner's problem: they judged that the programme might not be affected, and the issue might be cured or avoided by higher-temperature operation.[114]

Such discoveries threatened everything. Taken together, these effects appeared dire both to the engineers and the scientists—not just for Hanford's production, but for the entire wartime project and potential post-war careers in the field. Worthington, backed up by

[112] '100 Area facilities and operations', September 1945, Hagley 1957 Series III Box 58 folder 5.

[113] Worthington, Hood, 'Pile technology—effect of operation on graphite moderator (Wigner and Szilárd effects)—experience to August 1, 1945', memo to file, 13 September 1945, Hagley 1957 Series III Box 58 folder 4.

[114] Greenewalt, Crawford H., 'Manhattan Project Diary Vol. II', 1942–43, Hagley 1889, p. 295, 28 August 1944. The 'Szilárd complication'—not revealed to British colleagues, who were not permitted to visit Hanford—was the cause of the 1957 incident at Windscale which destroyed one of its two piles (Section 7.2.1).

Chicago, was pessimistic about the long-term use of the piles and, by implication, the sustainability of atomic energy as a viable technology:

> The metallurgical Laboratory is of the opinion that the hazard is quite real that prolonged operations at Hanford will render the pile inoperable. Such effects as extreme brittleness, change in volume, or dangerous loss of compressive strength cannot be disregarded as of negligible probability…The sudden release of…energy stored in a vast number of dislocations may conceivably occur…The Metallurgical Laboratory considers that the danger of an autocatalytic release of the stored energy with a period short (in comparison with the cooling period of the graphite) cannot be ignored.[115]

Du Pont engineers continued to operate the piles, but alerted the District Engineer of Manhattan Project about their pessimistic forecasts in February 1946, warning that, owing to the bowing of the tubes containing the uranium slugs, the operating life was certainly limited. The oldest unit, 100-B, consequently was reduced to 'stand-by', and another was shut down to ensure a back-up for plutonium and polonium production, both needed for future atomic bombs. The matter-of-fact Du Pont site history, written in the house style employed for all of their industrial plants, noted that 'the production facilities at Hanford that Du Pont turned over to General Electric had major operational problems'.[116]

In sum, the Hanford experience appeared unpromising as the launching pad for a new intellectual field. The contingency of the whole enterprise was appreciated by all: meticulous industrial planning but uncertain physical data; chronic pessimism amongst several of its key actors, notably Eugene Wigner and Leo Szilárd; and perturbing discoveries regularly along the way—xenon poisoning, the Wigner effects, the Szilárd complication. and, after just a few months of operation, the realization that the pile was aging quickly and unpredictably. The ongoing discussion between the engineers and scientists was an uncomfortable hodge-podge of deep research, urgent mitigation of known problems, and preparation of backups for further contingencies. Du Pont engineers, as much as the Met Lab scientists, were eager to end this bumpy collaboration as soon as wartime duties allowed. Post-war activities, it seemed, would have to be scaled down, slowed down, or abandoned.

Between late 1942 and the end of the war, then, the mixed teams of Met Lab and Du Pont personnel were seminal in defining American specialists in nuclear piles, just as ICI, NRC, and university personnel shaped the Anglo-Canadian cohort. Their interactions bred specialists unlike the pre-war models of physicists and industrial chemists. Clustered at sites of enduring significance—Chicago/Argonne, Oak Ridge, Hanford, and Montreal Laboratory/Chalk River—skills were acquired and disseminated through the first experiences with shared design, training, and reactor operation. And know-how was contained there by security rules. From these wartime roots, nuclear engineering was guided to grow in characteristic forms.

---

[115] Worthington, Hood, 'Pile technology—effect of operation on graphite moderator (Wigner and Szilárd effects)—experience to August 1, 1945', memo to file, 13 September 1945, Hagley 1957 Series III Box 58 folder 4.
[116] 'Hanford Story, chap 7, 8', Hagley 1957 Series V Box 50.

PART B

Incubation

'Take away your army generals, their kiss is death, I'm sure.
Every thing I build is mine; every volt I make is pure.
Oh, dammit! Engineering isn't physics, is that plain?
Take, oh take, your billion dollars, let's be physicists again'.
Arthur Roberts, 1946[1]

'Every time I learned something new...I'd tell myself, "nobody in history has ever worked on this equipment before, and here I am fooling around with it".'
Edward Jackson, 1948[2]

'Everyone could play the game of designing new nuclear power piles...we were like children in a toy factory'.
Alvin Weinberg, 1994[3]

---

[1] Roberts, Arthur, 'Take away your billion dollars (the Brookhaven song)', lyrics, UI box 35.
[2] Lang, Daniel, *From Hiroshima to the Moon: Chronicles of Life in the Atomic Age* (New York: Simon & Schuster, 1959), p. 34.
[3] Weinberg, Alvin M., *The First Nuclear Era: The Life and Times of a Technological Fixer* (New York: AIP Press, 1994), pp. 38–9.

4

THE ATOMIC NURSERY

If the war gestated atomic energy specialists, then the decade after it saw their birth and early development. They did not thrive unaided, however; the new experts were nurtured with copious resources, cosseted in secure environments, and isolated from contaminants. This period of incubation shaped their development and mature identity.[1]

4.1 Hesitant steps

It might be expected that, with their privileged knowledge of nuclear technologies, the trajectory of their subject was more easily forecast by participants in the Manhattan Project than for others. Yet, at war's end, the future was hard to predict. Despite deep engagement with particular scientific and engineering problems, few had an overall perspective of the goals, state of advancement of the project, or its timing. The lack of foresight about the potential for a post-war field was the result of the wartime compartmentalization of knowledge and organizational responsibilities.[2] As a result, the research and development centres that had grown with the Project had uncertain futures in August 1945. A Manhattan Engineer District report later explained this tentativeness in terms of the topsy-turvy hierarchies of wartime development:

> Manhattan District emergency wartime research was entirely subordinated to the primary objective of producing atomic bombs. It was necessary that production requirements take priority over 'academic', and often fundamental, research considerations [which] resulted in a relationship between project research and production activities uniquely different from the normal organizational procedures of scientific and engineering industry.[3]

---

[1] In terms of the sociology of the professions, Abbott's outline is apt here: internal competition occurs in particular occupational environments, and a profession's success depends as much on the situation of its competitors and on the system structure—in which, 'from time to time, tasks are created, abolished or reshaped by external forces'—as much as its own efforts [Abbott, Andrew D., *The System of Professions: An Essay on the Division of Expert Labor* (Chicago: University of Chicago Press, 1988), p. 33].

[2] For example, Col. K. D. Nichols, the District Engineer for Oak Ridge, had supervised the three major projects to produce fissionable material. The K-25 and Y-12 projects to separate U-235 by gaseous diffusion and electromagnetic separation were managed by unit chiefs, who had responsibility for both the production and associated research programmes. The researchers were managed by Area Engineers at the University of California, MIT, and Columbia University. By contrast, the plutonium project had separate production and research administrations: Hanford, under an Area Engineer, reported to Nichols; separate Area Engineers at Oak Ridge, Argonne, and a handful of other sites oversaw reactor research.

[3] DOE, OpenNet, 'History of the activities of the Manhattan District Research Division, October 15, 1945—December 31, 1946', 31 December 1946, DOE ON NV0714682, Chapter 1, p. 2.

The Manhattan Engineer District adapted, though, in the months after the Japanese bombings. It founded a new Research Division that October to cleave research and development operations from routine production—thereby annulling the forced marriage so painfully pursued by the Met Lab and Du Pont specialists. While it was recognized that this would be an interim measure until Congress legislated a civilian organization to replace Army control, the Research Division undertook relatively rapid actions to normalize scientific and engineering practices. The Division's priorities were to (a) declassify information; (b) institute a programme to distribute radioactive isotopes; and (c) encourage 'the many research and development programs in nuclear science which were being planned by other military agencies, non-military government agencies, academic institutions, and private and engineering industries'.[4]

Initial plans called for the production of some forty volumes of scientific information, for which 'a substantial percentage of the personnel of Metallurgical Laboratory [had] been accumulating data' over the final year of the war.[5] This openness would be tied to the publication of numerous patents. And the release of information was to be accompanied by an egalitarian culture at Argonne. During the second day of talks with the local Manhattan Project administrator, noted the minutes, 'a forum was held to discuss national lab policy. Agreed that visiting professors and students should have access for specific work, loaned by their universities', and that a directing committee (soon relabelled a Board of Governors) be appointed comprising three physicists, two chemists, one biologist, and one engineer from seven participating universities.[6] Nuclear knowledge, it appeared, would be shared and applied according to traditional Enlightenment ideals.[7]

But the emergence of a self-sustaining group of nuclear specialists was not assured. The 1945 Smyth report, the first military-sanctioned account of the bomb project, certainly suggested potential: the topics studied were 'not specific research problems as must be solved by a small team of scientists working for a few months but are whole fields of investigation that *might* be studied with profit for years'.[8] The technical specialists themselves were equally cautious. The spirit of the time and the fragility of the new expertise are suggested

---

[4] DOE, OpenNet, 'History of the activities of the Manhattan District Research Division, October 15, 1945—December 31, 1946', 31 December 1946, DOE ON NV0714682, Chapter 1, p. 3.

[5] Hilberry, Norman, 'Statement to the Advisory Committee on Future Argonne Laboratory Operation (draft)', memo, 24 November 1945, UI Box 134. See also Argonne Laboratory, 'Broad Policy on National Laboratories Recommended by General Groves' Advisory Committee on Research and Development' December 1945, UI Box 44.

[6] 'Minutes of conference with District Engineer and Metallurgical Laboratory', 5 April 1946, UI Box 19. This remarkably democratic organization eventually proved difficult to manage, owing to its complexity and competing interests. On the evolving Argonne–universities–AEC relationship see Greenbaum, Leonard, *A Special Interest: The Atomic Energy Commission, Argonne National Laboratory, and the Midwestern Universities* (Ann Arbor: University of Michigan Press, 1971).

[7] A clear model of post-war government strategy is given in Bush, Vannevar, 'Science, the Endless Frontier: A Report to the President by Vannevar Bush, Director of the Office of Scientific Research and Development', United States Printing Office, July 1945.

[8] Smyth, Henry D., 'A general account of the development of methods of using atomic energy for military purposes under the auspices of the United States Government 1940–1945', 1 July 1945, LAC MG30-E533 Vol. 1, p. VII-11 (emphasis added).

by the Foreword to *Elementary Pile Theory* (1950), the first published text for would-be nuclear engineers:

> During the war there was assembled at the University of Chicago a group of scientists and engineers who undertook the many complex problems associated with the planning and construction of nuclear-chain reactors which could be employed to produce plutonium and many other radioactive materials and radiations. This task reached its climax in 1944 when the reactors at Hanford were placed in operation. After this date most members of the group at Chicago dispersed to undertake other tasks at other sites of the Manhattan Project, most notably at Los Alamos. There remained at Chicago, however, a small group which was concerned with the 'stand-by' problems associated with the operating reactors and which devoted a fraction of its time to consideration of the longer-range aspects of reactor development. This group, which...looked to Professor Wigner for leadership, contained a number of men who intended to devote an appreciable fraction of their future life to the problem of reactor development. These individuals were interested in seeing established a plan whereby the science and technology of reactors could be projected continuously into the future once peacetime conditions would make it possible to broaden the basis of development to general as well as military problems.[9]

For this nucleus of workers—some subsequently populating the Argonne laboratory that sprang from the University of Chicago Met Lab, others manning the X-10 pilot-plant reactor at Oak Ridge from 1943, still others operating and studying the Hanford piles, and yet more decamping with Wigner to the Clinton lab after the war—careers dedicated to reactor development offered what mere weapons design did not.

But newly identified dangers corrected these first steps. With rising American concerns about Soviet expansionism and the threat to international and domestic security by cases of espionage and defection, the three allied governments continued to play an unprecedented role in directing nuclear science and engineering in the decade after the Second World War. The 'atomic secret' was to be preserved by the policies of the three countries that had contributed to it.[10] The political momentum in the USA shaped Senator Brien McMahon's Atomic Energy Act of 1946, leading to the exclusion of even Britain and Canada from future developments. Other early players—particularly France, whose nuclear specialists collaborating on the Anglo-Canadian project had already been repatriated to populate the Commissariat à l'Énergie Atomique, and the Soviet Union, whose scientists were striving to replicate and extend Manhattan Project results—were even more soundly locked out.[11]

---

[9] F. Seitz, in Soodak, Harry and Edward C. Campbell, *Elementary Pile Theory* (New York: John Wiley, 1950), p. v.

[10] On the construction of 'atomic secrets', see Herken, Gregg, '"A most deadly illusion": the atomic secret and American nuclear weapons policy, 1945–1950', *Pacific Historical Review* 49 (1980): 51–76. See also Kevles, Daniel, 'Cold War and hot physics: science, security and the American State, 1945–1956', *Historical Studies in the Physical and Biological Sciences* 20 (1990): 239–64 and Westwick, Peter J., *The National Labs: Science in an American System, 1947–1974* (Cambridge, Mass.: Harvard University Press, 2003), pp. 75–88.

[11] Early Soviet efforts were accelerated by espionage, but equally by strong scientific expertise and careful analysis of published material such as the Smyth report. According to contemporary CIA estimates, the Soviet project by 1950 involved some 10,000 engineers, technicians, and scientists. See Holloway, David, *Stalin and the Bomb: The Soviet Union and Atomic Energy, 1939–1956* (New Haven, Conn.: Yale University Press, 1994), Chapter 7 and p. 192.

Three factors are notable for the post-war decade. First, national programmes of atomic energy, and their diversity of goals, expanded dramatically. Second, this activity was cloistered within a small number of research centres under the scrutiny of their central governments. And third, security concerns and protective legislation isolated the three former allies, and their new laboratories were to mirror this traditional response to external threats.[12]

The post-war months, then, were marked by the faltering steps of atomic specialists, cautiously guided and redirected by their governments. In the USA, a new civilian organization and national laboratories were to be founded to redirect and expand development of atomic energy; in the UK, an urgent post-war atomic bomb project was approved, but its participants nurtured aspirations for developing nuclear power; and in Canada, the remaining members of the Chalk River project doggedly pursued their wartime focus on the possibilities of heavy-water reactors.

4.2 National Laboratories in post-war America

Of the three allies, the USA had by far the most established facilities, the greatest number of personnel trained in nuclear work, and the most tangible achievements. With the US Atomic Energy Commission (AEC), formed in 1946, came a transfer of responsibilities and planning for application of the new knowledge. The AEC was created by the McMahon Act, which defined the post-war nuclear terrain not only in the USA, but in Britain and Canada, too. According to the Act, the sharing of sensitive information with Britain and Canada would cease; possession of nuclear materials would become a government monopoly; and the new AEC would be a civilian organization (although the Commission remained closely associated with military goals and administration via its Military Liaison Committee).

The AEC inherited sites and personnel that were declining in number. From early 1946, these wartime facilities of the Manhattan Project began to transmute into a post-war core of American 'National Laboratories', although the outlook seemed clear for only one. Los Alamos, the third major research site of the Manhattan Project, and headed by Robert Oppenheimer during the war, was to remain dedicated to the design and testing of nuclear weapons after it. By contrast, the Clinton and Argonne Laboratories, both the progeny of Arthur Compton's Met Lab, had unclear goals in the early post-war months.[13] Yet these special labs were central to the birth and upbringing of new experts: they selected and fostered particular technical cohorts, and cloistered them through security rules.

In a perceptive study of the National Laboratories, historian Peter Westwick argues that they came to comprise a system peculiarly suited to the American context. Although

---

[12] On isolationism, see Shils, Edward, *The Torment of Secrecy: The Background and Consequences of American Security Policies* (Glencoe, Ill.: Free Press, 1956), pp. 86–9.

[13] Other National Laboratories were founded later. The Lawrence Livermore National Laboratory (1952), about 40 miles from San Francisco, was tasked with nuclear weapons design as a competitor to Los Alamos. The lab incorporated interdisciplinary teams of scientists and engineers in its work, the approach favoured by its founder Ernest Lawrence. Sandia Laboratory, located principally at Kirtland Air Force Base in New Mexico, became responsible for the non-nuclear portions of nuclear weapons. It was operated successively by the University of California, General Electric, AT&T, and then a subsidiary of the Lockheed Martin Corporation, and became a National Laboratory in 1979.

embodying state intervention in scientific research and development, he argues that they incorporated built-in competition through redundancy of goals—acting, in effect, like a subtle free-market in the production of nuclear expertise.[14] Although he does not draw upon other national experiences, the implication is that this American system was markedly different from the nuclear organizations in, for example, Britain, France, and the Soviet Union. I would interpret the growth of the National Laboratories somewhat differently. From the perspective of nuclear workers, rather than administrators in central government, I will argue that the evolution of the Labs was, on the one hand, disturbingly unplanned and economically (and perhaps intellectually) uncompetitive while, on the other, harnessed and steered securely towards national goals, with the state operating simultaneously as monopolistic funder and sole customer.

The existing personnel at these sites carried disciplinary momentum. In the first two years after the war, the AEC confirmed Los Alamos as a weapons laboratory; Argonne, under physicist Walter Zinn (1906–2000), was to claim sole responsibility for reactor work, and the Clinton Laboratory at Oak Ridge was directed to focus on isotopes, chemical reprocessing, and basic research in chemistry, physics, health physics, and biology (although this was soon challenged). And at Brookhaven National Laboratory—a third site vying to develop reactor proficiency—a fresh patch of intellectual terrain was carved out. The Lab was created from scratch in 1946 to serve prominent north-eastern universities, after lobbying of the Manhattan Engineer District by Columbia University and MIT. During the brief security thaw and its anticipated resumption of pre-war relations between science and technology, a scientific perspective initially prevailed.[15]

4.2.1 *Defining Oak Ridge specialists*

The AEC allocation of responsibilities proved particularly contentious at Oak Ridge. Engineering, rather than science, was dominant. As its later Director Alvin Weinberg recalled, 'the atmosphere of X-10, the first reactor to generate moderate quantities of plutonium, was more that of a Du Pont pilot plant than a University of Chicago research center'.[16] Overseen

---

[14] Westwick, Peter J., *The National Labs: Science in an American System, 1947–1974* (Cambridge, Mass.: Harvard University Press, 2003). See also Forman, Paul and José M. Sánchez Ron, *National Military Establishments and the Advancement of Science and Technology: Studies in 20th Century History* (Dordrecht: Kluwer Academic, 1996) and Seidel, Robert W., 'A home for Big Science: the Atomic Energy Commission's laboratory system', *Historical Studies in the Physical and Biological Sciences* 16 (1986): 135–75.

[15] Needell, Allan A., 'Nuclear reactors and the founding of Brookhaven National Laboratory', *Historical Studies in the Physical Sciences* 14 (1983): 93–122; Brookhaven NY had been chosen, after separate representations from scientists at Columbia University, MIT, and Harvard, as a compromise to serve as a resource for academics in the north-eastern US. Other sites had been rejected for reasons of cost or proximity to major urban centres. The name Brookhaven Laboratory was chosen 'because it was literally correct yet delightfully misleading. The site was in the town of Brookhaven but the name had a misleading association with quiet, shady streams which might make the laboratory site sound more attractive to potential new recruits than it actually was.' Ramsey, Norman F., 'Early history of associated universities and Brookhaven National Laboratory', *Brookhaven Lecture Series* 55 (1966): , p.7.

[16] Weinberg, Alvin M., *The First Nuclear Era: the Life and Times of a Technological Fixer* (New York: AIP Press, 1994), p. 48.

by Du Pont engineering staff, the laboratories had absorbed the company's operational culture. And the future appeared doomed to increasing reorientation towards chemical engineering: the University of Chicago had withdrawn from management of the Clinton Labs in the last month of the War, and was replaced by the Monsanto Chemical Company. Monsanto's industrial chemists and chemical engineers, argued AEC administrators, had relevant expertise for the site's remit.

Echoing the complaints of the wartime Met Lab physicists about working for a chemical corporation, the Clinton specialists were painfully aware of their low status. The backwater location seemed earmarked mainly for production and secondary research. Met Lab scientists were dismissive of the remote and unpromising location, and Du Pont staff had referred to the site as the 'gopher training school'.[17] With equal disdain, their Monsanto successors dubbed it 'Dogpatch' after the location popularized in the *Li'l Abner* comic strip that parodied a hillbilly community. Perceptions of the site threatened to indelibly label its nuclear workers, too (Figure 4.1).

So the nuclear specialists gestated at Oak Ridge were threatened by two transitions: a shift from physics to chemical engineering and from a goal-directed wartime industrial facility to more ambiguous post-war activities. The status of the site as a post-war incubator for the specialists was raised, though, by key administrators. With physicist Eugene Wigner as Technical Director sharing management of the facility with chemist James Lum of Monsanto as Executive Director, new directions were pushed. They expanded what had been conceived as a pilot plant into a laboratory for developing nuclear technologies: reactor design, radioisotope production, and chemical separation technologies.

The design of reactors was the plum activity for the Lab's scientists. Two projects initially vied for dominance. The first, supported by most of Clinton's scientists, particularly Wigner and Alvin Weinberg, was to design a reactor that would produce a high flux of neutrons—higher, indeed, than the Hanford reactors. Its purpose would be to further research on the effects of neutron irradiation and a first step towards producing a *breeder reactor* that could generate more fissile material than it consumed.[18] The other project, supported by physical chemist Farrington Daniels (1889–1972) who had been Director of the Met Lab at war's end, was to design a high-temperature gas-cooled reactor as the first step in investigating the feasibility of generating electrical power. He hoped to redirect post-war atomic energy research by demonstrating industrial involvement and application. Both projects enmeshed physics, chemistry, and engineering know-how into a new synthesis. But the Daniels pile—already relegated to the Clinton Lab as a non-urgent goal—would necessarily involve Monsanto engineers, and was resisted by the physicists as diverting them from scientific goals.[19]

---

[17] Johnson, Leland and Daniel Schaffer, *Oak Ridge National Laboratory: The First Fifty Years* (Knoxville: University of Tennessee Press, 1994).

[18] The high-flux reactor was also to be an American counterpart to the Canadian heavy-water reactor then in the planning stages, the NRX.

[19] See Hewlett, Richard G. and Jack M. Holl, *Atomic Shield, 1947–1952* (Berkeley: University of California Press, 1969), pp. 68–71; Alberty, Robert A., 'Farrington Daniels, March 9, 1889–June 23, 1972', *National Academy of Sciences Biographical Memoirs* 65 (1994): 109–21.

Fig. 4.1 Removing radioactive material from graphite reactor, Clinton Laboratories, Oak Ridge, 1946.[20] (Courtesy of US Department of Energy, Oak Ridge National Laboratory.)

The balance between scientific research and engineering development remained uneasy. Monsanto's tenure at Oak Ridge lasted only to 1947, when the AEC persuaded the University of Chicago to resume management. The Commission still favoured industrial management of its laboratories, though—a preference that cast its workers in a non-academic mould. Perhaps in a move to sweeten the connotations for the disaffected nuclear specialists, the AEC renamed the site the Oak Ridge National Laboratory (ORNL), and added 'National' to Argonne's title. For both sites, this apparently unsanctioned alteration subliminally boosted their occupational status.[21]

The aspirations of ORNL were further championed by Alvin Weinberg, successively Director of the Physics Division, Research Director, and overall Director between 1945 and 1974. About a third of its personnel, he argued, were engineers and scientists with recent experience

---

[20] DOE PHOTO 1870–6, 2 August 1946.

[21] Westwick, Peter J., *The National Labs: Science in an American System, 1947–1974* (Cambridge, Mass.: Harvard University Press, 2003), pp. 54–5. The key technology was also transmuted under AEC management, with 'piles' increasingly becoming 'reactors'.

in pile design and operation.²² In the face of the AEC relegation of the organization to chemical engineering, Weinberg interpreted the remit to include his pet scheme: homogeneous reactors. Homogeneous reactors were elegant in concept, dispensing with fuel rods, control rods, and complex mechanical structures. Instead, the reactor would consist of a slurry or solution of concentrated uranium in heavy water or molten salts. Neutron flux and thermal efficiency would be relatively high, and the scheme could be inherently self-controlling (i.e. reducing power output if the temperature rose). The recovery of fission products would also be simplified: instead of dismantling and dissolving solid fuel elements in acid, the liquid fuel would be more efficiently separated.²³ Weinberg argued that the research themes favoured at Argonne and Oak Ridge were thus complementary. The division of labour promised clear routes for nuclear specialists, low institutional conflict, and a better guarantee of national progress.

4.2.2 *Ascendancy at Argonne*

The aspirations at Argonne National Laboratory (ANL), however, initially threatened to overwhelm Oak Ridge reactor research. For its members, the end of the war seemed to signal not just a relaxation of security, but an outward momentum for their special expertise. The scientists at Argonne anticipated a return to pre-war intellectual hierarchies, but also a role for the national laboratory as the hub of a new government-directed field:

> the field of research will be greatly broadened by enabling universities within this region to utilize the facilities in Argonne...Although great progress had been made in the field of atomic bombs, little or no progress has been made in the commercial power field. A project of the magnitude of Argonne must be considered as a normal Government project, with the actual work done by a contractor with the Government providing administration and sites and the cooperation of the regional universities. With a program such as this and a scheme for visitation and guest work by professors of the regional universities, Argonne can be developed into the finest nuclear laboratory possible'.²⁴

The make-up of his new National Lab in early 1946 was unlike any pre-war laboratory: '50 scientists at Argonne (including their helpers); 90 scientists and helpers in New Chemistry; 200 personnel in Health and Safety; 200 scientists and technicians under Dr Ralph Lapp [Assistant Director]; 66 in patents and acting as writers; 150 guards; and 400 in administration, making a rough total of 1200'.²⁵ As the figures hint, however, the Lab cohort was rich with scientists and deficient in engineers, a make-up that was sustained by early policy.

---

²² Weinberg, Alvin M., *The First Nuclear Era: The Life and Times of a Technological Fixer* (New York: AIP Press, 1994), p. 48.

²³ Ibid., pp. 118–25. Despite its appeal, the concept of the homogeneous reactor—tested on several prototypes at Oak Ridge and elsewhere—proved impracticable to develop owing to excessive corrosion and the formation of gas bubbles, which destabilized the reaction.

²⁴ 'Draft Operating Policy of the Argonne National Laboratory', memo, 28 February 1950, UI Box 19.

²⁵ 'Minutes of conference with District Engineer and Metallurgical Laboratory', 5 April 1946, UI Box 19. By comparison, the comparable Soviet cadre concentrated on atomic energy at Igor Kurchatov's 'Laboratory 2' had grown to 650 employees in 1946 and was to rise to 1500 by May 1947 and 4000 by 1956 [Josephson, Paul R., *Red Atom: Russia's Nuclear Power Program from Stalin to Today* (New York: W. H. Freeman, 1999), p. 18].

Fig. 4.2 Walter H. Zinn and operator Fred Cokeing at the controls of the CP-3 reactor.[26] (Courtesy of Argonne National Laboratory.)

The Lab's developing culture was personified and channelled by Walter H. Zinn. A Canadian, Zinn obtained a first degree in mathematics from Queen's University, Ontario, and his PhD in nuclear physics from Columbia University in New York in 1934. He had worked with Leo Szilárd and Enrico Fermi at Columbia on early fission experiments, and then at the Met Lab through the war in developing the first reactors. When Fermi had left Chicago for Los Alamos, Zinn assumed direction of Argonne (Figure 4.2).[27]

At the 1946 meeting defining the remit of the new lab, Zinn made a plea for a monopoly on the field, stressing

---

[26] Argonne National Laboratory Neg. No. 201–663, May 1954.

[27] As discussed below, this uniquely trained specialist turned down Canadian overtures to recruit him to direct the post-war Chalk River programme, a role assumed by a Briton, W. Bennett Lewis. See Weinberg, Alvin M., 'Walter Henry Zinn', *Biographical Memoirs of Fellows of the National Academy of Sciences* 85 (2004): 365–74 and Fawcett, Ruth, *Nuclear Pursuits: The Scientific Biography of Wilfrid Bennett Lewis* (Montreal: McGill-Queen's University Press, 1994).

that the question of new piles was not only a question of increased research facilities but, much more, a question of new pile design and construction. He expressed the strong feeling that the objectives of the Laboratory should be new pile design and construction as well as fundamental research.

While the Board and General Nichols, the Manhattan Engineer District administrator with local responsibility, concurred in principle with this statement of objectives, Nichols warned against undertaking more than the available manpower would permit.[28] Even in this well-funded environment, it was apparent that experts were a scarce resource. Despite Zinn's successful salesmanship, Argonne would have to be steered away from nuclear physics towards, at best, science-based nuclear technologies.

A month later, Nichols again cautioned that funds 'will be more readily obtainable than man-power', and that Argonne 'will have as good a chance as the North East laboratory [Brookhaven] in getting additional piles', but that 'the decisions in each case must be based on the national situation as a whole'.[29] Although dampening his aspirations, the meeting clinched Zinn's acceptance of the directorship of Argonne, just weeks after an offer to head the Canadian Chalk River establishment.

His proprietary urge to concentrate expertise at Argonne was bolstered by pragmatism. Reactors were rising in demand. In 1948, the Board discussed the proliferation of interest in atomic piles: both the Army and Navy had an interest in reactors, and the Air Force was asking assistance; General Electric was beginning to explore power generation possibilities, with a close tie to the pile research and development group at Argonne; the Navy was 'thinking of power piles tactically but as yet know nothing about atomic energy'. As Glenn Seaborg later recalled, the new AEC 'had to rely almost entirely on Zinn and Argonne for its reactor development program', and consequently supported it well.[30]

Nevertheless, the chairman of the AEC, David Lilienthal, told reporters candidly that 'the first commercially practical atomic power plant is not just around the corner, but around two corners' and that estimates were merely an educated guess, 'and no one's guess is very educated'.[31] Board members were concerned that so much reactor research was being undertaken by so few experts, one observing drily that 'there are only enough competent men in the country to staff work on two piles let alone four or five'.[32] Zinn proposed a threefold solution: to find men to share the responsibility; to work on relatively few proposals, and to 'build a power pile at another location, say in a Naval shipyard'. Even more unsettling was the lack of policy regarding how industry would fit into the picture. One Board member summarized that

> companies such as Westinghouse, Allis-Chalmers etc have agreed to come in on Zinn's terms, this being the only present solution which makes sense... Dr Zinn would be happy to have industry send in people to the Laboratory for training in the field; he would expect their assistance in determining the feasibility

---

[28] 'Minutes of conference with District Engineer and Metallurgical Laboratory', 5 April 1946, UI Box 19.

[29] 'Minutes of Board Meeting, ANL', 6 May 1946, UI Box 19.

[30] DOE, OpenNet, 'Remarks by Seaborg at the 25th anniversary of Argonne National Laboratory, Argonne Illinois (AEC-S-12-71)', 19 June 1971, DOE ON NV0712412, p. 5.

[31] 'Commercial atomic power: how soon?' *Discovery* 8 (11) (1947): 331.

[32] Dr Speddings, 'Minutes of special meeting of ANL Board of Governors', 4 January 1948, UI Box 19.

of designs and in some cases industry could build the item after the feasibility of the design is determined.[33]

Even for the most prosperous of the wartime allies, then, limited experience throttled down the post-war expansion of atomic energy.

Rising demand for the restricted expertise in reactor design thus threatened Walter Zinn's promotion of ANL as the locus of American nuclear engineering. Eugene Wigner—chief designer of most of the first generation of reactors (the X-10 pilot plant, CP-3 heavy water reactor, and Hanford plutonium production reactors, as listed in Table 5.1)—had been translated to ORNL as Technical Director. There, his team continued to explore novel reactor designs. Even worse, ORNL embodied a distinct complement of experts having an engineering slant, and practicing and preaching their own brand of proficiency. ORNL competition further threatened to usurp Argonne's role as centre of the new field by attracting influential government-sponsored students on the training courses it began to offer.

The sharing of reactor design between Argonne and Oak Ridge was consequently troublesome. The internal competition was not seen as productive, and so in 1948 the AEC decided to centralize the reactor programme at Argonne and to relocate the Clinton Lab staff. The hierarchical division of expertise is captured by a Christmas carol sung by Oak Ridge physicists during the Christmas of 1947 to the tune of 'Deck the halls':

> Pile research is not for us'ums, Fa la la la la, la la la la
> Leave it for our Argonne cousins, Fa la la la la, la la la la
> Engineering is for us'ums, Fa la la la la, la la la la
> We're a bunch of dirty peons, Fa la la la la, la la la la[34]

Zinn nevertheless opposed the decision, partly because he believed the best Oak Ridge workers would not move, but also because the institutional cultures were distinct. Salary grades of scientists were markedly different, and the few engineers on the Argonne staff, he suggested, would create a potential division. In effect, he argued that the two segregated sites were too dissimilar to be successfully integrated.

A special meeting of the ANL Board of Governors to address schemes for integrating the two labs revealed concerns about diluting the expertise of 'pure research' with 'pile engineering personnel'. The minutes record that 'Dr Zinn had attempted to avoid class distinction in the outright establishment of two groups in the Laboratory, those doing pure research

---

[33] 'Minutes of special meeting of ANL Board of Governors', 4 January 1948, UI Box 19. See also Hewlett, Richard G. and Jack M. Holl, *Atomic Shield, 1947–1952* (Berkeley: University of California Press, 1969). Westinghouse was contracted in 1948 to develop Argonne's concepts for the reactor for the first nuclear submarine. General Electric later assumed similar responsibility for the reactor powering the first nuclear aircraft carrier with Oak Ridge designers, with both projects funded by the US Navy. The Air Force equivalent was the Nuclear Energy for the Propulsion of Aircraft (NEPA) project from 1946, which enrolled companies such as Fairchild, Convair, and Pratt–Whitney. Early expertise in reactor technologies thus flowed from the national labs to American industries.

[34] Stacy, Susan M., 'Proving the Principle', http://www.inl.gov/proving-the-principle/, accessed 12 June 2009, p. 44.

and those doing directed work'. Others feared that an influx of engineering-oriented nuclear workers would drive pure researchers to Brookhaven.[35] The Deputy Director of Argonne, Norman Hilberry, consequently proposed a month later to expand the 'pile engineering groups' with 'a group of top-flight physicists [to] provide the leading influence'. A kind of disciplinary eugenics, he seemed to reason, would assure viable laboratory stock.[36]

These early efforts to shape Argonne's skill-set were countered by rapidly expanding goals. The responsibility given to Zinn's ANL was heavy. The AEC asked him to draft a national reactor plan, although it relented to permit ORNL to continue certain reactor researches. At ANL, ongoing design of a fast-neutron-breeder reactor continued (EBR-1), and two ORNL projects were incorporated into his management portfolio: its high-neutron-flux Materials Test Reactor (MTR) became a joint project with Argonne, and a project to develop a pressurized water reactor for submarine propulsion (the Submarine Test Reactor, or STR) was moved to Chicago in the summer of 1948.

Even with this concentration of development projects under its control, an engineering orientation at Argonne remained unpalatable for the administrators. Despite Zinn's aspirations to centralize national expertise in reactors at Argonne, accommodation of engineering and commercial talent was elusive. He reported to the Board of Directors in 1949 on the Lab's concept of a central station reactor based on natural uranium, admitting that 'an engineering group has not yet been organized but as people in the laboratory are relieved of other duties they will be put on this work'. The transcript captures his dogged promotion of reactor research:

> It is unfortunately true that one needs to be a politician as well as technically competent in this field... We are trying to introduce stability into this program. We have had some violent oscillations that have settled down in recent months. From here on we need a year or so on construction of reactors and a year for thinking about reactors that should be coming along. We hope we will get some activity started on homogeneous reactors. For the foreseeable future we will try to hold our course, to get something built, and think about the next step so that we don't have to go through another siege.[37]

In practice, however, Argonne found a solution in further segregation: ANL would move its engineering development to a separate site in Arco, Idaho, from 1949 (later known as the National Reactor Test Station, and later still as the Idaho National Engineering Laboratory).[38]

---

[35] 'Minutes of special meeting of ANL Board of Governors', 4 January 1948, UI Box 19. Incidentally, the Board Minutes during these crucial early years were remarkably candid and, in some cases, augmented by full transcripts of the discussions.

[36] Hilberry, Norman to F. Daniels, letter, 5 February 1948, UI Box 19. To bolster reactor work, it was also planned to expand the Metallurgy Division (responsible for testing suitable construction materials) and the Process Chemistry Division (tasked with extracting suitable products such as radioactive isotopes and plutonium for reactor fuel). In addition, it was hoped that Health Physics for 'some fundamental research' and Theoretical Physics, with 'the procurement and operation of an Eniac', would support the goal-directed reactor programme and 'accommodate the new problems in this field introduced by individual differences in reactor design'.

[37] 'Minutes of Board Meeting, ANL', 7 March 1949, UI Box 19.

[38] The official history of the Idaho National Laboratory is Stacy, Susan M., 'Proving the Principle', http://www.inl.gov/proving-the-principle/, accessed 12 June 2009.

As Zinn put it, 'the testing station will be a place to build reactors and get experience in their operation, rather than to do experiments such as will be done on the research reactor at Argonne by the physicists, chemists and biologists. The reactors at Idaho will provide space for engineering types of experiments'. Regardless of where the 'reactor farm' was located, he insisted, 'the research and experimental work required will be done at the Laboratory'.[39] By farming out the less appealing tasks, then, the young US National Labs actively bred distinct flavours of nuclear expert but ineffectually defused animosities between them.

Despite ongoing contentions between the nuclear specialists at American sites, the early discussions between the Atomic Energy Commission and Argonne National Laboratory established principles that were to endure: management of national laboratories by universities or university consortia; responsibility there for training and provision of facilities to students and practising technical workers; and, most significantly, a conscious aspiration to move away from military development and towards scientific research at most sites. Even so, these aims contained the seeds of conflicts that were mirrored at a higher administrative level. The AEC sought to ensure both weapons development and civilian applications of nuclear energy, and the underlying technologies were secret, at present; the universities, by contrast, emphasized education and open-ended research.

A sense of the participants' misgivings about governmental largess—and its conditions—is suggested by physicist Arthur Robert's 1946 song 'Take away your billion dollars', also known as 'The Brookhaven song':

> Upon the lawns of Washington the physicists assemble,
> From all the land are men at hand, their wisdom to exchange.
> A great man stands to speak, and with applause the rafters tremble.
> 'My friends,' says he, 'you all can see that physics now must change.
> Now in my lab we had our plans, but these we'll now expand,
> Research right now is useless, we have come to understand.
> We now propose constructing at an ancient Army base,
> The best electronuclear machine in any place, – Oh
>
> It will cost a billion dollars, ten million volts 'twill give,
> It will take five thousand scholars seven years to make it live.
> All the generals approve it, all the money's now in hand,
> And to help advance our program, teaching students now we've banned.
> We have chartered transportation, we'll provide a weekly dance,
> Our motto's integration, there is nothing left to chance.
> This machine is just a model for a bigger one, of course,
> That's the future road for physics, as I hope you'll all endorse.'
>
> And as the halls with cheers resound and praises fill the air,
> One single man remains aloof and silent in his chair.
> And when the room is quiet and the crowd has ceased to cheer,
> He rises up and thunders forth an answer load and clear:
> 'It seems that I'm a failure, just a piddling dilettante,
> Within six months a mere ten thousand bucks is all I've spent.

---

[39] 'Minutes of Board of Governors' Meeting, ANL', 2 May 1948, UI Box 19.

> With love and string and sealing wax was physics kept alive.
> Let not the wealth of Midas hide the goal for which we strive – Oh
>
> Take away your billion dollars, take away your tainted gold,
> You can keep your damn ten billion volts, my soul will not be sold,
> Take away your army generals, their kiss is death, I'm sure.
> Every thing I build is mine; every volt I make is pure.
> Take away your integration; let us learn and let us teach,
> Oh, beware this epidemic Berkeleyitis, I beseech.
> Oh, dammit! Engineering isn't physics, is that plain?
> Take, oh take, your billion dollars, let's be physicists again.[40]

The song, pointedly published in the first volume of the post-war American journal *Physics Today*, lampooned the terrain becoming visible for Big Science: physicists would be *directed* by government via the military; they would be *funded* copiously from government coffers; they would be part of a movement of *integration*, in which engineering goals were paramount; and they would put pure research and teaching on the backburner for the duration.[41]

Identity was being shaped, too, by international events. The sense of freedom to choose research goals was abruptly curtailed by news of the successful Russian test of its first atomic bomb.[42] Following announcement by President Eisenhower of the detection of the Soviet weapons test in September 1949 and his administration's policy to build an American H-bomb, Walter Zinn informed representatives of the Participating Institutions of the reorientation of American reactor development:

> there have been considerable changes in the emphasis of the work being done in some of the divisions of the Laboratory. The Laboratory administration has asked the AEC what effect the President's announcement will have on its reactor programs. The answer given was that when a reactor has military value effort should be made to expedite this reactor. It is very difficult to define 'military value'.

---

[40] Roberts, Arthur, 'Take away your billion dollars (the Brookhaven song)', lyrics, UI Box 35 and *Physics Today* 1 (1948), 17. Significantly, a decade later—after the period of cloistered secrecy—Roberts provided a new version that surrendered to the seduction of funding, having the final verse: 'Give us back our billion dollars, better add ten billion more/If your budget looks unbalanced, just remember this is war./Never mind the Army's shrieking, never mind the Navy's pain./Never mind the Air Force projects disappearing down the drain./In coordinates barycentric, every BeV means lots of cash,/There will be no cheap solutions,—neither straight nor synchroclash./Oh, if we outbuild the Russians, it will be because we spend./Give, oh give, those billion dollars, let them flow without an end.' [Roberts, Arthur, 'Ten Years Later', 1956].

[41] On the wider context of the role of science in the post-war USA, see Wang, Jessica, *American Science in an Age of Anxiety: Scientists, Anticommunism, and the Cold War* (Chapel Hill, NC: University of North Carolina Press, 1999); Leslie, Stuart W., *The Cold War and American Science: The Military-Industrial-Academic Complex at MIT and Stanford* (New York: Columbia University Press, 1993); Galison, Peter and Barton J. Bernstein, 'Physics between war and peace', in: M. R. S. Mendelsohn and P. Weingart (eds.), *Science, Technology, and the Military* (Dordrecht: Duckworth. 1988), pp. 47–86; Boyer, Paul, *By the Bomb's Early Light: American Thought and Culture at the Dawn of the Atomic Age* (New York: Pantheon, 1985).

[42] Discussed further in Section 4.4 below.

Fig. 4.3 Hanford 'F' pile complex, beside the Columbia River; the building housing the pile is between the water towers at right.[43] (Courtesy of US Department of Energy, Declassified Document Retrieval System.)

Zinn noted that development of a reactor for submarine propulsion, sponsored by the Navy, was to assume the highest priority, with over a hundred ANL staff already dedicated to the programme.[44]

Corporate administration and changing state goals proved equally unsettled at the more frankly industrial sites. In Hanford, Washington, Du Pont management was replaced by the General Electric Company in September 1946 (Figure 4.3).[45] This again represented a shift in disciplinary and occupational identity, but this time from chemical plant to engineering systems. This oscillation of expertise at the post-war Manhattan Project sites illustrates the tentative embedding of the new knowledge and sensitivity to local context. Not surprisingly, then, national trajectories for atomic energy also proved remarkably different.

---

[43] DOE DDRS N1D0030555, 22 December 1944.

[44] Council of Representatives, 'Minutes of Meeting of Council of Representatives, Participating Institutions, ANL', 2 May 1950, UI Box 44.

[45] General Electric, 'Four years at Hanford', report, 1951, DOE DDRS D19803392.

4.3 Chalk River for Canadians

A distinct nurturing environment was provided in Canada. At war's end, the Anglo-Canadian project at Chalk River offered the most likely site for continuity of research. As *Little Boy* and *Fat Man* were dropped on Japan, the first Canadian pile was in its final stages of preparation and on the brink of its first criticality. Nevertheless, its expert personnel were disappearing more rapidly than at the Manhattan Project sites. The Montreal Laboratory was closed in June 1946, with all remaining employees reallocated to Chalk River. Alongside the sudden departure of the French workers at the end of the war went the gradual staged transferral of most of the British staff.

Most of those British workers went to the new Atomic Energy Research Establishment at Harwell directed by John Cockcroft.[46] Cockcroft had been sounded out about accepting the new post in April 1945, and accepted the Directorship that October. His sudden recall to Britain led to an anxious search for a Canadian replacement. As C. J. Mackenzie confided to the favoured candidate, Walter Zinn at Argonne:

> We are particularly anxious to get a Canadian-born director, as the project is going to be completely Canadian in every respect. We will probably have in the future British teams of scientists who come to us as visitors, but there will be absolutely no administrative control or direction from Britain and the teams will be at Chalk River as guests and we hope teams from the United States will be there in the same capacity.[47]

In reality, Mackenzie was aware of haemorrhaging staff levels at Chalk River and the difficulty of engaging competent replacements owing to salary levels lower than in the US. He also fumed privately about 'the atrocious start we made under the original promoters and directors' and the lack of senior direction for his young theoreticians.[48]

As discussed above, Zinn declined the offer from Mackenzie in favour of building an American centre of reactor expertise, despite assurances that 'the Chalk River project will be completely divorced from petty political interference, and the staff is not under the Civil Service Commission nor its control vested in any department of Government'. The problems of attracting new blood were not merely financial or ambition, though: Zinn intimated that 'the Americans put terrific pressure on him, pointing out that he was the only man with experience in designing and operating medium-sized piles, that he had been in the American show from the start, knew all the inside dope, and had a responsibility, particularly as he had become a naturalised citizen'.[49] John Cockcroft, in turn, judged Zinn an 'Americanized Canadian' who might not maintain British interests. Another

---

[46] The government decided to fund Harwell in October 1945, and by the following April the Harwell airfield and buildings were being converted for the new laboratories. For the official history of the British post-war project, see Gowing, Margaret and Lorna Arnold, *Independence and Deterrence: Britain and Atomic Energy*, Vol. I: *Policy Making, 1945–52* (London: MacMillan, 1974) and Gowing, Margaret and Lorna Arnold, *Independence and Deterrence: Britain and Atomic Energy*, Vol. II: *Policy Execution, 1945–52* (London: MacMillan, 1974).

[47] Mackenzie, C. J. to W. Zinn, letter, 17 April 1946, LAC RG77 Vol. 283.

[48] Mackenzie, C. J. to J. Chadwick, letter, 12 February 1946, LAC RG77 Vol. 283.

[49] Mackenzie, C. J. to J. Cockcroft, letter, 1 October 1947, LAC RG77 Vol. 283.

Canadian candidate, Robert L. Thornton, similarly rejected the Directorship in favour of resuming his position at the Lawrence Radiation Lab at the University of California.[50] Instead, Mackenzie chose Cockcroft's preferred successor as Director of Research, W. Bennett Lewis, a British nuclear physicist who had directed radar work at the Air Ministry Establishment during the war.

Despite the repatriation of many of the non-Canadian participants, Mackenzie was impressed by how easily workers at the National Research Council, and its Chalk River staff, made the transition from wartime to peacetime activities (Figure 4.4).[51] This continuity, aided by the heterogeneous but cooperative profile of committed scientific and engineering staff, undoubtedly gave the Canadians an early post-war advantage over their British and even American counterparts.

Chalk River, under the wing of the National Research Council, remained the home of the Canadian 'Atomic Energy Project'. It corresponded to C. J. Mackenzie's desire to model post-war Canadian research on wartime British and American models. As Chair of the War Technical and Scientific Development Committee in 1943 he had argued that 'Canada should have strong research groups tied in to the related industries which also should maintain research establishments', noting that

> the UK appreciates the value of research and has established a large number of research stations under the Admiralty, the Ministry of Aircraft Production, the Dept of Scientific and Industrial Research, etc and has appropriated very large sums for their activities. In the US also very large amounts are being spent on scientific research.[52]

Nevertheless, the post-war Atomic Energy Project began to dominate the National Research Council budget and administration. The completion and evaluation of the first pilot reactor ZEEP (1945) was quickly followed by the much more ambitious NRX (1947). In the following years, research blossomed without a clear organizational goal; the site appeared ripe for transition from research to a more immediate application. Mackenzie noted that 'atomic energy developments are at the stage where venture money will pay the same sort of dividends as did radar and our other war activities'.[53] As a result, in 1952 a new Crown Corporation, Atomic Energy of Canada Limited (AECL), took over the responsibilities from the NRC of shepherding these activities. Mackenzie shifted roles to lead AECL until his retirement in 1953. The Canadian government took the opportunity to further consolidate activities: the new President of AECL was W. J. Bennett, Director of Eldorado Mining and

---

[50] Bothwell, Robert, *Nucleus: The History of Atomic Energy of Canada Limited* (Toronto: University of Toronto Press, 1988), p.80. Robert Lyster Thornton (1908–1985) was born in Bedfordshire but raised in Canada, gaining a PhD in atomic spectroscopy at McGill University. Moving to the Radiation Laboratory to work with Ernest Lawrence in 1933, he worked during the war on the Lab's calutron project to separate uranium isotopes via electromagnetic separation, becoming assistant director of the Y-12 plant at Oak Ridge.

[51] Mackenzie, C. J., *The Mackenzie–McNaughton Wartime Letters* (Toronto: University of Toronto Press, 1978), p. 144.

[52] Mackenzie, C. J., 'War Technical and Scientific Development Committee 19th meeting', 2 July 1943, MG30 B122 Vol. 3.

[53] Mackenzie, C. J. to C. D. Howe, letter, 15 August 1952, LAC MG30-B122 Vol. 3.

Fig. 4.4 Chalk River nuclear worker, 1947.[54] (Chris Lund, National Film Board of Canada, courtesy of Library and Archives Canada.)

Refining, which had been nationalized as a Crown Corporation in 1944 and was later to merge with AECL itself.

Thus—unlike the American contentions fomented by competing sites and competing goals—the Canadian experience during these early years was increasing integration at a single site aimed at exploration.

4.4 The British atomic bomb and beyond

In the UK, an organization comparable to those in America and Canada was slower to develop, but it provided distinct opportunities to negotiate a space for nuclear specialists. The Tube Alloys Directorate within the Ministry of Supply mutated in 1946 into the Division of Atomic Energy (DAE). The Division's first centre of activity was the Atomic Energy Research Establishment (AERE), directed by John Cockcroft. Established in late 1945 at a former Royal Air Force base (RAF Harwell, Berkshire, some sixteen miles from Oxford),

---

[54] Photographer Chris Lund, National Film Board of Canada, LAC MIKAN 3377.

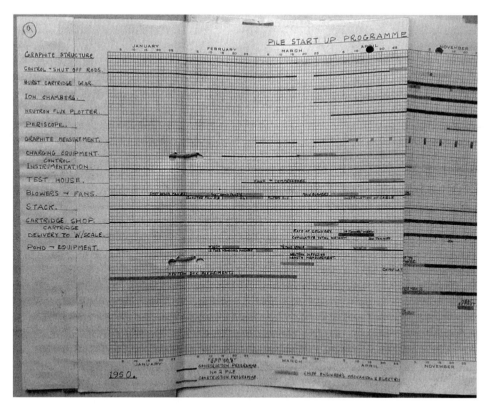

Fig. 4.5 Portion of management chart for completing the first British plutonium-generating pile, Windscale Pile 1, 1950.[55] (Reprinted with permission of National Archives (UK).)

it hosted the first British piles, GLEEP (1947) and BEPO (1948), and developed separation processes.

Within six months a second centre was created to house a new Industrial Group under Christopher Hinton (1901–1983), a former ICI manager responsible for the British munitions filling factories during the war. Hinton's organization was to be based at the former headquarters of Ordnance Supply in Risley, Lancashire. By 1950 the Group had designed and built uranium purification plants to supply the piles at Capenhurst, Cheshire, factories to produce their fuel elements at Springfields, Lancashire, and put its first production pile at Windscale, Cumbria, into operation (Figure 4.5). A plant at Windscale was soon separating plutonium from the spent fuel elements.[56]

The third branch of the Atomic Energy Division was the Atomic Weapons Research Establishment (AWRE) begun in 1950 at another former military base, RAF Aldermaston under the direction of William Penney, a British mathematician who had been a member of

---

[55] NA AB 62/42.

[56] The Capenhurst diffusion plant was used to produce the slightly enriched uranium fuel for the Windscale piles, but also the nearly pure U-235 required for Hiroshima-type uranium bombs and for later fast reactors [Bertin, Leonard, *Atom Harvest: A British View of Atomic Energy* (San Francisco: W. H. Freeman, 1957)].

the Los Alamos team during the war. By late 1952, using the Windscale plutonium, it had designed and tested the first British plutonium bomb.

These post-war sites promised employment for a new breed of nuclear specialists. Hinton, and to a lesser degree Cockcroft, were to create a distinct and enduring British identity for the workers. As in the USA, inadvertent features also shaped identity: the rapid expansion and ambitious plans for the organizations, for example, created severe shortages of competent participants.[57] As a result, this money-rich but expertise- and resource-poor post-war environment defined enduring characteristics of nuclear engineers.

As with Zinn's Argonne, industrial benefits were an early aim. The Division of Atomic Energy hoped initially 'to sub-delegate whole sections of the design and supply of plant to the industry'.[58] The civil servants managing the programme, as historian Margaret Gowing summarized, felt that an atomic pile 'did not fit easily into any of the existing industrial classifications', but judged that the best match for the complex of plants was the chemical plant industry.[59] On the other hand, this was not to be a civilian operation like the pre-war chemical industry. As a result, industrial collaboration proved difficult to attract. ICI had pursued research in atomic energy as a wartime duty; having enthusiastically identified peacetime commercial power generation in the 1940 MAUD report, the company declined to participate in the post-war British bomb programme. Smaller British firms after the war were even less enthusiastic, usually identifying the required skills as too demanding for their over-stretched resources. Part of the firms' reticence was likely attributable to an impression that continued Ministry work would be relatively unprofitable, secretive, and hamstrung by red tape. Moreover, the urgency of development provided 'far less certainty than any industrial firm would accept'.[60] Even within the Ministry of Supply views were diffident. The planned industrial programme placed uncomfortably heavy demands on the capacity of the chemical engineering and heavy electrical industries, both of which were vital to the revival of the export trade. The atomic energy programme worked against the demands of market economics. Industrialists were, in general, unenthusiastic about participating in the atomic energy work.[61] As a result, the management of the post-war project would have to continue under government jurisdiction, much like wartime research and

---

[57] Amenities and housing were also in short supply at new centres such as Harwell and Chalk River. Administrators faced the additional problems of attracting and retaining workers at the stark and isolated sites. See, for example, Behrens, D. J., 'Life at Harwell', *Atomic Scientists' News* 2 (1953): 173–6 and Kinsey, Freda, 'Life at Chalk River', *Atomic Scientists' Journal* 3 (1953): 18–23.

[58] Gowing, Margaret and Lorna Arnold, *Independence and Deterrence: Britain and Atomic Energy*, Vol. I: *Policy Making, 1945–52* (London: MacMillan, 1974), p. 168.

[59] Gowing, Margaret and Lorna Arnold, *Independence and Deterrence: Britain and Atomic Energy*, Vol II: *Policy Execution, 1945–52* (London: MacMillan, 1974), p. 155.

[60] Owen, Leonard, 'Nuclear engineering in the United Kingdom—the first ten years', *Journal of the British Nuclear Energy Society* 2 (1963): 23–32 and 296–8.

[61] '1953–1955 Industrial representatives attached to atomic energy establishments', NA AB 16/1324. Employees of the British Chemical Plant Manufacturers' Association, Rolls Royce, General Electric, and English Electric had been paid by their firms to work at AERE, but the Division of Atomic Energy noted that 'there was no intention to subsidize industry to enable them to acquire a knowledge of Atomic Energy techniques'.

development. During this period, then, post-war British nuclear specialists worked in a familiar wartime context.

4.5 Shaping secret programmes

These contexts shaped the post-war specialists. The three distinct national programmes amounted to separate experiments in the creation of a new intellectual field.[62] And because of the insulating effects of post-war secrecy, these pockets of expertise grew in distinct flavours—what American radiochemist Glenn Seaborg later described as 'small islands of technical information sealed off from the rest of society'.[63] Security proved seminal to their professional identity. Sociologist Edward Shils observed at the time that secrecy is a means of controlling suspected conspiracies.[64] Indeed, conspiracies—being actions organized to achieve an end sought by a group having common interests—have much in common with professional activities; by attempting to control the former, one is likely to stifle the latter.[65]

The security implications of nuclear knowledge, particularly in the American context and during the Cold War, have been the focus of studies by Shils during the mid-1950s, and Daniel Moynihan from the perspective of the post-Soviet period. Both distinguished *functional* secrecy—pragmatically necessary control of sensitive knowledge—from *symbolic* secrecy, identified as an ideological extremism having deleterious effects for a nation and, by extension, to social groups within it.[66] The nuclear context of secrecy, which amplified and routinized such symbolic secrecy, has been addressed insightfully by research from the top-down perspective of policy studies,[67] but seldom in relation to its influence on technical

---

[62] On analyses of distinctions between engineers on the national scale, see the thematic issue and opening essay of Chatzis, Konstantinos, 'Introduction: the national identities of engineers ', *History and Technology* 23 (2007): 193–6.

[63] DOE, OpenNet, 'Remarks by Seaborg at the 25th anniversary of Argonne National Laboratory, Argonne Illinois (AEC-S-12-71)', 19 June 1971, DOE ON NV0712412, p. 6. Seaborg (1912–1999), a University of California, Berkeley, scientist who was responsible for plutonium chemistry at the University of Chicago Met Lab, discovered 10 transuranic elements, gaining the 1951 Nobel Prize in Chemistry for the work.

[64] Shils, Edward, *The Torment of Secrecy: The Background and Consequences of American Security Policies* (Glencoe, Ill.: Free Press, 1956), pp. 27–33.

[65] As he notes, 'Since secrecy is so damaging to solidarity, the mere possession of a secret gives rise to the suspicion of disloyalty' [p. 35]; 'Within professions and professional societies, within occupational groups…the emergence of an alleged national crisis attenuates autonomy and enfeebles the will to autonomy' [p. 45].

[66] Shils, Edward, *The Torment of Secrecy: The Background and Consequences of American Security Policies* (Glencoe, Ill.: Free Press, 1956) and Moynihan, Daniel P., *Secrecy: The American Experience* (New Haven, Conn.: Yale University Press, 1998).

[67] E.g. Krige, John, 'Atoms for Peace, scientific internationalism and scientific intelligence', *Osiris* 21 (2006): 161–81; Seidel, Robert W., 'Secret scientific communities: classification and scientific communication in the DOE and DoD', in: M. E. Bowden, T. B. Hahn and R. V. Williams (eds.), *Proceedings of the 1998 Conference on the History and Heritage of Scientific Information Systems* 1999), pp. 46–60; Kevles, Daniel, 'Cold War and hot physics: science, security and the American State, 1945–1956', *Historical Studies in the Physical and Biological Sciences* 20 (1990): 239–64; Galison, Peter and Barton J. Bernstein, 'Physics between war and peace', in: M. R. S. Mendelsohn and P. Weingart (eds.), *Science, Technology, and the Military* (Dordrecht: Kluwer Academic, 1988), pp. 47–86. On the framing of secrecy for policy purposes, see Herken, Gregg, '"A most deadly illusion": the atomic secret and American nuclear weapons policy, 1945–1950', *Pacific Historical Review* 49 (1980): 51–76.

identity, i.e. from the grassroots. Previous studies of secrecy in the nuclear field have focused on those vocal and visible members of the 'atomic scientists movement' who actively promoted internationalism and progressive ideals during the post-war period,[68] or on contemporary weapons designers.[69] By contrast, the effects of secrecy on their cousins the engineers, technicians, and other skilled nuclear specialists, who arguably were even more affected, kept them relatively voiceless and unexamined.

Because nuclear specialists were managed in new isolated environments, their intellectual products were classified for restricted dissemination; their political activities and labour representations were scrutinized; their collective identity was shaped by pre-existing institutional and industrial affiliations; and their training, job categories, and disciplinary labels were assigned largely by their respective governments. All of these aspects of working life were dissimilar to prevailing pre-war scientific and industrial contexts.

Secrecy operated in three spatial domains: the international secrecy operating between nation-states, the domestic secrecy between different atomic energy sites, and the interactional secrecy required of individual nuclear workers. Central to each was intellectual isolation. The term *cloister* is apt, an analogy identifying atomic energy establishments as akin to monastic sites, and the association of their dedicated workers with enclosed religious orders. Like mediaeval monasteries, wartime and post-war national laboratories in the USA, UK, and Canada promoted distinct regional variations; they were not founded primarily with economic motivations, but on locally nuanced and isolated intellectual foundations; they combined idealism with pragmatic duties; and they served a strong central authority. Although a limited analogy, it suggests the peculiar environments for nuclear workers during their early years, and the combination of personal, institutional, and intellectual containment.[70]

---

[68] E.g. the sympathetic study by Smith, Alice Kimball, *A Peril and a Hope: The Scientists' Movement in America, 1945–47* (Chicago: University of Chicago Press, 1965), and Strickland, Donald A., *Scientists In Politics: The Atomic Scientists Movement, 1945–46* (Lafayette: Purdue University Studies, 1968) on its factionalism and ideological conflict. Both focus on the two post-war years during which the participants had the greatest public exposure and influence. For a layman's view of the Federation of Atomic Scientists, see Lang, Daniel, 'The unscientific lobby', in: Lang, Daniel, *From Hiroshima to the Moon: Chronicles of Life in the Atomic Age* (New York: Simon & Schuster, 1959), pp. 53–65. On the differentiation of technical specialists, see Whalley, P. and S. R. Barley, 'Technical work in the division of labor: stalking the wild anomaly', and Perlow, L. and L. Bailyn, 'The senseless submergence of difference: engineers, their work, and their careers', both in: S. R. Barley and J. E. Orr (eds.), *Between Craft and Science: Technical Work in U.S. Settings* (Ithaca: ILR Press, 1997), pp 23–52 and 230–43, respectively.

[69] E.g. Parfit, Michael, *The Boys Behind the Bombs* (New York: Little and Brown, 1983); Rosenthal, Debra, *At the Heart of the Bomb: The Dangerous Allure of Weapons Work* (Reading, Mass.: Addison-Wesley, 1990). Along the lines of this section, but focusing on a different occupational group, is Gusterson, Hugh, *Nuclear Rites: A Weapons Laboratory at the End of the Cold War* (Berkeley: University of California Press, 1996), Chapter 4, an anthropological study which discusses the role of secrecy in shaping the identities of American weapons scientists.

[70] Donald Strickland similarly suggests a religious caricature for the Federation of Atomic Scientists as 'Jesuits in the Imperial Chinese Court—fascinating the rulers with their technology, implanting their doctrine in attenuation, attracting both the admiration and the criticism of their co-religionists' [Strickland, Donald A., *Scientists In Politics: The Atomic Scientists Movement, 1945–46* (Lafayette: Purdue University

Table 4.1 Initiatives and events relating to atomic energy secrecy.

Year	USA	UK	Canada
October 1939	Briggs Advisory Committee on Uranium formed		
April 1940		MAUD Committee formed	
December 1940			National Research Council experiments
July 1941	S-1 Uranium Committee formed		
July 1941		MAUD reports released Tube Alloys Project begun	
August 1942	Manhattan Engineer District formed		
August 1943	Quebec Agreement	Quebec Agreement	Quebec Agreement
August 1945	Smyth Report		
January 1947	McMahon Act		
March 1946			Alan Nunn May spy case
February 1950		Klaus Fuchs spy case	
May 1950	Julius and Ethel Rosenberg spy case		
August 1950		Bruno Pontecorvo defection	
September 1950	McCarran Internal Security Act		
June 1952	McCarran-Walter Immigration and Nationality Act		
December 1952			NRX reactor accident
August 1954	Revised Atomic Energy Act		
August 1955	(First) International Conference on the Peaceful Uses of Atomic Energy (Switzerland)		

Three distinct temporal regimes of security are also relevant, as suggested by the brief chronology listed in Table 4.1. The first was the wartime secrecy imposed by the Manhattan Project and precursor committees in the USA and UK, described in Chapter 3. The second comprised nearly a decade after the 1946 McMahon Act in the USA, which had limited the circulation of knowledge between the former allies and during which security measures ebbed and flowed in response to perceived foreign and domestic threats of espionage in each country. And a third, considerably more relaxed, period followed the 1954 'Atoms for Peace' initiative of the Eisenhower administration and especially its consolidation by the subsequent UN-sponsored Conference on the Peaceful Uses of Atomic Energy in Geneva in August 1955, to be discussed in Chapter 6.

In brief, then, the wartime Manhattan Project secrecy, followed by a decade of selective and varying security measures, were succeeded by a new openness from the mid-

---

Studies, 1968), p. 64]. Taking another tack on the analogy, Alvin Weinberg mused on the need for a 'nuclear priesthood' to tend the dangerous technology of nuclear power [Weinberg, Alvin M., 'Social institutions and nuclear energy', *Science* 177 (1972): 27–34].

1950s. These three security regimes did not evolve towards complete transparency, however: military secrecy was to give way to rising commercial secrecy. In short, the enduring nature of technical specialists in the nuclear field owed much to their cloistered origins.

4.5.1 *Compartmentalization*

The initial regime of security had been instigated by the wartime Manhattan Project, during which the US, UK, and Canada had an on-again, off-again collaboration. The wartime investigations of atomic energy were undertaken in environments appropriate to a potential military secret, and management focused largely on restricting and channelling the flow of information about the very existence of the project and its technological trajectory. The principal feature of security, from the standpoint of the technical participants, was intellectual isolation within and between the handful of sites developing reactor technology. The post-war role of governments in linking science to national goals is well known.[71] Paul Forman has highlighted the unusual compartmentalization of post-war American national labs, with rapidly growing knowledge ring-fenced against the threat of espionage. As wartime installations were expanded into national laboratories focusing on distinct intellectual terrain, site segregation—a direct consequence of security concerns—was important for breeding the first cohorts of nuclear specialists.

The fitful collaboration between the UK, USA, and Canada had shaped their respective programmes early on. American access to Anglo-Canadian information had been unfettered through 1944 and 1945. The 'Special Secret Committee' established in February 1945, allowed 'the Trust, US Military Intelligence and the like to have direct information on what was being done in Canada... In addition, secret reports on the work done in Canadian laboratories on the development and improvements in analytical techniques and so on have been circulated'.[72] By contrast, no British or Canadian workers had been permitted to visit the Hanford site—the most secret of American installations—and the design principles and practicalities of its plutonium-producing piles were learned piecemeal and second-hand. Not only was expertise still secret; it had to be reinvented at each national site and passed on by unformalized routes.

The Manhattan Project hierarchy was reiterated by a secret cable immediately after the Japanese bombings, which defined post-war US censorship policy:

> Nothing may be written, discussed or used in any media of publication on the following.
> - Specific processes, formulas and mechanics of operation.
> - Stocks, location of stocks, procurement of stocks and stock consumption.
> - Quality and quantity of production of active material.
> - Physical characteristics of the weapon and methods of using it.
> - Speculation in the future development of the processes for military purposes.

---

[71] See, for example, Galison, Peter and Bruce William Hevly, *Big Science: The Growth of Large-Scale Research* (Stanford, Calif.: Stanford University Press, 1992); Seidel, Robert W., 'A home for Big Science: the Atomic Energy Commission's laboratory system', *Historical Studies in the Physical and Biological Sciences* 16 (1986): 135–75; Weinberg, Alvin M., *Reflections on Big Science* (Cambridge, Mass.: MIT Press, 1967).

[72] Thomson, L. R. to C. J. Mackenzie, letter, 1 October 1946, LAC RG77 Vol. 283.

- Information as to the relative importance of the various methods or plants or of their relative functions or efficiencies.

The policy censored more than bomb-making; in effect, it capped the fragile field of nuclear engineering as a whole. In the light of wider political factors and the anticipated benefits of continued collaboration with the Americans, C. J. Mackenzie and John Cockcroft recommended that Canada and the UK follow the same policy.[73]

Security inhibited and became ingrained for many participants. When General Groves ruled that French contributors must leave Chalk River at the end of the war, those who had left the Anglo-Canadian project to return to France instituted similar security consciousness. Key participant Lew Kowarski reported that 'Chatillon has its guards, as you in Brookhaven have yours; we are responsible for some of the secrets and our people are bound by certain, not very explicit but very binding, secrecy rules'.[74] What had been understood as sensible wartime security precautions for nuclear engineering were thus straightforwardly exported to other contexts. This combination of administratively decreed and personally adopted segregation helped to create proto-disciplines along clearer national lines. As suggested by the post-war French adoption of security apparatus, some of the features of wartime secrecy—segregated environments, cosseted isolation, and careful monitoring of safety procedures—were amplified over the following decade to manage new perceived threats. Other conditions important in shaping technical identity—the filtering and categorization of personnel—developed in distinct national contexts.

At the new American National Laboratories, by contrast, security concerns during the first post-war years were relatively unproblematic. Arthur Roberts' lyrics for 'Take away your billion dollars' said nothing of secrecy, which was not yet demanding great attention among his colleagues at Brookhaven. As during the war, security concerns were identified as an operational nuisance rather than as a fundamental issue affecting the nature of science and engineering development. Indeed, the creation of the US Atomic Energy Commission initially meant that the military organization that had been imposed by the Manhattan Project was somewhat relaxed. The Argonne National Lab, for example, merely asked technical staff to sign a loyalty oath:

I _____ do solemnly swear

A. That I will support and defend the Constitution of the United States against all enemies, foreign and domestic; that I will bear true faith and allegiance to the same; that I take this obligation freely without any mental reservation or purpose of evasion.

B. That I am not a Communist; that I do not advocate nor am I a member of any organization that advocates the overthrow of the Government of the United States by force or violence; that this oath (or affirmation) will remain in effect during the period that I am a holder of an Atomic Energy Commission fellowship.[75]

---

[73] 'Statement on US censorship policy', memo, 13 August 1945, LAC RG77 Vol. 283.

[74] Kowarski, L., 'Atomic energy developments in France', *Atomic Scientists' News* 2 (1948): 16–21. Zoé, the first French reactor, was built at Fort de Châtillon, near Paris, in 1947.

[75] 'Draft Operating Policy of the Argonne National Laboratory', memo, 28 February 1950, UI Box 19. See also Brown, Ralph S., *Loyalty and Security: Employment Tests in the United States* (New Haven, Conn.: Yale University Press, 1958).

Ritualistic avowals of loyalty, like the recitations of matins prayers in monasteries, thus carried inordinate weight in working life.

Walter Zinn at Argonne casually excused the 'startlingly large number' of security personnel to the University of Chicago manager as 'required due to the multiplicity of locations and due to the insistence of the Atomic Energy Commission that all areas be under guard'.[76] On the other hand, the now publicly known atomic energy facilities raised new security concerns about espionage, and worked against such a relaxation.

The physical isolation of secure sites bred intellectual isolation, too. As Peter Galison has discussed, the classification of knowledge—military secrets, trade secrets, inventive expertise—accelerated during the twentieth century and especially after the Atomic Energy Act of 1946—'the founding document of modern secrecy'—with profound consequences.[77] Governments had been cautious to contain the circulation of nuclear knowledge that could have military application. Thus, a Committee on Declassification set up by General Groves in 1945 had generally opted to withhold both 'facts of nature' and engineering data.[78] But the ostensibly civilian AEC interpreted 'restricted data' just as conservatively to include not merely 'all data concerning the manufacture or utilization of atomic weapons', but also 'the production of fissionable material [and its use] in the production of power' and other applications.[79]

But the 1946 McMahon Act further compartmentalized information and personnel by ending cooperation with the new British and Canadian programmes: henceforth there would be no collaboration on atomic energy between the former allies, although some was nevertheless available by barter and special pleading. The post-war security aimed to accomplish three things: protect secrets of military value, particularly from domestic and foreign communist espionage; isolate the public from the potential dangers of radioactivity; and to sustain the hard-won technological advance of the wartime work for other national or commercial advantage.

It was a similar story in Britain. But while the mutual American–British concealment was motivated by economic as much as military confidentiality, the perceived threat of Soviet espionage dominated the following decade. Indeed, the Division of Atomic Energy was sometimes more cautious than its American counterpart. For example, in reviewing a proposal by the AEC Director of Classification, Christopher Hinton admitted that the current secrecy was 'unhealthy and undesirable', but argued that more than 'the technology of advanced reactors and the design of atomic weapons' should be classified; the British view was that 'instrumentation, health physics and control techniques in connection with reactors' should also not be considered open fields.[80]

Canadian research—even though eschewing military applications—absorbed many of the security attributes of American and British sites. C. J. Mackenzie complained to Canadian Minister C. D. Howe that information flow continued to be in one direction only: the

---

[76] Zinn, Walter to W. B. Harrell, letter, 4 February 1948, UI Box 19.

[77] Galison, Peter, 'Removing knowledge', *Critical Enquiry* 31 (2004): 229–43; on classification of nuclear knowledge, see pp. 232–5.

[78] Manley, John H., 'Secret science', *Physics Today* 3 (1950): 8–11.

[79] Argonne National Laboratory, 'Subcontractor contract template', July 1946, UI Box 44.

[80] 'Secrecy in nuclear engineering: comments on a paper by Dr J.G. Beckerley, USAEC 1952 11/1/7/2(17)', NA AB 6/1063.

Americans released information to the Canadian company Eldorado about uranium processing in exchange for badly needed raw ore, but 'in spite of a strong case and an active campaign for cooperation on the part of the American group on reactor design, up to the present time we have had no concessions whatsoever'.[81] Canadian access to American sites was also restricted. Beginning in 1948, requests by Canadian scientists to attend training courses on isotopes at Oak Ridge had to be channelled from Chalk River consecutively to the Department of External Affairs, the Canadian Ambassador in Washington, and the US State Department, which then sought security clearance from the FBI, because the Americans mistrusted Canadian security procedures.[82]

Security concerns also limited staff appointments at Chalk River. The McMahon Act discouraged Americans from joining foreign establishments, and Canadians were deterred by restrictions on dissemination: as the Director W. B. Lewis complained, 'there are a few on the staff here who are working on subjects which they are free to publish and are very likely to leave if publication were not possible'.[83]

But in a professional atmosphere soured by the McMahon Act, self-censorship also inhibited nuclear workers. An academic at the University of Chicago, for example, hesitated to take a post at McMaster University in Canada because he was advised by a colleague that he could be convicted of divulging US atomic secrets to his students. Administrators at McMaster and C. J. Mackenzie, while dismissing such fears, could offer only diffident assurances that

> there is not the slightest chance of... being prosecuted for carrying on ordinary teaching and research work at any university, providing always that he observes what he is morally and ethically bound to do; that is, not to pass on directly, and as such, specific technological information which he has obtained as a member of one of the super-secret projects in America.[84]

As Galison argues, nuclear security, while having few guiding principles, bore a strong resemblance to long-established trade secrecy.[85] Manhattan Project administrators had feared British commercial interests in atomic energy, and post-war British access to American information was even more seriously curtailed. Despite the embedding of ICI's working culture in the new atomic factories, Christopher Hinton found technical information difficult to obtain even from ICI, his former employer and manager of Tube Alloys.[86] This did not concern merely nuclear secrecy, but also old-fashioned commercial confidentiality. Hinton later recalled:

> In most design offices, when the plant flow sheet has been drawn, it is possible to take, from the files, drawings which have been used for other plants and to modify these or use them as a basis for the design of some

---

[81] Mackenzie, C. J. to C. D. Howe, letter, 15 August 1952, LAC MG30-B122 Vol. 3.

[82] 'Visits to scientific establishments—Oak Ridge Institute ', 1948, LAC RG25-B-2 Vol. 2143. The September 1945 revelations by Igor Gouzenko, a Soviet cipher clerk in Ottawa—which implicated Alan Nunn May (1911–2003) of Chalk River and Klaus Fuchs (1911–1988) and led to the arrest of over three dozen suspects in Canada—were an important precedent for the McMahon Act and the Canadian reaction to regain American confidence.

[83] Lewis, W. B. to D. A. Keys, 24 September 1947, LAC RG77 Vol. 283.

[84] Mackenzie, C. J. to H. G. Thode, letter, 29 December 1947, LAC RG77 Vol. 283.

[85] Galison, Peter, 'Removing knowledge', Critical Enquiry 31 (2004): 229–43, pp. 238–9.

[86] NA SUPP 10.

of the vessels that are needed... the trouble did not lie in the fact that we were incapable of doing this work when starting from a clean sheet of paper, it lay in the fact that this took time, and the experienced engineers (of which we were desperately short) had to do work which could otherwise be done by draughtsmen. I asked my old division of ICI if they would let me have designs of typical vessels; they refused, claiming that their processes were secret. Considering that I had asked for drawings of vessels that might be used on any chemical plant and which had no secret features, I thought this was singularly unhelpful.[87]

The ramifications of US security on Canadian and British workers rebounded on American workers as well. At the Argonne National Laboratory, for example, dedicated to non-military applications of atomic energy from 1946, security had been relatively light. Guards were required at all buildings, and visitor access was controlled. But from 1950, with the rising national concerns of atomic spies, FBI checks became mandatory. The nuclear workers protested against such restrictions, leading the Associate Director of ANL to announce a public relations initiative:

Until recent months it was very difficult and often impossible to receive scientific visitors who did not hold an AEC 'Q' clearance. This deprived the Laboratory of the normal exchange of scientific ideas so necessary in research. It also placed us at some disadvantage in the recruitment of scientific personnel.

Now our staff members in Buildings 202, 310 and 180–181 may receive US citizens as scientific visitors with almost the same freedom as they would enjoy on a university campus. In nearly three quarters of the rest of the Laboratory, subject to a preliminary AEC check and certain simple but necessary internal precautions, scientific visitors under escort may be received on unclassified visits for useful and approved purposes... I do not think curiosity will provoke any flood of visitors since no one need receive an unwanted visitor...[88]

Even so, the newly 'open' labs were Biology, Meteorology, and Effluent Studies—none of them concerned directly with nuclear engineering—and could be visited without security formalities only by 'US citizen visitors coming for scientific purposes' and with prior appointments.

For each of the British atomic energy establishments, security was an early and continuing concern. A reporter on a press tour of the still-incomplete Harwell in 1948 noted that

when scientists started research on atomic bombs they took a path which was bound to lead away from the free scientific world in which knowledge knew no national frontiers, a path which was bound to lead, in the absence of international control of atomic energy, into a strange murky country of guarded laboratories and secrecy regulations... our first sight of Harwell [was of] the ten-foot high barbed wire fence around the establishment. The first people we met were uniformed War Department Police and plain-clothed security officials of the Ministry of Supply.

Nevertheless, he concluded,

It would be wrong to leave readers with the impression that Harwell swarms with security officials; a covey of visiting newspapermen would of course induce an artificial swarm of these conscientious 'secret servants'. Security precautions there have to be; at Harwell they appear to be completely adequate, but they are not excessive and I am quite sure that the average scientist working at Harwell does not find them irksome.[89]

---

[87] IME Hinton papers A.3, p. 152.
[88] Bryce, J. C., 'Admission of uncleared scientific visitors', memo, 30 September 1953, UI Box 44.
[89] Dick, William E., 'The hangars hide uranium piles', *Discovery* 9 (9), 1948: 281–5; quotations p. 281 and 285.

4.5.2 Scrutiny

The view from the ground was different. While reporters may have found such measures necessary and reasonable, post-war organizations of nuclear workers—including substantial numbers of engineers as well as scientists—addressed the disadvantageous national, professional, and personal consequences of government security policies.

The Association of Oak Ridge Engineers and Scientists (AORES), founded in 1946, combined several bodies active immediately after the war: the Association of Oak Ridge Scientists (working at the Clinton Laboratory), the Atomic Production Scientists of Oak Ridge (employees of the Tennessee Eastman Corporation), and the Atomic Engineers of Oak Ridge (employees of the Carbon and Carbide Corporation). Their counterparts at other Manhattan Project sites were the Atomic Scientists of Chicago, the Association of Los Alamos Scientists, and the Association of Manhattan Project Scientists, who collectively formed the Federation of Atomic Scientists (FAS) in October 1945.[90] In the UK, the Atomic Scientists Association (ASA) also formed in 1946 to represent the technical, social, and policy views of engineers in the British programme, many of whom had been involved in Tube Alloys and the British Mission.

The organizations provided an early opportunity to demarcate scientific and engineering political identities. The prominence of scientists over engineers in these organizations is notable, and reflected the hierarchy of post-war atomic energy. As Richard Hewlett observes, the new American AEC was disproportionately top-heavy with scientists, appointing only a single engineer among its first board, and under-representing engineers both publicly and within the organization.[91]

Nuclear scientists had associated first at Chicago and then Clinton Laboratories, but remained relatively aloof of production scientists and engineers at the uranium separation plant sites, arguing that, as action groups, they anticipated that difficulties in reaching agreement would be multiplied by larger numbers and diversity of members. The Hanford site showed few stirrings of interest, and, owing to job segregation, there were dramatic differences in levels of activity and opinion that later were labelled with political connotations. As John Marshall, a Du Pont production physicist there recalled,

> there wasn't any organized opposition to the use of the bomb. We were concerned, and hoped it wouldn't be used against cities. There were relatively few people there who knew there was a bomb, just the scientific staff and some management. There was a discussion among a few of us about what would be done with the bomb but we didn't try to exert pressure on anybody'.[92]

But engineers in both the UK and USA remained a more invisible constituency during this period. Reflecting the reticence of British engineers, the Secretary of the Federation of

---

[90] See Smith, Alice Kimball, *A Peril and a Hope: The Scientists' Movement in America, 1945–47* (Chicago: University of Chicago Press, 1965), especially Chapter 3.

[91] Hewlett, Richard G., 'Beginnings of development in nuclear technology', *Technology and Culture* 17 (1976): 465–78. More recently, Pap Ndiaye has highlighted the downgrading of engineers in nuclear historiography [Ndiaye, Pap A., *Nylon and Bombs: DuPont and the March of Modern America* (Baltimore: Johns Hopkins University Press, 2007), Chapter 4].

[92] Sanger, S. L. and Robert W. Mull, *Hanford and the Bomb: An Oral History of World War II* (Seattle: Living History Press, 1990), p.129. On these post-war activities, see 'The Federation of Atomic Scientists', *Bulletin of the Atomic Scientists* 1 (1945): 2 and Smith, Alice Kimball, *A Peril and a Hope: The Scientists' Movement in America, 1945–47* (Chicago: University of Chicago Press, 1965), especially Chapter 3.

Atomic Scientists proffered that American engineers 'are mostly apolitical, but when they do have views they are of a stereotyped conservative nature', a view contested by historians of American engineering David Noble and Edwin Layton.[93]

The reputed connection between political viewpoint and reliability affected nuclear workers at every level. In the UK, the ASA, seeking 'to keep the public informed about developments in atomic energy', steered an awkward thirteen-year course between representing official views and those of its independent members. Among its early policy recommendations were calls for international control, and for the use of atomic energy only for peaceful purposes. Its attention to security matters made the ASA more vulnerable. Its first meeting had agreed to supply technical advice for the defence in the espionage trial of Alan Nunn May. The Association of Scientific Workers (AScW), founded in 1918, was an early ally, offering association with the ASA.[94] A close link was rejected, though, given the ASW's history of championing labour and progressive causes that were increasingly suspect in the post-war political environment.[95] Even so, the forthright stances of the Atomic Scientists Association appeared dangerous to some. Sir Wallace Akers, the ICI manager of the wartime Tube Alloys project, cautiously asked for guidance about the company's attitude, noting that he 'imagined that ICI couldn't prevent people joining in a private capacity'. Senior management at ICI, in fact, 'suggested that it is all to the good for reasonably minded people to become members'.[96] But government support was more negative: John Cockcroft, a Vice President of the ASA, argued that the Civil Service discouraged its employees from holding office in such an association, with the result that 'the initiative is apt to pass into the hands of the very left wing scientists'.[97] The personal dimension of security was well characterized by British physicist Neville Mott: 'Anyone entering a government establishment will know that for the rest of his career his political life and associations will be under observation, and his advancement and perhaps his job will depend on them. This will not add to the attraction of a career in government service.[98]

The views of Cockcroft and Mott reveal the caution of scientists to voice opinions that might deviate from the primary funder in the atomic energy field, the government. While members of the Division of Atomic Energy did, indeed, become officers, they complained about some of the public positions adopted by the Association. For example, the *Atomic Scientists' News*—modelled on the Federation of Atomic Scientists' *Bulletin of the Atomic Scientists* in America—argued that the Division of Atomic Energy was excessively cautious

---

[93] Meier, Richard L., 'The origins of the scientific species', *Bulletin of the Atomic Scientists* 7 (1951): 169–73, quotation p.171. On the political context of American engineers, see Noble, David F., *America By Design: Science, Technology, and the Rise of Corporate Capitalism* (New York: Knopf, 1977), which identifies a largely conservative stance allied to corporate interests, while Layton, Edwin T., *The Revolt of the Engineers: Social Responsibility and the American Engineering Profession* (Cleveland, Ohio: Case Western Reserve Press, 1971) identifies a more socially committed progressivist current.

[94] 'The Association of Scientific Workers and Atomic Energy', *Atomic Scientists' News* 1 (1947): 37–8.

[95] The AScW had been founded as the National Union of Scientific Workers immediately after the First World War, and was renamed in 1927. It amalgamated with the Association of Supervisory Staffs, Executives and Technicians in 1968, and reorganized eight years later to become the Association of Scientific, Technical and Managerial Staffs.

[96] Akers, Wallace to M. W. Perrin, letters, 29 and 31 August 1946, NA AB 16/52.

[97] Cockcroft, John to R. E. France, letter, 6 September 1951, NA AB 27/6.

[98] Mott, N. F., 'The scientist and dangerous thoughts', *Atomic Scientists' News* 2 (1949): 171–2.

about security criteria; R. M. Fishenden of Harwell retorted that this was 'exaggerated nonsense', and that the BEPO reactor held secrets that had to be safeguarded according to the UK, USA, and Canadian agreement on classification.[99] The ASA found itself criticized, too, for holding a conference on radiation hazards behind closed doors—what a science magazine condemned as '*public* enlightenment by means of *private* conferences!'.[100] At the same time, an internal debate raged about the representation and political role of the ASA. One of its Vice Presidents, Prof. H. W. B. Skinner of Liverpool University, argued that the Association was unrepresentative of the 'genuine profession of Atomic Scientists' employed at the government research centres: 'therefore in most ways the real Atomic Scientists have to be represented by others—partly ex-Atomic Scientists and partly people who have never had anything to do with this type of work'.[101]

Thus, security concerns questioned the authority, allegiances, and political affiliations of nuclear scientists, and even more thoroughly muzzled their engineering colleagues.

4.5.3 *Filtering*

Towards the end of the 1940s further events interpreted as destabilizing to American interests seemed to vindicate the policy of rising security. In September 1949, the Soviet Union tested its first atomic bomb; a month later, with the end of the Chinese civil war, Mao Zedong proclaimed the People's Republic of China, and signed a mutual assistance pact with Joseph Stalin; and, in mid-1950, North Korea invaded the south.

These developments, overtly provocative to Western eyes, were mirrored by unsettlingly covert ones, too. Espionage was discovered repeatedly within the nuclear programmes. The first atomic spy scandals, involving scientists Alan Nunn May (made public in March 1946) and Klaus Fuchs (February 1950), and the defection of University of Liverpool scientist Bruno Pontecorvo to the Soviet Union (August 1950), had links directly to the origins of the three atomic energy programmes.[102] The American espionage case of Julius and

---

[99] Hawes, Lewis, 'Far too secret a secret', *Atomic Scientists' News* 3 (1949): 7–19; Fishenden, R. M. to Editor of *Atomic Scientists' News*, letter, 19 August 1949, NA AB 16/52. On the decline of the Federation of Atomic Scientists after 1949, see Wang, Jessica, *American Science in an Age of Anxiety: Scientists, Anticommunism, and the Cold War* (Chapel Hill, NC: University of North Carolina Press, 1999), pp. 82–4.

[100] Skinner, H. W. B., 'Atomic energy and the public interest', *Discovery* 12 (9), 1951: 269–72; quotation p. 270.

[101] Ibid., quotation p. 271.

[102] The activities of Alan Nunn May, Klaus Fuchs, Bruno Pontecorvo, and the Rosenbergs unsettled administrators of the British programme. Their loss to those programmes was also significant. Fuchs headed the Theoretical Physics Division at post-war Harwell. And of Pontecorvo, an associate wrote, 'It is a very bad business in itself, and also for us at Liverpool since he will be impossible to replace. Nothing whatever has been heard of him, and it is horrible to think that people one knows can drop out suddenly in this way' [Skinner, H. W. B. to W. B. Lewis, letter, 5 December 1950, Kingston, Ontario, QU Box 1 File 13]. For a contemporary British account reflecting the paranoiac contemporary concerns, see Newman, Bernard, *Soviet Atomic Spies* (London: Robert Hale, 1952). On Fuchs, see Williams, Robert Chadwell, *Klaus Fuchs, Atom Spy* (Cambridge, Mass.: Harvard University Press, 1987). See also Turchetti, Simone, 'Atomic secrets and government lies: nuclear science, politics and security in the Pontecorvo case', *British Journal for the History of Science* 36 (2002): 389–415 and Agar, Jon and B. Blamer, 'British scientists and the Cold War: The Defence Research Policy Committee and information networks, 1947–1963', *Historical Studies in the Physical and Biological Sciences* 28 (1998): 209–52.

Ethel Rosenberg—arrested in 1950 and executed in 1953—challenged the dependability of lower-tier nuclear workers, and provided ammunition for Senator Joseph McCarthy (1908–1957) to prosecute a campaign alleging communist subversion throughout American institutions.[103] For a period of four years, during most of which the Korean War exacerbated paranoia, McCarthyism heightened national security through narrowly conceived patriotism. Subsequent spy convictions and Senator Joseph McCarthy's hearings aggravated such problems. During the highly publicized Rosenberg trial, for example, Hanford staff members were cautioned not to elaborate on any technical information disclosed in the courtroom.[104]

For individual engineers and scientists working on the national atomic energy programmes there was a more personal relevance of security: its influence on employment and career progression. Given the security context, it is not surprising that the political backgrounds of potential staff in each country were increasingly vetted from 1950. This scrutiny proved seminal for the first cohort of specialists. They worked for a single employer: their government. It meant that a politically filtered generation was admitted to the new field.

Loyalty oaths, for example, were replaced by mandatory FBI checks at Argonne in 1950, which could also be instigated by anonymous reports or suspicions of existing staff members.[105] AEC workers suspended under the Atomic Energy Act after an FBI investigation could find themselves denied due process by the Personnel Security Review Board, unable to learn the charges against them or to cross-examine witnesses, because the Commission was unwilling to reveal its sources of information.[106] Even though specialists denied posts at the national labs often found employment at firms or universities engaging in non-classified work, the risk of career consequences was considerable: government-sponsored research was growing rapidly at industrial research establishments and universities, and it appeared that political liberalism was being identified increasingly as disloyalty.[107]

This scrutiny of political views was more contested in the UK and Canada, which had not embraced the same degree of communist suspicion as in the USA. But by the early 1950s, the filtering of project personnel was increasing. When Klaus Fuchs, Head of Theoretical Physics at Harwell who had worked at Columbia University for the Manhattan Project, was convicted of espionage, security measures in Britain increased further—so much so that the Atomic Scientists Association issued a statement hinting at Stalinesque excesses in 'the Civil Service purge of persons who are members of extremist political organizations and likely to weaken

---

[103] Indeed, 'none of the famous scientist spies—Fuchs, Nunn May and Pontecorvo—were American; none of the American spies—Greenglass, Sobel, Gold or Rosenberg—were higher than technicians or engineers' [Shils, Edward, *The Torment of Secrecy: The Background and Consequences of American Security Policies* (Glencoe, Ill.: Free Press, 1956), p. 186]. Rosenberg was an electrical engineer employed during the war by the Army Signal Corps Engineering Lab in New Jersey, who obtained Manhattan Project information via his brother-in-law David Greenglass, a machinist.

[104] Shaw, D. F., 'Forthcoming espionage trial', memo, 7 March 1951, DOE DDRS D0983254.

[105] 'The New Government Employee Security Program', *Bulletin of the Atomic Scientists* 9 (1953): 175. See also Hewlett, Richard G. and Jack M. Holl, *Atomic Shield, 1947–1952* (Berkeley: University of California Press, 1969), pp. 88–95.

[106] 'Security procedures in the USA', *Atomic Scientists' News* 1 (1948): 162–4.

[107] 'American scientists involved in security investigations', *Atomic Scientists' News* 2 (1948): 49–54.

national security'. The Atomic Scientists Association urged the British government not to extend security to other institutions such as industry and universities.[108]

National filtering was also increased. American Senator Pat McCarran (1876–1954) subsequently sponsored two bills to limit communist influence on American research institutions (Table 4.1). The Internal Security Act (1950) and Immigration and Nationality Act (1952) limited entry to the USA by foreign scientists, technologists, and educators. The visa restrictions imposed by the State Department isolated Americans from their international peers.[109] The British government was equally wary in the months after Bruno Pontecorvo's defection. When the passport of physicist E. H. S. Burhop was revoked after he had accepted an invitation to visit Moscow in 1951, a flurry of protest to the 'island arrest' led to its reinstatement, with the caution that Burhop consult the Foreign Office before considering any visit to a country in the Russian orbit.[110]

For American nuclear workers there were three forms of fallout from the political atmosphere: an increased demand for urgent nuclear development and production, more arduous loyalty checks, and rising levels of secrecy. Their counterparts in Britain and Canada, hoping to regain contact with their American peers, adopted similar security measures. Chalk River consolidated its status as a closed site accessible only to cleared visitors. As perceived international and domestic threats grew, its borders became more immiscible: the Canadian press was refused access, and the editor of *Nucleonics*, the first American periodical in the field, was even more summarily dismissed.[111] The clamp-down apparently surprised and insulted British colleagues, too: David Keys, Vice President of the project, was compelled to assure the irritated John Cockcroft that the clearance procedures applied equally to British, Americans, and Canadians visiting Chalk River.[112] But Americans, too, chafed at the heightened security. Even though a USAEC liaison officer was based at Chalk River from 1950, at least one colleague at Hanford urged that current restrictions on exchange of information be liberalized for American benefit.[113]

---

[108] 'The purge in Britain', *Discovery* 11 (6) (1950): 202. The ASA admitted that its Council was divided about the need for atomic security that involved individuals to be penalized for associations that were not prescribed by law, but urged the government not to extend security to other institutions such as universities and industry. The same month (April, the month after Fuchs' conviction), the French government dismissed Frédéric Joliot as High Commissioner of Atomic Energy in France owing to his communist associations. Overtly political, rather than engineering, decisions may have been behind the subsequent French adoption of gas-graphite rather than heavy-water reactors favoured by Joliot [Hecht, Gabrielle, 'Political designs: nuclear reactors and national policy in postwar France', *Technology and Culture* 35 (1994): 657–85, p. 664].

[109] For a special issue devoted to the visa problems created by this 'American paper curtain', see *Bulletin of the Atomic Scientists* 8, 1 (October 1952).

[110] 'Freedom in science', *Atomic Scientists' News* 1 (new series) (1951): 2.

[111] Keys, D. A. to Editor of *Nucleonics*, letter, 23 April 1951, LAC MG30 B59 Vol. 4.

[112] Keys, D. A. to J. Cockcroft, letters, 17 July and 8 August 1951, LAC MG30 B59 Vol. 4.

[113] 'Atomic energy—Canada—General—US Atomic Energy Commission liaison officer at Chalk River', LAC RG25 Vol. 6675; Mather, R. E., 'Report of visit to Chalk River Ontario', memo, 22 January 1951, DOE DDRS D4193933. According to one staff member there, the roles of the liaison officer at Chalk River and contacts at the US national laboratories included espionage, revealing 'an undercurrent of distrust between the two projects' [Wallace, Philip R., 'Atomic energy in Canada: personal recollections of the wartime years', *Physics in Canada* 56 (2000): 123–31, quotation p. 128].

This ratcheted security thus had operational drawbacks. Secrecy was becoming increasingly complex to manage; cross-checking how secrets could inadvertently be revealed through innocuous publications was becoming arduous. At the same time, declassification was being superseded by the rise of independent nuclear research conducted in other countries. After the Soviet Union detonated its first atomic bomb, the American committee on declassification decided to fully declassify some simple reactor designs. The continuing strategy, however, was to trade off *security through secrecy* by *security through achievement*, i.e. to control nuclear secrets in a way that optimized the advance of the three former Allies.[114]

For each of the three countries, then, the post-war decade provided special occupational environments. National laboratories fostered the nuclear specialists born during the war, favouring dissimilar traits and defining distinct goals. Their expertise was valued so highly that security became the guiding principle. Protection of their secrets from espionage and sabotage isolated the experts from each other and further filtered their ranks. And within those cosseted environments—Argonne, Oak Ridge, Hanford, Chalk River, Harwell, and Windscale—the nature of their special knowledge of reactor technology was worked out.

---

[114] Beckerley, J. G., 'Declassification of low-power reactors', *Nucleonics* 8 (1951): 13–16; Manley, John H., 'Secret science', *Physics Today* 3 (1950): 8–11.

5
'LIKE CHILDREN IN A TOY FACTORY'

During the war, theoretical ideas had been translated with dizzying rapidity into immense industrial enterprises. The concentration of wartime experience at a handful of centres was crucial in nurturing the new specialists. But the drive for the new field was provided by glimmers of technological possibilities and the opportunities that they might provide to build a new discipline. As described by Abbott, new technologies create new intellectual tasks and potential jurisdictions; these can be appropriated by technical groups that devise appropriate 'paths to abstraction', rendering particular skills into more generic academic attributes.[1]

While bomb development had continued at Los Alamos, contributors to the Manhattan Project at the Met Lab in Chicago had had more time on their hands from the spring of 1944. Largely free of urgent tasks, they were planning further experimental reactors. Alvin Weinberg (1915–2006), at the time a young physicist working in Chicago with Eugene Wigner, later evoked a sense of the novelty and mystery of this new domain:

> As I think back on Wigner's design of the Hanford plutonium-producing reactors, I can think of only one analogy: Mozart, who would compose a piano concerto in a few days. The whole thing was accomplished in about four months…the entire conception was mind-boggling – a chain-reacting pile producing 500 megawatts at a time when no one had operated a critical pile at any power…we can hardly imagine the sense of strangeness that pervaded the enterprise in 1942.[2]

During the last year of the war, he recalled, 'everyone could play the game of designing new nuclear power piles…Crazy ideas and not-so-crazy ideas bubbled up, as much as anything because the whole territory was unexplored—we were like children in a toy factory'.[3]

The Manhattan Project had collected and amalgamated expertise during the war, and the post-war centres retained a critical mass of specialists eager to explore the technologies of atomic energy further. As a number of studies in historical sociology have demonstrated,

---

[1] Abbott, Andrew D., *The System of Professions: An Essay on the Division of Expert Labor* (Chicago: University of Chicago Press, 1988), pp. 92–3, 98–108; quotation p. 93.

[2] Weinberg, Alvin M., *The First Nuclear Era: The Life and Times of a Technological Fixer* (New York: AIP Press, 1994), p. 26. Weinberg (1915–2006) had obtained a PhD in mathematical biophysics before the war and, like the others of his generation, acquired and shaped the new field of nuclear engineering through his wartime activities at the University of Chicago and then Oak Ridge.

[3] Weinberg, Alvin M., *The First Nuclear Era: The Life and Times of a Technological Fixer* (New York: AIP Press, 1994), pp. 38–9.

technologies can form the heart of new technical collectives and disciplines.[4] The new national laboratories at Argonne, Oak Ridge, Chalk River, Harwell, and Brookhaven, alongside the production facilities at Hanford and Windscale, became centres of the new field of atomic energy, focused on its new technology: the nuclear reactor. In retrospect, these laboratories, driven by the enthusiasms of their creators, seem to have been almost free-running during the period 1945–1953. Site by site, their aim was to construct and evaluate reactor designs conceived largely during the war. Such facilities generated products and effects on a scale never before envisaged: vastly more radiation than produced by the world supply of radium, which itself had been discovered less than half a century earlier; new artificially produced elements, by-products of the reactors, having a plethora of unknown properties; and phenomena that challenged engineering knowledge and practice.

With distinct histories, administrators, and institutional goals, they also defined variants of the new discipline of nuclear engineering.[5] For the physicists, engineers, chemists, and biologists involved, the nuclear reactor was a generator of copious phenomena and, as many soon realized, enough to build a career on. Many did just that.

5.1 The core of a new discipline

5.1.1 Early explorations

The Smyth report, the official American summary completed weeks before the Hiroshima and Nagasaki bombings, had summarized forecasts of possible directions for post-war research and development:

> The possible uses of nuclear energy are not all destructive, and the second direction in which technical development can be expected is along the paths of peace...a multitude of suggestions [come] from men on the various projects, principally along the lines of the use of nuclear energy for power and the use of radioactive by-products for scientific, medical, and industrial purposes. While there was general agreement that a great industry might eventually arise, comparable, perhaps, with the electronics industry, there was disagreement as to how rapidly such an industry would grow; the consensus was that the growth would be slow over a period of many years...there is a good probability that nuclear power for special purposes could be developed within ten years and that plentiful supplies of radioactive materials can have a profound effect on scientific research and perhaps on the treatment of certain diseases in a similar period.[6]

---

[4] Abbott, Andrew D., *The System of Professions: An Essay on the Division of Expert Labor* (Chicago: University of Chicago Press, 1988), pp. 91–4; Galison, Peter, *Image and Logic: A Material Culture of Microphysics* (Chicago: University of Chicago Press, 1997); Turner, Stephen, 'What are disciplines? And how is interdisciplinarity different?' in: P. Weingart and N. Stehr (eds.), *Practising Interdisciplinarity* (Toronto: University of Toronto Press, 2000), pp. 46–65, p. 60; Joerges, Bernward and Terry Shinn (eds.), *Instrumentation: Between Science, State, and Industry* (Dordrecht: Kluwer, 2001).

[5] By contrast, the weapons research and development centres at Los Alamos and Aldermaston had different concerns and employed distinct technologies; their expertise diverged increasingly from the other sites.

[6] Smyth, Henry D., 'A general account of the development of methods of using atomic energy for military purposes under the auspices of the United States Government 1940–1945', 1 July 1945, LAC MG30-E533 Vol. 1.

Even scientists who had not participated in the Manhattan Project identified the potential for new occupations: the Hungarian-American aerodynamicist and military consultant Theodore von Kármán (1881–1963) suggested that, like radio and radar, 'atomic engineering' could be a new field opened by physicists to the benefit of engineers.[7]

Yet clear goals were elusive. Only Britain conceived a programme for power reactors in these early post-war years, but relegated it to the background as atomic weapons took priority. Indeed, a remarkable 1947 secret memo by John Cockcroft laid out a clear and prescient plan for development of British nuclear research, nuclear power, and a nuclear industry.

SECRET

Power Development Programme

The following seems likely to be the principal steps in a power development programme:-

(1) gas-cooled piles built primarily for plutonium production but generating power as a by-product

It is at present intended that UK should follow up the construction of two or three air-cooled piles with a programme of construction of gas-cooled piles. These would use aluminium sheathed natural uranium and would produce an emergent gas temperature of about 350°C. This gas would be used to generate steam for power production.

On the present schedule the first gas-cooled pile might operate in 1951/2.

(2) A pilot high temperature natural uranium pile built primarily for power production

The next step would be to carry design studies and experimental engineering work to the stage when reasonable estimates of performance and cost can be made. The following are the major items of work:-

(i) Studies to determine temperature of emergent gas required to obtain a reasonable thermodynamic efficiency of 25–30 per cent.
(ii) Comparison of steam turbine and gas turbine.
(iii) Determination of optimum gas flow arrangement.
(iv) Check of heat transfer by experimental work.
(v) Mechanical design of pressure envelope.
(vi) Decisions whether to use sheathing and, if so, what. Beryllium, refractory? Consequent metallurgical work.
(vii) Estimates of radioactivity likely to be deposited on pumps, etc., if no sheathing used—this will involve an experimental programme.

(cont.)

---

[7] von Kármán, Theodore, 'Atomic engineering?' *Journal of Applied Physics* 17 (1946): 2–3. He argued that 'the development of atomic power...is primarily of the nature of engineering', and that engineers should receive more scientific training just as medical doctors 'should have sufficient education to follow the development of biology' [p. 3].

> SECRET *Cont.*
>
> (3) <u>Pile for district heating</u>
>
> We should make a study of the economics of providing heat for a district heating system from a natural uranium pile, taking as a basis the Egerton report which recommended a primary circulation of hot water—250 deg C.
>
> It might be found that a Super-Harwell pile would provide the basis for such a scheme.
>
> (4) <u>Experimental high temperature fast burning plutonium slow reactor</u>
>
> The principal objective of this stage is to obtain information for a slow breeder programme. The first plutonium for such a reactor might be available in Canada in 1948 and in the UK in 1950. Allowing a two-year construction period, a UK design should therefore be completed in 1948. The design should be based on a realistic estimate of the plutonium which might be obtained for the pile, and I suggest that at this stage we take a target of not more than 5 kilos with a maximum of about 10 kilos.
>
> In view of the fact that effective breeding requires high burning rates, and therefore high neutron fluxes, the pile must provide experience on the behaviour of materials under the latter conditions. We may perhaps hope to get advance information from the NRX pile but will probably have to take a decision as to the moderator before this is available.
>
> The following points require to be determined:-
>
> (i) choice between beryllia and graphite as moderator. The behaviour of beryllia under high thermal stresses and under irradiation are the principal unknown. UK should press on with the first and ask Chalk River to provide the answer to the second.
> (ii) Study of fission product dispersal and removal.
>
> (5) <u>Fast reactor</u>
>
> We should aim at the construction of a fast reactor by 1951, the objective being to provide basic information and operating experience for a fast breeder programme.
>
> We might take as a basis a design which has already been developed. Fuel requirements for this are estimated at 15–20 kilos.
>
> (6) <u>Slow reactors using two-fold enriched 235</u>
>
> We have recommended the construction of a UK diffusion plant to provide in its low enrichment stages about 1.2 kilos per day of about two-fold enriched 235. This material is not likely to become available before 1952.
>
> J. D. Cockcroft
> 7th May 1947[8]

---

[8] Cockcroft, John D., 'Power Development Programme', 7 May 1947, LAC MG30 B59 Vol. 4. For a similarly goal-oriented perception, see Oliphant, M. I. E., 'How the atom can be harnessed for peace: power production by nuclear methods; with diagrammatic drawings by G. H. Davis', *Illustrated London News*, 207 13 October 1945: 399–401. On the fast reactor, see Section 5.2 below.

The Cockcroft memo illustrates a clarity of rational planning that was pursued effectively in Britain despite a civil-service dispersal of responsibilities and a frenetically varied research programme. Cockcroft's clear statement set the course for national development of atomic energy and, implicitly, its specialist workers.

At Chalk River, Canadian staff gave their all to reactor design. As NRC director Mackenzie underlined to James Chadwick, 'our government has suggested that they are not interested in work on the bomb, and we in Canada have never received a particle of information in connection therewith'.[9] The shunning of military applications and sponsorship placed the Canadian Atomic Energy Project in a more precarious but still favoured position—with no clear short-term remit but protected by the national status afforded by the new field and by the seductive promises of future applications. Indeed, this period with little governmental interference and 'scientific self-government' was referred to by some Canadian participants as 'the Golden Age'.[10]

The Anglo-Canadian ZEEP (Zero Energy Experimental Pile) achieved criticality weeks after the war ended. Conceived in 1944 as a small-scale pilot reactor, it provided the experience necessary for a much larger heavy-water reactor, the NRX (National Research Experimental, 1947), which became the most intense source of neutrons into the 1950s and so an important resource for the British and American workers, too.

Similarly, at Harwell, the GLEEP (Graphite Low Energy Experimental Pile, 1947) and BEPO (British Experimental Pile zero, 1948) research reactors were test-beds for studying technoscientific properties, a domain also being explored in the USSR and France.[11] And by 1950, Britain had two plutonium production piles at Windscale, Cumbria, equivalent to the Hanford facility but using air-cooling instead of water. Argonne National Laboratory, with the Navy as client, began testing a submarine prototype reactor at about the same time, an association that led eventually to the first American power reactor at Shippingport, Pennsylvania. And, a year later, the National Reactor Testing Station in Arco, Idaho, began operating the more innovatory EBR-1 (Experimental Breeder Reactor).[12] See Table 5.1.

---

[9] Mackenzie, C. J. to J. Chadwick, 28 January 1946, LAC RG77 Vol. 283.

[10] Redhead, Paul A., 'The National Research Council's impact on Canadian physics', *Physics in Canada* 56 (2000): 109–21; quotation p. 112.

[11] If the Soviet programme initially lagged behind Canada's Chalk River, it quickly outpaced that of Britain: the first Soviet reactor went critical on Christmas Day 1946, about 8 months before GLEEP, and the first plutonium production reactor on June 8, 1948, some 16 months before Windscale Pile 1. Key participants included Nikolai Dollezhal, chief designer of the production reactor, who worked with Igor Kurchatov, director of the bomb project, and oversaw the design of the first reactor for generating electrical power at Obninsk in 1954. By that time, a heavy-water reactor had also been built by Abraham Alikhanov and, as in the USA and UK, design studies of a variety of other reactor systems were under way [Holloway, David, *Stalin and the Bomb: The Soviet Union and Atomic Energy, 1939–1956* (New Haven, Conn.: Yale University Press, 1994), Chapter 16; Josephson, Paul R., *Red Atom: Russia's Nuclear Power Program from Stalin to Today* (New York: W. H. Freeman, 1999), pp. 22–3]. Similarly, the scientists at Montreal who returned to revitalize the French programme contributed to ZOE (Zero power, Oxide fuel, Eau lourde moderator) the first French reactor, started up in December 1948.

[12] The National Reactor Testing Station, established in 1949, had seventeen operating reactors within a decade, and another six being built ['The National Reactor Testing Station', *Nuclear Energy Engineer* 13 (1959): 200].

Table 5.1 Early research reactors in the USA, Canada, and UK.[13]

Year	University of Chicago/Argonne Laboratory, Illinois, USA	Clinton Laboratory/Oak Ridge, Tennessee, USA	Chalk River, Ontario, Canada	Atomic Energy Research Establishment, Harwell, UK	National Reactor Testing Station, Arco, Idaho, USA
1942	CP-1				
1943	CP-2	X-10			
1944	CP-3				
1945			ZEEP		
1946					
1947			NRX	GLEEP	
1948	STR	LITR		BEPO	
1950	ZPR-1	BSR			
1951		ORR			EBR-1
1952					MTR
1953	ZPR-2	TSF			BORAX-I
1954				DIMPLE, ZEPHYR	BORAX-II
1956				LIDO, DIDO	
1957			NRU, PTR	PLUTO	ETR

As this suggests, early British and American explorations of the field were strategically pursued to achieve short-term military aims, but were also opportunistic in piggy-backing research on them. Canadian nuclear workers were relatively unencumbered, and more diffident about specific applications in their early predictions. As the Vice President and most senior scientific advisor to the project, David Keys damped down the enthusiasm of one correspondent in 1951:

---

[13] This list omits the eight plutonium production reactors at Hanford, Washington, started up between 1944 and 1955, two at Sellafield, Cumbria, UK, in 1950, and five at Savannah River, Georgia, between 1953 and 1955, as well as two research reactors at Los Alamos (LOPO (1944), a low-power homogeneous reactor, and CLEMENTINE (1946), a fast neutron breeder reactor). From 1956, the first generation of power generating reactors came online. As indicated, Argonne (which managed the Idaho Test Station and designed the Hanford Reactors) and Oak Ridge dominated American reactor design. Following the postwar fashion for calculating machines, most of the early names were acronyms. BEPO: British Experimental Pile '0'; BORAX: Boiling Reactor Experiment; BSR: Bulk Shielding Reactor; CP: Chicago Pile; EBR: Experimental Breeder Reactor; EBWR: DIMPLE: Deuterium-moderated Pile Low Energy; Experimental Boiling Water Reactor; ETR: Engineering Test Reactor; GLEEP: Graphite Low Energy Experimental Pile; LITR: Low Intensity Testing Reactor; MTR: Materials Testing Reactor; NRX: National Research Experimental; NRU: National Research Universal. ORR: Oak Ridge Research Reactor; PIPPA: Pile for Producing Plutonium and Power; STR: Submarine Thermal Reactor; TSF: Tower Shield Facility; ZEEP: Zero Energy Experimental Pile; ZEUS: Zero Energy Uranium System; ZPR: Zero Power Reactor. The CP-1, CP-2, X-10, ZEEP, GLEEP, and BEPO were air-cooled, graphite-moderated natural uranium reactors; CP-3, NRX, NRU, DIMPLE, and PLUTO were heavy-water moderated natural uranium reactors; EBR was the first reactor to usefully generate electricity, and it and ZEPHYR pioneered the use of plutonium as a fuel and demonstrated the possibility of breeding; ZPR-1 was the first prototype submarine power unit.

I believe it will be many years before power will be developed by such a process for commercial uses. When such plants are constructed, they will probably find application in special places where it is difficult to obtain power by any of the usual methods...My personal opinion is that nuclear power will be achieved but will supplement rather than replace any of our conventional sources.[14]

A year later, C. J. Mackenzie, still at the helm of the Canadian activities, mused that it was a matter of confidence, pragmatism, and politics as much as technological trajectory:

> Living in a young country where we are inclined to be optimistic, we feel that even with our present piles we are getting valuable operating experience every day, and by the time we have five more years' experience on our production piles and have available the results of development work now under way, we should know a great deal about power units a few years after the first one starts to operate.
>
> I do not believe it is of fundamental importance to try and set the date at which we can say we will enter the atomic power age. In my opinion, any such date can never be identified. The whole development will be a gradual one. In 1952 the most important thing, in my opinion, is to get a power production pile into operation at the earliest possible moment. If our statesmen and politicians get the idea that the useful application of atomic power is still half a century away, it will make it very difficult to get the financial support we need right now.[15]

But Mackenzie showed pessimism, too, about a new technical profession: to the Dean of the University of Toronto's Faculty of Applied Science and Engineering he wrote a couple of months later, 'At the moment I can't see how the atomic age—whatever that means—has much to do with the fundamentals of engineering science'.[16]

5.1.2 *Chain-reactor potential*

In their distinct ways, the early reactor programmes in the United States, Canada, and Britain explored the potential of reactors and consolidated the growing skills of their nationally segregated development teams. As Mackenzie's thoughts suggest, the new field's balance between comfortably familiar conventional engineering and novel science was contested repeatedly in different working environments.

For Canadian administrators and the nuclear specialists themselves, the NRX reactor led to renewed confidence in the interdisciplinary technical workforce at Chalk River. Project VP David Keys noted that

> realizing that the pile was a very large scale physical instrument constructed by engineers in which the tolerances permitted were much less than that usually required in engineering practice, a special pile start-up committee of physicists and engineers was appointed...The success attending the activities of the pile start-up committee which brought research, engineering and operational personnel into such close co-operation has been so outstanding that the pile operating committee has been reconstituted so as to keep this unification of effort in continuous operation.
>
> ...it is very gratifying to record that all the procedures in connection with the pile operation have been carried out with a precision and confidence worthy of a group of operators and scientists possessing previous experience in such matters, which is an indication of the high quality of staff engaged at this

---

[14] Keys, D. A. to C. C. Cook, letter, 21 January 1948, LAC MG30 B59 Vol. 4.
[15] Mackenzie, C. J. to F. E. Simon, letter, 4 December 1952, LAC MG30-B122 Vol. 3.
[16] Mackenzie, C. J. to K. F. Tupper, letter, 24 April 1952, LAC MG30-B122 Vol. 3.

project. The fact that the pile is now in daily operation has produced a visible increase in enthusiasm on the part of all those working at the plant.[17]

American workers, on the whole, vaunted the novelty of the nuclear domain, and were equally quick to identify new expertise based upon it.[18] But, at other sites—particularly at the Risley headquarters of Britain's Industrial Group of the Division of Atomic Energy—reactor design could be, and was, undertaken with relatively straightforward extrapolations of engineering practice. Although this judgement was driven by the pragmatic difficulty of recruiting experienced nuclear workers, it also reflected the engineering outlook of influential senior British engineers. The argument for the novel basis of nuclear engineering is consequently important to explore here.

Reactor development gradually became the core of a recognized form of expertise applied to specified ends. By the early 1950s three purposes for nuclear reactors had been defined: research, production, and power. Research centred on radioactivity—particularly neutron irradiation—and its transformative effects. Research reactors such as the Canadian NRX consequently were designed to yield the highest possible flux of neutrons. By irradiating materials to note their degradation over time, designers gained crucial information for building subsequent reactors, which were expected to be perpetually off-limits for maintenance or alteration owing to their high-radiation environments. Nuclear physicists, materials engineers, and metallurgists became important clients of these facilities. Irradiation also transformed materials to yield *radioisotopes*. The planning, production, and application of these isotopes to new uses—particularly medical treatments and as tracers for biological and mechanical engineering studies—became an important and novel subfield with links to pre-war radium therapy; radiochemists and biologists were thus frequent users.

By contrast, production reactors were designed for different purposes and clients; they consequently engendered a distinct administrative hierarchy and enrolled different professionals. Plutonium was the primary product of the first generation of production reactors, such as those at Hanford, Windscale, and, soon afterward, the Savannah River Works, as discussed in Section 5.5. After the war, scientists were thin on the ground at such sites. Indeed, production reactor sites and their management were usually modelled on chemical plants, so its workers adopted, or adapted to, largely familiar job functions and responsibilities.

The third class of reactor, intended for generating electrical power, was conceived on the back of these two pre-existing categories. From 1945, power reactors had been mooted optimistically and freely: they might be portable, for use in cars, trains, or even aircraft; they might be suitable for untended, energy-starved locations, such as arctic sites; or they might be compact and self-contained units to supply power on military campaigns. Such hopes were often remarkably naïve. A 1948 General Electric internal memo suggested that

---

[17] Keys, D. A., 'Monthly report', March 1954, LAC MG30 B59 Vol. 5.

[18] E.g. Condon, E. U., 'Physics gives us—nuclear engineering', *Westinghouse Engineer*, 5 November 1945: 167–73; Kramer, Andrew W., 'Atomic energy is here', *Power Plant Engineering*, 49 September 1945: 74–7.

if the Hanford works could employ an 'atomic power plant' to generate 25 MW of electrical power, it would save some $300,000 per year.[19]

By the early 1950s, however, most such applications were recognized to be expensive, unreliable, and potentially dangerous. The worldwide stock of uranium was also estimated to be very limited, perhaps obviating any possibility for 'burning' uranium to generate power. The concept of a power reactor also raised distinct design problems and operational difficulties. In order to generate electricity affordably, for example, a reactor would have to mimic a conventional coal- or oil-fired power plant: the reactor would have to operate at as high a temperature as possible, in order to yield efficient production of electricity from steam turbines. Research and production reactors shared the need to produce heat, but only as a by-product of neutron and plutonium production. But optimizing reactor design for thermal production was both complex and dangerous. Such reactors had to generate as much heat at as high a temperature as possible; this process inevitably increased radiation in a similar proportion. This irradiation was beyond engineering experience, and the effect on materials was largely unknown. An additional constraint was that a power reactor, to be economically viable, would have to operate for at least a decade or two, and probably without the option of modifying or repairing its radioactive core. These interdependent technical and economic factors consequently shaped the power reactors' associated experts. Power reactors merged nuclear workers with engineers from the electricity industry, who embodied expertise in disparate fields. It also demanded new skills from power-station workers.

Together, these three types of reactor defined new types of technical experts. Reactor design itself grouped several existing scientific and engineering specialties. As one UK Atomic Energy Authority physicist recalled, 'we were all there essentially to support the reactor, and a reactor problem might have several manifestations... One of the great things was that there was this continuum, and not people in little silos'.[20] The context of secrecy surrounding the early field nevertheless isolated these silos and encouraged local definition of such experts. Not surprisingly, then, nuclear workers were dissimilarly categorized at each site.

Following the test of Britain's first atomic bomb in 1952, the personnel of the Industrial Group, like Harwell, could begin to adopt a new remit focused on refining the art of nuclear reactors. Administration of the British programme changed in 1954 and, with it, the level of secrecy. The Atomic Energy Authority (UKAEA) was formed to transfer the direction of nuclear research and production from the Ministry of Supply (Figure 5.1).[21] The new UKAEA was modelled on the American AEC, changing the British nuclear programme from what had been a centrally controlled military endeavour under the former Labour government to a nominally civilian organization under Winston Churchill's Conservative

---

[19] Scott, E. E., 'Atomic energy power generation', memo, 14 February 1948, DOE DDRS DA02779406. On analyses through the early 1950s, see Bormaier, R. J., 'Reports on atomic energy in relation to power', memo, 26 October 1953, DOE DDRS D8471637.

[20] Gregory, Colin to S. F. Johnston, interview, 19 September 2007, Thurso, Scotland, SFJ collection.

[21] On the evolving organization of the nuclear programme and its industry in Britain, see Millar, R. N., 'Future power programme in Britain', *Nuclear Engineering* 1 (1956): 304; Central Office for Information, *Nuclear Energy in Britain* (London: HMSO, 1969); and Pocock, Rowland Francis, *Nuclear Power: Its Development in the United Kingdom* (London: Unwin Brothers, 1977).

Fig. 5.1 Old and new: Crest adopted by the UK Atomic Energy Authority, 1954, with stylized fuel rods and radiation as symbols of its new expertise based on nuclear reactors. (Reprinted with permission of the UK Atomic Energy Authority.)

administration. Indeed, the chair of the UK Atomic Energy Authority could state in 1956, 'in the coming decades...the nuclear power industry will make an increasing contribution to, and provide a continuing stimulus for, the rest of British industry. The role of the Atomic Energy Authority in this will be to lead the way—*to design, build and operate new types of reactor*'.[22]

Beyond this three-fold division of reactor purposes, there was also a wide spectrum of reactor concepts to explore. Even though the basic principle of the chain reactor can be readily described, it was understood by the end of the war that a rich assortment of materials and nuclear effects could be combined to yield reactors having new, and probably improved, properties.

The first piles in the USA, Canada, and Britain had adopted distinct design principles. The Chicago CP-1 pile (1942) opted for the use of pure graphite as a moderator and natural uranium as the fuel. Graphite blocks were stacked with interspersed cylinders of uranium oxide. The rate of the chain reaction could be controlled by beryllium-coated rods inserted through the pile to absorb neutrons. The larger Clinton X-10 pile (1943) generated more heat, so the uranium pellets were arranged along channels that could be cooled by forced air. The even larger Hanford piles (1944–5) relied on water flow—some five million gallons per day—to cool the core, which consisted of uranium rods encased in aluminium cans, again in channels within the graphite mass. The sole purpose of the Hanford reactors was production of plutonium: some of the irradiated uranium was transmuted into plutonium, which was then chemically separated for use in weapons. Radiation and heat—some 250 megawatts—were merely the undesired and dangerous by-products. By contrast, the X-10 pile also served as a source of neutrons for post-war studies of the biological and material effects of radiation. Each of these designs, as we have seen, was a compromise between scientific elegance, available materials, and engineering feasibility, and relied on the merging of scientific and engineering perspectives.

While graphite proved an effective moderator, Anglo-Canadian work at Montreal and Chalk River during the war focused on heavy water as a moderator. The ZEEP and NRX reactors used heavy water and uranium oxide, cooled by normal (light) water and using

---

[22] Plowden, Edwin in Jay, Kenneth, *Calder Hall: The Story of Britain's First Atomic Power Station* (London: Methuen, 1956), p. v (emphasis added).

control rods fabricated from boron powder in steel tubes. The large post-war Windscale piles, like the smaller Clinton X-10, used a graphite pile, uranium oxide fuel rods, and forced air cooling.

All these reactors relied on moderators to reduce the speed of neutrons so that they could be captured efficiently by uranium nuclei to cause fission. But such reactors remained inefficient and massive, usually requiring tonnes of uranium (some 230 tonnes in the case of the Hanford designs). It was evident from the late 1940s, however, that known stocks of uranium were too small to support an extensive reactor programme for producing plutonium or for generating power. A new idea was the fast neutron reactor. This used no moderator; instead, the fast neutrons released by fission, which can sustain a chain reaction only if they are much more numerous, were obtained from either enriched uranium (containing a higher proportion of U-235 than naturally available uranium), or from plutonium, or from a combination of natural uranium and plutonium. The excess neutrons could also be captured by a 'blanket' material, where they would transmute some of it into fissile material for use in another reactor, thus breeding more fuel than is consumed. As with the simpler thermal neutron reactors (usually referred to simply as *thermal reactors*), fast reactors had a large spectrum of design variables. The blanket of tubes could contain non-fissile U-238 or thorium. The small CLEMENTINE reactor (Los Alamos, 1946) used plutonium as fuel and mercury as a coolant, and the EBR-1 (National Reactor Facility, Idaho, 1951) employed plutonium fuel and liquid sodium/potassium as coolant. The first British fast reactor (Dounreay, 1959) used enriched uranium and liquid sodium-potassium coolant, and the second (Dounreay, 1974) was cooled by liquid sodium; both were designed for electrical power generation.

Other schemes employed liquid lead or helium gas as coolants. More radically, as mentioned in Section 4.2.1, the *homogeneous fast reactor* scheme investigated at Oak Ridge Laboratories consisted merely of a suspension of another uranium isotope (U-233) in heavy water and surrounded by a blanket of thorium (Th-232); the simple system was described by its proponent, Alvin Weinberg, as 'a pot, a pump and a pipe'.[23] Such conceptual brevity was seductive, and encouraged the new specialists to embody their ideas in hardware.

Within this broad spectrum of materials and nuclear processes were numerous other variants. Power plant designs being developed for submarines generally opted for reactors that were moderated and cooled by light water.[24] Similar plants for surface ships proved uneconomic, though, and serious American investigations toward a nuclear-powered aircraft were eventually scuttled by the inescapable problems of shielding the reactor and in

---

[23] Weinberg, Alvin M., *The First Nuclear Era: The Life and Times of a Technological Fixer* (New York: AIP Press, 1994), pp. 114–22. Neutrons emitted by the U-233 transmuted thorium (Th-232) into more U-233. The first aqueous homogeneous (slow) reactor was LOPO, which went critical at Los Alamos in May 1944. During the 1950s, Oak Ridge investigated several combinations of fluid-fuel reactors.

[24] Walter Zinn, Director of Argonne National Laboratory, had envisaged the prototype submarine reactor as a model for a land-based power station and a good use of limited resources: 'if it becomes possible to build the Navy reactor, a number of the central station power problems will simultaneously have been solved' [Zinn, Walter, 'Minutes of Board Meeting, ANL', letter, 2 May 1948, UI Box 19]. The reactor design of the *Nautilus* nuclear submarine consequently provided lessons for the first American nuclear power station at Shippingport, Pennsylvania.

containing a crash.²⁵ Thus the purpose of the reactor dramatically shaped its engineering characteristics. It could be optimized for plutonium production (needed for bombs); for high neutron flux (important for scientific experimentation); for heat generation (required for electrical power generation); or for breeding more fuel (essential, if the estimates of uranium reserves that circulated during the early 1950s proved accurate). In short, the chain reactor had mutated into a spectrum of devices having distinct purposes.

In Britain this rapid succession of ideas was telescoped into fewer years than the American programme. And, just as plutonium generators, power stations, and fast breeders were being laid out on draftsmen's tables, the possibility of nuclear *fusion* was mooted. This would release energy by fusing nuclei in conditions of extremely high temperature; in effect, it was a controlled thermonuclear reaction. In 1956, Harwell experimented with ZETA (Zero Energy Toroidal Assembly), a large apparatus intended to heat gas into plasma by pulses of electric current and to contain it via magnetic fields long enough for a fusion reaction to occur. Groups at four American sites—Oak Ridge, Los Alamos, Livermore, and Princeton University—also began fusion research under the aegis of the AEC during the early 1950s, but pursued different concepts that only gradually came to overlap.²⁶

Yet another divisive technological factor working against a disciplinary identity was the choice between chain reactors and particle accelerators as the paradigm apparatus of the field.²⁷ During the decade most important for disciplinary coalescence, the 1950s, particle accelerators increasingly demarcated physicists from engineers, although both technologies relied on a merging of know-how. Accelerators could generate beams of particles having well-defined energies and geometries but relatively low intensity. The results of collisions with targets could be observed and analysed with increasing precision. Nuclear reactors, by contrast, often generated a much more intense but chaotic shower of particles that had a different utility: transmuting elements into other radioactive species of value for medicine; degradation studies to determine the effects of radiation of materials of engineering importance; and biological studies to determine the effects of radiation on genetics.

---

²⁵ Early estimates had put the problems of nuclear propulsion of aircraft as 'a whole order of magnitude more difficult' than for submarines [Prentice, B. R., 'Development of atomic energy for subsonic plane', memo, 10 December 1950, DOE DDRS DA02114307; see also Lacy, C.E., 'Report of the Technical Advisory Board to the Technical Committee of the Aircraft Nuclear Propulsion Program', 4 August 1950, ANP-52].

²⁶ Westwick, Peter J., *The National Labs: Science in an American System, 1947–1974* (Cambridge, Mass.: Harvard University Press, 2003), pp. 202–4. See also Pocock, Rowland Francis, *Nuclear Power: Its Development in the United Kingdom* (London: Unwin Brothers, 1977), pp. 79–80. Despite ardent publicity during 1956–7, ZETA proved to be less successful than hoped, and nuclear fusion influenced the British nuclear programme only temporarily.

²⁷ As Peter Galison demonstrated in his study of particle physics [*Image and Logic: A Material Culture of Microphysics* (Chicago: University of Chicago Press, 1997)], scientific apparatus became a means of discriminating between physicists reliant on imaging detectors (such as cloud chambers and spark chambers) and those devoted to logic devices (e.g. coincidence detectors), a choice that determined not merely their occupational competences and working culture but also, to a degree, their conceptualization of nature.

For nuclear physicists, while reactors presented novelty and new phenomena, particle accelerators had a longer lineage. Key players in the Manhattan Project such as John Cockcroft in England and Ernest Lawrence in the USA had designed particle accelerators during the interwar period as sources of energetic particles. Indeed, the first minute quantities of plutonium had been accumulated as the output of Lawrence's cyclotron.

As Director of the Montreal Group in late 1944, Cockcroft outlined plans for a post-war British research establishment that would include not only a graphite reactor but also an accelerator for fundamental research. His Anglo-Canadian team of scientists emphasized their orientation towards fundamental research:

> Nuclear physics is in its infancy and there is no reason to expect that the phenomenon of nuclear fission is the only means of obtaining nuclear energy. Hence it is necessary that pure research should be pursued on a broad front. It would be a grave mistake to concentrate too much effort in the field of nuclear fission.[28]

Similarly Walter Zinn, the ardent promoter of Argonne as the centre of reactor expertise, was as interested in high-voltage particle accelerators as in nuclear reactors. He sought to balance the 'applied' science of nuclear reactor research at ANL with the 'pure' science of accelerators. Until his departure in 1956, however, Argonne was to serve primarily as the reactor development centre for the AEC, while still favouring a physics-based approach over engineering methods. Zinn resigned as Director of Argonne over the issue of research priorities there, after physicists baulked at his reticence to pursue particle accelerator development in the direction demanded by Midwest universities. Zinn the physicist had transmuted into Zinn the engineer: he left ANL to become president of the General Nuclear Engineering Corporation in Dunedin, Florida, established to provide engineering and advisory services for the design, construction, and operation of pressurized-water power reactors.[29]

While Zinn and Cockcroft were comfortable with the industrial-scale engineering of both chain reactors and large atom smashers, other scientists were ill at ease with the apparatus for nuclear energy centred on reactors. Accelerators had initially been conceived as of value solely for physicists and scaled up from laboratory apparatus. Reactors, from their earliest demonstrations, were a shared product with engineers. And as a means to an engineering end as well as sources of new phenomena, reactors carried industrial connotations still mistrusted by some scientists. Unable to be appropriated for physicists alone, the new reactors increasingly were linked to engineering identities.[30]

Within a single decade, then, nuclear engineering took shape around the planning and implementation of a bewildering range of technologies in two branches: the wartime uranium, plutonium, and post-war hydrogen (thermonuclear) bombs, which rapidly became

---

[28] Chalk River scientists to J. Cockcroft, 27 December 1945, CCFT25/20 quoted in Hartcup, Guy and T. E. Allibone, *Cockcroft and the Atom* (Bristol: Adam Hilger, 1984).

[29] See Hewlett, Richard G. and Jack M. Holl, *Atoms for Peace and War, 1953–1961* (Berkeley: University of California Press, 1969), pp. 258–60.

[30] Only from the 1970s were particle accelerators employed much for biological experiments and medical treatments.

the preserve of bomb designers still bottled up at a handful of sites;[31] and the thermal reactor, fast reactor, and fusion reactor, leaking expertise beyond the original government laboratories. Even then, the theoretical and practical foundations—unfamiliar except to a privileged few—overlapped only partially. The technologies behind such schemes extended beyond the developing experience of budding nuclear engineers and threatened to destabilize the nascent profession.

Nuclear engineering, then, incorporated expertise on the still largely unexplored nuclear properties, chemistry, and metallurgy of solids, liquids, and gases, in both elemental and compound forms; on the phenomena surrounding irradiation for materials and biological systems; and clever engineering combinations that exploited newly discovered phenomena.[32] And—as more gradually understood through the 1950s—it also relied intimately on skills in modelling and controlling rapid and inherently unstable physical processes, and in maintaining complex systems in environments that were not accessible to human operators.

5.1.3 *Engineering mysteries*

There was an urgency to investigate the design implications of these new phenomena and interactions. Fundamental characteristics of the materials of the nuclear pile were still inadequately understood. For example, as discussed in Chapter 3, the phenomenon of *Wigner growth* or the *Wigner disease* had been discovered by Du Pont engineers responsible for the Hanford reactors: after months in the high-radiation environment of the pile, the graphite blocks began to swell perpendicular to the direction in which they had been extruded, warping to distort the closely engineered dimensions of the pile.[33] The effect had been foreseen by Eugene Wigner, who predicted a bloating of the graphite blocks, but its magnitude and operational consequences were beyond prediction: engineering experiments were needed. Upon learning of the Hanford experiences in 1948, the Windscale pile designers at Risley incorporated design features that ensured that the fuel channels would not narrow enough to prevent the removal of fuel rods. However, British scientists at Harwell discovered a year later that their own samples of graphite behaved quite unlike the American material, expanding in all directions when irradiated, having consequences for the developing Windscale designs. In the midst of frantic late design changes, Canadian scientists at Chalk River reported a few months later that their own measurements of graphite expansion were only one-fifth of what the Americans had reported.[34] Thus fundamental physical science and large-scale empirical engineering ran neck and neck to produce rapidly evolving reactor designs.

---

[31] Los Alamos, Aldermaston, and Lawrence Livermore. According to the *Nuclear Weapons Databook* (Cambridge, Mass.: Ballinger Publishing, 1987), Hanford produced 121 kg of plutonium between 1944 and the end of 1945—enough for 19 Nagasaki-type bombs.

[32] On the rapid growth of analytical chemistry owing to atomic energy, see Bushey, A. H., 'The less familiar elements of the atomic energy program', report, 12 October 1953, DOE DDRS D198149819; on the comparable benefits for civil engineering, see Pilkey, O. H., 'The civil engineer and atomic energy', draft for American Society of Civil Engineers Conference, 22 February 1962, DOE DDRS DA615027.

[33] Crane, P. W. to J. M. Tilley, memo to file, 20 February 1945, Hanford, DOE DDRS D8255334.

[34] Arnold, Lorna, *Windscale 1957: Anatomy of a Nuclear Accident* (London: Macmillan, 1992), p. 12.

More serious again was the related phenomenon that Hanford engineers had dubbed *Wigner release* or *Szilárd's complication*. Knowledge of Wigner release was rudimentary during the first months of operating the Hanford reactors at the beginning of 1945;[35] as a consequence of the McMahon Act, they were still unknown in Britain when the two Windscale piles went critical in October 1950 and June 1951, respectively. The bombardment of graphite by neutrons causes the lattice structure of the atoms to be gradually disrupted, leading not only to Wigner growth but also to the accumulation of potential energy stored in the displaced atoms making up the lattice defects. These defects can relax spontaneously and *en masse*, releasing energy and causing heating which allows a cascade of further relaxations. The result was an unplanned, unpredictable, and uncontrollable heating of the reactor core. As a wartime internal Hanford report warned,

> If the water cooling system fails even in a portion of the system, a disaster may occur because the... stored energy which can be released below 1000°C will be freed. This can occur even though the power in the pile has been turned off. This... could readily start an explosion if the operating temperature of the pile were above 125°C... The magnitude of the disaster would depend, of course, on the magnitude of the stored energy which can be released.[36]

Such an energy release, even on a small scale, would be exacerbated in the British piles: cooled by air rather than water, they could allow the hot fuel rods to burst their jackets and burn along with the graphite itself. Spontaneous heating events indeed occurred within a year of the second Windscale pile's completion and initially went unexplained.[37] Once understood, the two piles were periodically annealed to avoid such occurrences, i.e. heated to allow controlled and gradual relaxation of defects. As the original secret Hanford report had suggested,

> it is possible that this release could be carried out by heating the graphite near the control rods to a temperature above 200°C during a shut-down and thereby start a minor explosion. The flow of water could be maintained in all tubes so that they do not overheat.[38]

However, the British piles had not been designed to facilitate this process. The heat was provided by the reactor core itself, the cooling was via forced air instead of water, and there were too few temperature sensors to adequately monitor the heating process. In October 1957, during the ninth such annealing procedure, fuel cartridges in one area of the core overheated and ruptured; the exposed uranium oxidized and caught fire and, fanned by the powerful air blowers, released radioactive particles through the cooling chimneys to the surrounding environment. As UKAEA historian Lorna Arnold noted, 'the accident was by hindsight inevitable... in consequence of the over-loading of organization and staff, and insufficient knowledge and understanding of complex phenomena'.[39] This was not merely

---

[35] Section 3.2.5.
[36] Seitz, F., 'Report on Wigner disease', memo to file, 3 February 1945, DOE DDRS D4763348, p. 3.
[37] Fanner, A. A. and J. M. Hill, 'Incidents in the life of the Windscale piles', 26 May 1953, NA AB 7/18254; 'Analysis of lost production time, Windscale Piles, by J. L. Phillips 1 October 1957', NA AB 7/3234. By contrast, the water-cooled Hanford piles were operated at a higher temperature to promote continuous annealing.
[38] Seitz, F., 'Report on Wigner disease', memo to file, 3 February 1945, DOE DDRS D4763348, p. 4.
[39] Arnold, Lorna, *Windscale 1957: Anatomy of a Nuclear Accident* (London: Macmillan, 1992), pp. 17 and 154.

a British problem. Following a post-accident visit, General Electric engineers noted that 'some very nice fundamental work' was proceeding at Harwell, with 'about 25 very capable technical people working on graphite problems at Windscale and an equal number at Harwell', but 'not enough is known by either the UK or the US concerning the fundamental processes... This information would be helpful in extrapolating present experience to future Hanford operations'. Further *engineering* research and closer collaboration would benefit the nuclear experts in both countries.[40]

As highlighted by the Windscale accident a particular problem for nuclear reactors, compared to other power sources, was the density of heat generation and the use of materials in unfamiliar engineering contexts. The core may be only metres across, but can generate tens of megawatts of heat in this volume. The transfer of heat away from the core presented novel challenges. The coolant—gas or liquids such as helium, air, water, or liquid metals—must not react with, or erode, the moderator or fuel, or absorb neutrons that would stifle the chain reaction. The fissile fuel itself generates considerable heat even when the chain reaction is stopped by control rods. Fast reactors, which produce a much higher heat density, raise additional questions of unanticipated problems of materials interactions, maintenance, and practicalities of a closed cycle of burning, breeding, recovering, and further burning of fissile materials.[41] Exposed to a much higher flux of high-energy neutrons, fast reactors also demanded much further research into the radiation damage and aging of materials.

Physics and industrial chemistry conceded ground to nuclear and chemical engineering. For each of these operational mysteries, scientific knowledge had to merge seamlessly with craft skills and tacit expertise. As one of Dounreay's first chemists noted, 'Converting uranyl nitrate into uranium metal is what I call a witchcraft and sorcery process. It depends upon phases of the moon, [and whether] there is an r in the month; it's an art, it is not a science'.[42]

5.1.4 *Engineering values*

The various trade-offs relied on understandings beyond the ken of conventional engineers. Engineering decisions had profound effects on design, but were based on locally constructed values. For example, the Director of Chalk River, W. Bennett Lewis (1908–1987), promoted 'neutron economy' as the fundamental design principle of its reactors.[43] He argued that, given the expense and possibly limited supply of uranium, it was necessary to employ every available neutron for useful production of fissions, heat, and electrical power. Nevertheless, the Canadian design choice of individual pressure vessels to carry cooling

---

[40] Curtiss, D. E. and R. E. Nightingale, 'Trip report US-UK Graphite Conference, London, England, December 16–20, 1957', memo to file, 9 January 1958, DOE DDRS D4763348.

[41] Indeed, investigating the technical feasibility of fast breeder reactors occupied a generation of nuclear specialists between the 1950s and 1990s, with national projects in the USA, Britain, France, and the USSR seeking to identify potential problem areas ['International Atomic Energy Agency, International Symposium on Design, Construction and Operating Experience of Demonstration Liquid Metal Fast Breeder Reactors (Bologna, Italy 10–14 April 1978)', Dounreay Superarchive Box/SR shelf/FRDC/FEWP/P(78)20, IAEA-SM-225/4-1 FRFPDC(84)P5-1].

[42] Lillyman, Ernie to S. F. Johnston, interview, 18 September 2007, Thurso, Scotland, SFJ collection.

[43] Fawcett, Ruth, *Nuclear Pursuits: The Scientific Biography of Wilfrid Bennett Lewis* (Montreal: McGill-Queen's University Press, 1994), pp. 94–5.

water over individual fuel rods, for example, meant a lower efficiency for the reactor because of neutron absorption, and was taken because a single pressure vessel was deemed impracticable. Mixing frugality with pragmatism was an old trick in power engineering.[44]

Engineering designs were also influenced critically by economics. The feasibility of a British power-reactor programme was judged, for example, by its cost relative to coal-generated electricity. If nuclear power could be forecast as being marginally cheaper or having a longer availability than coal, then power-plant design, construction, operation, and maintenance were judged worthwhile. The relative cost included considerable uncertainty, though: materials in nuclear reactors were being operated in novel conditions of temperature and irradiation. Unfamiliar dangers—such as degradation of materials or accidents of radiation release—were recognized but as yet largely unpredictable. And a significant adjustable parameter in that economic equation was the manufacturing cost and subsequent economic value of the plutonium produced by the reactor. For the American and British programmes, plutonium was the primary goal, and its value for military strategy, and indirectly for national status, was crucial but incalculable. Such factors, calculable or not, nevertheless left the future status of nuclear engineers precariously balanced on a decimal point.

Ironically, the economic value of plutonium—a waste product in a country that eschewed military applications—was readily calculated even in Canada. C. J. Mackenzie, first as President of the NRC and subsequently as President of AECL, secretly negotiated a price for the 'plutonium credit' with the American government. His staff designed their NRU reactor specifically with the intention of funding its operation and, it was hoped, the entire Canadian programme, by plutonium sales.[45] Awash with research money, Canadian nuclear workers

---

[44] 'Efficiency engineering' had become fashionable with industrial managers after the turn-of-the-century work of Frederick W. Taylor and especially during and after the First World War, when government-supervised factories encouraged 'scientific management' [Wilson, John F., *British Business Histor1720–1994* (Manchester: Manchester University Press, 1995), pp. 162–5; Maier, C. S., 'Between Taylorism and Technocracy: European ideologies and the vision of industrial productivity in the 1920s', *Journal of Contemporary History* 5 (1970): 27–61]. On its role in legitimating a professional group, see Abbott, Andrew D., *The System of Professions: An Essay on the Division of Expert Labor* (Chicago: University of Chicago Press, 1988), pp. 194–5.

[45] The original deal negotiated agreed that the USA would 'buy all plutonium produced at Chalk River at a price between $170,000 and $180,000 per kilogram'. C. D. Howe briefed the Canadian Finance Minister that 'a price of $145,000 per kilogram will permit the government to amortize the new plant over ten years, and...a price of $175,000 per kilogram would allow the Government to amortize past expenditure as well as future expenditures over the same period' [Howe, C. D. to D. C. Abbott, letter, Ottawa, Ontario, LAC RG29-F-2 Vol. 5361]. Owing to its American commitments 'for all fissile material with the exception of the amount we require for our own use', Mackenzie was hesitant to supply the British programme with 2 kg of plutonium as they requested [Mackenzie to Cockcroft, letter, 26 May 1952]. He saw plutonium as a solution to long-term funding of Canadian work: 'I am quite sure, however, that plutonium will always be valuable, and the next plant we plan, after our present production unit [NRX] will be a power production unit with two main products—power and fissile material. On this basis we think the economics will eventually work out satisfactorily' [Mackenzie, C. J. to F. E. Simon, letter, 4 December 1952, LAC MG30-B122 Vol. 3]. Britain adopted the same course with its Calder Hall and Chapelcross power reactors, which replaced the two Windscale piles as plutonium producers.

remained ignorant of the supporting economics.[46] In Britain, by contrast, the factor was a hidden variable: there was no sense of a tangible price for plutonium, but the benefit of producing plutonium for the nation's military and for raising national esteem could be estimated. Thus not just economics, but politics, too, were key ingredients for nuclear engineering design.

5.2 Walter Zinn and nucleonics at Argonne

The intellectual potential of atomic energy was recognized and constructed differently in each country's context of classified knowledge, government funding, and technical cultures. These novel arrangements were tentative and coalesced differently at each site. The cloistered workers and division of labour fostered by the Manhattan Project made the subjects unusually malleable by the most experienced and enthusiastic of the new specialists. As explored in Chapter 4, within four months of the Japanese bombings plans for state-supported science were developing as a final thrust of the Manhattan Engineer District. The draft 'Plan for the continuation of the Argonne Laboratory', in particular, set its ultimate objective as 'the formation of a government corporation created and financed by the permanent federal agency established to supervise development in the atomic energy field'.[47] Thus its participants argued that both a new field, and a new research and development regime to undertake it, were the remit of the University of Chicago's Met Lab.

There had already been thought given to the new discipline, and the social and political background in which it would be practised. Arthur Compton had authorized a committee under Zay Jeffries of General Electric to draft a report entitled 'Prospectus on Nucleonics' between July and November 1944 which, amongst many other topics, mooted the possibility of 'cooperative laboratories' for large-scale post-war research.[48] And the Met Lab personnel, in particular, came to favour the term *nucleonics* introduced in the report's title for their nascent field. The Jeffries report chose the term—an early example of specialist vocabulary demarcating the new subject—from a list of possibilities. The authors explained

---

[46] Because it contradicted Canadian policy of abstaining from nuclear weapons work, the plutonium economy of the Chalk River programme was confidential. MacKenzie noted, 'I am particularly anxious that all correspondence on the policy level, covering such things as the exchange of Plutonium, barter arrangements of major kind, and any other matters which I should discuss with the officials of the Government or the Control Board, should be sent to my Ottawa office. I do not wish such sensitive matters of policy to get into the general records at the plant' [Mackenzie, C. J. to J. Cockcroft, letter, 23 May 1952, LAC MG30-B122 Vol. 3].

[47] Hilberry, Norman, 'Plan for continued operation of Argonne Laboratory', 5 December 1945, UI Box 44.

[48] Reproduced in Smith, Alice Kimball, *A Peril and a Hope: The Scientists' Movement in America, 1945–47* (Chicago: University of Chicago Press, 1965), pp. 539–59. The Committee members were Enrico Fermi (having transferred from Chicago to Los Alamos), Charles A. Thomas of Monsanto Chemical Company, and three Met Lab personnel: James Franck, the associate director of Chemistry; Thorfin R. Stone, head of the health division: and Robert S. Mullikan, a physicist acting as information director. Comments were sought from Manhattan Project group leaders. The resulting report included seven sections covering the development of atomic energy and new methods, the expected impact on international relations and social order, and the organization of nucleonics in the USA. Communicated in broad outline to the Met Lab staff, the report remained classified until 1957.

that, 'reflecting the modern trend toward close correlation between science and industry, and following the lead of "electronics", we propose that the word "nucleonics" shall refer to both science and industry in the nuclear field'.[49]

The earliest post-war documents revealed their aim to be national provision of 'research and development in the field of nucleonics':

> the nation must carry ahead a strong program in research and development in the field of nucleonics. This program will involve improvement in methods of production of fissionable materials, the development of methods of power utilization and the development of the entire field of the production and use of artificial radioactive materials. It will also involve a longer term program concerned with methods for the release of nuclear energy other than utilization of the phenomena of fission...It is essential therefore that the understanding of a new chain reacting system be translated from the field of the research scientist into the field of industrial operation.

The internal memo, drafted under the direction of Walter Zinn, identified the nature of nucleonics as part science, part engineering:

> In the field of research there are those two major areas, that of pure research and that of 'process' research. The pure research program will be concerned in large extent with the long term investigations. It will be primarily involved in obtaining the new concepts and understandings upon which any future advances in the field of the release of nuclear energy must be based. On the other hand the 'process' research will be concerned chiefly with the shorter term problems involved in obtaining new understandings of chain reacting systems. Both of these tasks are primarily research jobs although both may involve much in the way of engineering for their successful accomplishment. Moreover, the delineation between the two fields is not sharp and the discoveries in each tend to support the progress of the other.[50]

The term *nucleonics*, in fact, remained closely associated with the Chicago group. General Electric instituted a Nucleonics Division in 1946 and a Nucleonics Department at the Hanford works that it now managed; a magazine with that name was founded that year, and was later succeeded by *Nucleonics Week*. In the decade after the war, when small firms were springing up to supply the atomic sites, the term became most closely associated with instrumentation for nuclear studies. By analogy to the term *electronics*, the root *nucleon* refers either to the neutron or proton, the principal constituents of atomic nuclei. Radioactivity—in the form of neutrons, alpha particles (a neutron–proton pair), beta

---

[49] Jeffries, Zay, 'Prospectus on Nucleonics', report, University of Chicago Metallurgical Laboratory, 13 November 1944, reproduced in: Smith, A. K., *A Peril and a Hope: The Scientists' Movement in America: 1945–47* (Chicago: University of Chicago Press), Appendix I, p. 540. While the report was being drafted, nucleonics was the label most frequently used by the Met Lab staff to describe the new subject. In July 1944, Arthur Compton offered Enrico Fermi a job at Chicago after the war to study nucleonics. In the same vein, the University of Chicago was then discussing 'National requirements for research and development in the field of nucleonics' [Hilberry, Norman, 'Statement to the Advisory Committee on Future Argonne Laboratory Operation (draft)', memo, 24 November 1945, UI Box 134] and undertaking 'the operation of a laboratory at Argonne for cooperative research in nucleonics under a contract with the United States Government' [Harrell, W. B. to A. H. Frye Jr, letter, 9 March 1946, UI Box 19].

[50] Hilberry, Norman, 'Statement to the Advisory Committee on Future Argonne Laboratory Operation (draft)', memo, 24 November 1945, UI Box 134.

particles (electrons) and gamma-rays—was a ubiquitous product of reactors and of materials irradiated within them, so nucleonics was a convenient label for a fertile field.[51]

But as Zinn's definition suggests, nucleonics equated to reactor science and engineering in the short-term. Argonne physicists were equally interested in particle accelerators, however, which would provide beams of energetic subatomic particles, more precisely and in greater variety than reactors could, for experiments in nuclear physics. The label *nucleonics* related closely to the directions pursued at Argonne, but was increasingly ill-fitted to the directions taken by other institutions and the later national nuclear industries. For example, another variant of the field, focusing on the production and applications of radioisotopes, was of more interest at Oak Ridge National Laboratory, Chalk River, and AERE (Harwell). Henry Seligman, head of AERE's Isotope Division, proposed dubbing the field 'isotopics', but failed to win adherents.[52] This mismatch of interests from site to site illustrates the ambiguities of defining the new field and its specialists.

Nevertheless, confident of the key attributes of atomic energy research, Zinn's vision of nucleonics also stressed that the expertise should be nurtured at Argonne:

> It is particularly in these research fields that the training of new personnel is a matter of imperative national necessity. This need for training must be considered as a primary element in the development of any over-all national plan for the field.[53]

A few months later, he could summarize, 'this laboratory is to continue the development of the uranium chain reaction. Its efforts are to be devoted to those phases of the atomic energy development which are not directly connected with weapons research, but which are of fundamental importance to the field of nuclear energy'.[54] As such documents indicate, the Chicago scientists and engineers played a seminal role in defining the post-war field, its goals, and personnel.

5.3 Anti-discipline: Christopher Hinton and the British nuclear worker

5.3.1 *Post-war legacies*

The nuclear specialists returning to Britain were shaped differently by wartime experiences than were their American and Canadian counterparts. An amalgam of chemical, electrical, mechanical, and civil engineers coalesced alongside physicists under the direction of government ministries both during and after the war. But rather than forming a proto-profession around an intellectual centre such as nucleonics, British nuclear workers found themselves pigeon-holed into existing scientific and engineering categories. Their jurisdictional skirmishes for recognition and status were unequally matched and repeatedly lost.

---

[51] The term rapidly fell from use during the 1950s; an atypical survival is the term 'bionucleonics' used in courses at Purdue University as late as 1959 ['Bionucleonics at Purdue', 2 December 1959, UI Box 13].

[52] Herran, Néstor, 'Spreading nucleonics: the Isotope School at the Atomic Energy Research Establishment, 1951–67', *British Journal for the History of Science* 39 (2006): 569–86.

[53] Hilberry, Norman, 'Statement to the Advisory Committee on Future Argonne Laboratory Operation (draft)', memo, 24 November 1945, UI Box 134.

[54] Zinn, Walter, 'memo to file', memo, 26 July 1946, UI Box 134.

Imperial Chemical Industries was an important influence, seeding the management and engineering personnel of the Tube Alloy project and being the training ground for many of the post-war project's engineers. As related in Chapter 3, from 1940 the MAUD committee in Britain had balanced high science alongside engineering pragmatism by combining university researchers with their counterparts in the chemical industry. The ICI engineers played a crucial role in assessing viability and in conceiving and implementing the wide-ranging experiments.

Less comfortably, the sites of the Division of Atomic Energy were to be associated intimately with leading-edge research. The need to manage a government-mandated technical industry had been tried before, but only during time of war. The atomic factories, closely linked to the munitions factories of the Second World War, had a more pertinent First World War counterpart: the organization by the British Ministry of Munitions of government factories for explosives production. In the munitions factories, chemists from around the British Empire had mingled with production staff. Bemused civil servants overseeing both the wartime factories and Hinton's Industrial Division likely saw them (in historian Margaret Gowing's words) as a 'strange experiment in which civil service machinery was used to carry out a large industrial enterprise'.[55]

The diffident government organization of the post-war British nuclear project was mirrored by the awkward pigeonholing of its early specialist workers. The Tube Alloys project had placed an early reliance on chemical engineers, although they were a little-recognized profession in Britain during and immediately after the war.[56] At Chalk River, where 'a small group of chemical engineers [or chemists, depending on the source] had built a primitive pilot plant in a tar paper covered tower on the banks of the Ottawa river', the nucleus of the Harwell chemical engineering division was established.[57] There, in the shadow of the American wall of security on nuclear information to its former allies, the British contingent developed a flow sheet (i.e. a chemical engineering recipe) for industrial separation of uranium and plutonium fission products. A critical mass of the Chalk River Division moved to Britain in 1950 and eventually grew into two branches concerned respectively with chemical engineering operations (e.g. heat transfer and hydrodynamics) and process technology (e.g. technology for purifying and handling

---

[55] Gowing, Margaret and Lorna Arnold, *Independence and Deterrence: Britain and Atomic Energy*, Vol. II: *Policy Execution, 1945–52* (London: MacMillan, 1974), p. 67.

[56] Divall, Colin and Sean F. Johnston, *Scaling Up: The Institution of Chemical Engineers and the Rise of a New Profession* (Dordrecht: Kluwer Academic, 2000). Just as Eugene Wigner, the paramount reactor designer in the USA, had a chemical engineering background, so too did Nikolai Dollezhal at Obninsk, chief engineer for the first Soviet plutonium production reactor, designer of submarine power plants and, eventually, Chernobyl. Trained in mechanical and thermal engineering, he rose in the hierarchy of Soviet chemical engineering and, thereafter, nuclear engineering [Josephson, Paul R., *Red Atom: Russia's Nuclear Power Program from Stalin to Today* (New York: W. H. Freeman, 1999), pp. 21–5].

[57] Franklin, N. L., 'The contribution of chemical engineering to the U. K. nuclear industry', in: W. F. Furter (ed.), *History of Chemical Engineering* (Washington: American Chemical Society, 1980), pp. 335–66; Rae, H. K., 'Three decades of Canadian nuclear chemical engineering', in: W. F. Furter (ed.), *History of Chemical Engineering* (Washington: American Chemical Society, 1980), pp. 313–34; quotation from Cockcroft, John, 'Interpretation of scientific knowledge', *Chemical Age* (1949): 613–14, p. 613.

graphite, chemical recovery of fissionable materials such as plutonium, and the fixing of undesired fission by-products in glasses).[58]

The Industrial Division nurtured a distinct complement of engineering specialists. This was largely due to the continuity of organizational culture imposed by Christopher Hinton, responsible for designing and managing the factories of the post-war nuclear programme.[59] Having studied mechanical engineering at Cambridge and becoming an exceptionally young chief engineer of the Alkali Group at ICI in 1930, Hinton had briefly supervised explosives production and then, as already noted, directed the munitions filling factories for the Ministry of Supply for the duration of the war.

The post-war nuclear programme in Britain involved a labyrinth of committees responsible for discrete technical and administrative problems, but struggled to find enough competent staff.[60] Most of Hinton's first generation of staff had gestated in wartime chemical factories. His background powerfully shaped the nature of his post-war nuclear workers. Having been offered a post equivalent to his pre-war status, Hinton scavenged ICI for associates who had been offered similarly 'paltry jobs' after the war, bringing to the nascent nuclear programme personnel who had been influential in the wartime Royal Ordnance Factories (ROF).[61] The core group that was to direct the industrial programme over the next decade was made up of long-time associates of Hinton.

Not surprisingly, nuclear production also was centred on such wartime sites, and inherited some of their character. The headquarters of Hinton's industrial group at Risley, Lancashire, had been the temporary headquarters of Ordnance Supply, and was a former Ministry of Supply shell filling factory. The first uranium fuel plant (erected at Salwick, Lancashire, renamed Springfields, and gradually taken over from the ICI caretaker staff) and the first nuclear reactor and associated chemical separation plant for military plutonium production (at Sellafield, Cumbria—renamed Windscale) had been wartime ordinance factories. Hinton recalled:

---

[58] White, A. S., 'Chemical engineering research at A.E.R.E.' *The Chemical Engineer* (1962): A66–A73; Nichols, C. M. and A. S. White, 'Chemical engineering and atomic energy', *Chemistry and Industry* (1963): 51–5; Davis, W. K. and W. A. Roger, 'The chemical engineer and nuclear energy', in: W. T. Dixon and A. W. Fisher Jr (eds.), *Chemical Engineering in Industry* (New York: American Institute of Chemical Engineers, 1958), pp 88–97.

[59] See Section 4.4. Knighted for his work in 1951 and made a life-peer in 1965, Hinton was a focus of public attention for the programme. For the largest collection of his unpublished materials, see IME Hinton papers A.4–A.6; for contrasting admiring and critical views, see Bertin, Leonard, *Atom Harvest: A British View of Atomic Energy* (San Francisco: W. H. Freeman, 1957) and Pringle, Peter and James Spigelman, *The Nuclear Barons* (London: Joseph, 1982).

[60] 'Production Piles—Air Commodore Rawley's History', NA AB 8/557.

[61] Charles J. Turner, for example, who had co-designed the first polythene plant at ICI before the war, was responsible for the first chemical separation plants of the British nuclear programme; Henry Gethin Davey (1911–1960), who had taught chemical engineering at the Trefforest School of Mines before the war, became superintendent of ROF Drigg and then superintendent of works at the first nuclear pile at Windscale; Robert Alexander, from ROF Drigg and ROF Sellafield, became assistant works manager at UKAEA Capenhurst; Stanley F. Hines, employed as a chemist at ROF Pembrey and then as manager of the Drigg TNT factory, transferred to Capenhurst where he became works General Manager. For Davey see ICE Membership Applications 2187 and 7508; for Alexander (1915–), see ICE MA 2232 and 7988.

> The initial atomic energy establishments had been represented to me as munitions factories for the manufacture of fissile material for bombs... and, having learned by experience that the strategic location of factories for the next war is normally based on experience in the last, I guessed that the most acceptable factory sites would be in the North West.[62]

Representations based on wartime understandings shaped the post-war engineering field. The working environment was largely familiar, too. Hinton's first plans for a production reactor site envisaged it as an upgraded Royal Ordnance Factory, just the type of organization that he had directed during the war:

> It is proposed that each of the main process areas should, in fact, be a little factory; self contained except in so far as it is serviced from the central administration and services area. It will be rather like a "Group" in a Filling Factory with its Group offices. Headquarters will act in the same way as the corresponding services and Administration Groups on a Filling Factory.[63]

The continuity of purpose, working culture, and disciplinary orientation was thus firmly reinforced.

5.3.2 *A British workforce*

The industrial roots distinguished the early British reactor programme from its American counterpart, and determined its intellectual labelling. To Hinton and most of his colleagues with experience of wartime production, an atomic pile was not a wholly new or even particularly complex creation. Its environment of high temperatures, unfamiliar materials, and biological dangers had analogues in the long-standing chemical industry. To his workers, the new danger of radiation represented merely a variant of the century-old hazards of chemical works.

Given this assessment and the shortage of experienced specialists, Hinton surrounded himself with practically trained men who lacked formal engineering credentials. The nucleus of his organization—a dozen men brought from the munitions and filling factory organization in 1946—included two having university degrees and only one with direct experience with atomic energy. The civil service pay scale forced a further dichotomy on workers, defining them as various grades of either 'engineer' or 'chemist'.[64] In Hinton's design office for chemical plants at Risley, nuclear reactor facilities and chemical separation processes were designed mainly by those labelled 'engineers'. Even so, their skills were developed practically, not academically. Harold Disney, for instance, the chief designer of the first large gaseous diffusion plants and later Director of Engineering of the Industrial Group, had worked at railway wagon building works and taken evening classes to Higher

---

[62] IME Hinton A.2, p.127.

[63] 'Factory planning for production pile 1947', NA AB 7/284.

[64] Staffing shortages were exacerbated by the restrictions of salary scales and job categories maintained by the Ministry of Supply. These were eased with the creation of a separate Authority, when research engineers were graded on the more inclusive category of 'Scientific Officers'. Owen, L., 'Nuclear engineering in the United Kingdom—the first ten years', *Journal of the British Nuclear Energy Society* 2 (1963): 26; P. N. Rowe, interview with Johnston, 10 March 1998; 'Careers in Nuclear Engineering at Harwell and Winfrith 1960 December' NA AB 17/234.

16. Mr. Christopher Hinton, F.R.S., and senior members of the United Kingdom Atomic Energy Industrial Group. *Seated (left to right)*: Mr. R. E. France, O.B.E., former assistant secretary, Department of Atomic Energy; Mr. R. W. Preston, first general manager of the ore-processing factory at Springfields; Mr. W. L. Owen, C.B.E., deputy director; Sir Christopher Hinton; Mr. D. A. Shirlaw, C.B.E., director of administration; Mr. D. W. Cole, asst. controller, production; Mr. L. Rotherham, in charge of research and development. *Standing:* Mr. H. H. Bannister, O.B.E.; Mr. S. F. Hines, O.B.E., general manager, Capenhurst uranium enrichment factory; Mr. H. V. Disney, designer and builder of Capenhurst; Mr. J. W. Kendall, designer and builder of the Windscale piles; Mr. H. G. Davey, Windscale general manager; Mr. C. S. Turner, designer and builder of the Windscale chemical plant.

Fig. 5.2 Christopher Hinton (centre, seated) and his team, with Harold V. Disney and James W. Kendall standing to his right and left, respectively.[65] (Courtesy of W. H. Freeman & Co.)

National Certificate (HNC) level before the war; James Kendall, responsible for Pile Design at Risley, had managed a building firm before joining the wartime filling factory group.[66] Despite Kendall's lack of 'any recognisable practical or theoretical training in engineering', Hinton described him as 'a born engineer' and made him responsible for the first Harwell and Windscale reactor designs (Figure 5.2).[67]

Backing up this motley if highly competent team were the industrial chemists of ICI General Chemicals at Billingham who had studied uranium purification and gaseous diffusion, respectively, during the war.[68] By contrast, Harwell (initially, at least) employed mainly 'chemists' and few 'engineers'.

As his key staff choices attest, Hinton favoured practical engineering experience over technical training. When in 1950 a letter from an associate in the Division of Atomic

[65] Bertin, Leonard, *Atom Harvest: A British View of Atomic Energy* (San Francisco: W. H. Freeman, 1957), p. 200a.

[66] Hinton, Christopher, *Engineers and Engineering* (Oxford: Oxford University Press, 1970), pp. 3, 5, 39; Gowing, Margaret and Lorna Arnold, *Independence and Deterrence: Britain and Atomic Energy*, Vol. II: *Policy Execution, 1945–52* (London: MacMillan, 1974), pp. 22, 31. Nevertheless, this mixing of what was called in Germany 'school culture' and 'shop culture' (science- versus practically-based expertise) was not unusual in engineering professions of the period; see, for example, Jarausch, Konrad, *The Unfree Professions* (Oxford: Oxford University Press, 1990) and Gispen, Kees, *New Profession, Old Order: Engineers and German Society, 1815–1914* (Cambridge: Cambridge University Press, 1989).

[67] Hinton, Christopher, Hinton A.3, Institution of Mechanical Engineers Archives.

[68] Design groups at Risley typically comprised one or two designers, five to ten draughtsmen and an occasional chemical advisor. 'Plans for staff organization, 1946–1950', NA AB 19/56; IME Hinton papers A.3.

Energy suggested replacing the departing physicist with a 'chemical engineer with some knowledge of atomic energy', Hinton instead proposed an ex-ICI dyestuffs mechanical engineer.[69] Similarly, he hesitated to appoint Frank Kearton, a young ICI 'chemist and not an engineer' who had been responsible for work on diffusion plant during the war, because Hinton felt he could have lowered morale and delayed the design progress of his older associates, despite his recognized technical competence.[70]

Given the chronic lack of manpower, it is surprising that Hinton's organization made little use of the nuclear researchers/designers at Harwell through the early 1950s, particularly Arthur S. White's new Chemical Engineering Division.[71] The large and expensive laboratories there had been intended to develop new chemical separation processes but, given the self-sufficiency of the Industrial Division, turned instead to research on solvent extraction and novel designs for reactors. Other long-term problems of nuclear engineering, particularly the corrosion of pipes and vessels aggravated by high temperatures, erosion, and radiation effects, were tackled less systematically. This competition for intellectual space paralleled American activities at Oak Ridge and Argonne.

5.3.3 *A British specialism*

The dearth of recently trained university engineers was not entirely due to the perspective Hinton carried away from ICI. Workers were reluctant to enter the security-bound, militarily-oriented, and underpaid civil service immediately after a long war. As a result, a high proportion of the professional appointments of Hinton's group were of men already in the civil service, whom the team knew personally, or who 'had been put in touch with us by old friends'.[72]

The problem was one of extending as well as reinventing knowledge now denied to them by the McMahon Act. Hinton later commiserated with Sir Leonard Owen, his former deputy from ICI, that 'every day, we tried to be expert in three completely different fields of engineering, the mass production of the diffusion plant, the chemical engineering of the uranium and separation plant, and the heavy mechanical and civil engineering of the piles. And in all of them we were pioneering'.[73] Key physicists directing the programme were cautious in scouting new terrain. Hinton recalls that William Penney, heading the bomb-design facility at Aldermaston, provided practically useful information prefaced with 'we don't really know, but—'.[74] On the other hand, he saw John Cockcroft—with whom a close working relationship

---

[69] Letter from Eric Welsh to Hinton, 1 March 1950 NA AB 19/66.

[70] Hinton, Christopher, 'Unpublished memoirs Chap XIV: The Diffusion Plant', IME Hinton A.4. Kearton (1911–1992) was later to become Director and Chairman of Courtaulds.

[71] The head of the Chemical Engineering Division for twenty years, White (1903–1987) had neither a research degree nor wide research experience, having obtained his chemical engineering diploma from UCL in 1934 after losing his job during the Depression, and then working at ICI Dyestuffs in Manchester. Yet 'Harwell', said an associate, 'bristling with whizz kids and academic protégés, needed a sound and experienced engineer to keep their feet on the ground'. Rowe, P., 'Obituary: Arthur Southan White', *IChemE Diary & News,* April 1987: 3; interviews, Johnston with P. N. Rowe (10 March 1998) and Sir Frederick Warner (19 December 1997); ICE Membership Application 3300.

[72] Hinton, Christopher, 'Unpublished memoirs Chap XVI: The Reasons Why', IME Hinton A.4, p. 231.

[73] Hinton, Christopher, 'Unpublished memoirs Chap XIV: The Diffusion Plant', IME Hinton A.4.

[74] Hinton, Christopher, 'Unpublished memoirs Chap XI: Risley Organization', IME Hinton A.3.

on nuclear engineering problems was essential—as 'a benevolent sphinx', who never gave unequivocal answers. Cockcroft's biographers note that he and Hinton were introduced in January 1946 by Wallace Akers and that 'the relationship between scientist and engineer, although at this stage reasonably cordial, proved in the long run to be far from harmonious'.[75] In contrast to Cockcroft's university education and apprenticeship as an electrical engineer, Hinton's complementary career arc—a school apprenticeship at the Great Western Railway works in Swindon, followed by the Mechanical Engineering Tripos at Cambridge— equipped him for years of experience as Chief Engineer of an ICI Division. For both men, the notion of 'applied physics' mapped poorly onto 'engineering'.

Besides the diffident guidance of Cockcroft and Penney, Hinton's new Industrial Group in the Division of Atomic Energy of the Ministry of Supply, excluded from touring American facilities, benefited from visits to Chalk River. Hinton's initial visit in October 1947 provided his first view of a functioning reactor; he was followed that year by many of his staff, including W. Leonard Owen, his de facto deputy and Director of Engineering at Risley; Charles Turner, later responsible for the plutonium separation chemical plants; Donovan Cole, in charge of production and commissioning of the atomic factories; and, Tom Tuohy, later Works Manager of the Windscale piles.[76] Owen, an ex-ICI engineer like Hinton, argued for the importance of versatile and experienced engineers:

> I will have nothing said against the scientists. But the people this country needs most today are the men who can understand what the scientist's idea is and put it into practice in brick, mortar, iron and steel... We are always trying to get more engineers of the sort who can discuss all the different aspects of the many specializations, the operations side .., the development side .., the scientific side, and can chat with the secretarial and administrative types about Treasury considerations and with the doctors about health. Your design engineer takes all this into account'.[77]

Of even more importance for building British expertise was the handful of engineers who had gained wartime experience in the Montreal Group and at Chalk River. Foremost among them was Ronald Newell, Engineering Division Head at Chalk River and early advisor to Hinton on the siting of the British production piles.[78] Newell had fourteen years' prior experience in chemical plant design, and had been a member of the Montreal Laboratory from its foundation in 1942.[79] Dennis Ginns (1912–), who had worked on the Chalk River designs, was the first to explain the nature of reactors to the original dozen members of the Industrial Group collected in February 1946.[80] Both Newell and Ginns were ICI men like

---

[75] Hartcup, Guy and T. E. Allibone, *Cockcroft and the Atom* (Bristol: Adam Hilger, 1984), p. 140.

[76] Visitors to Chalk River from Harwell, the Atomic Energy Council (which directed the early Atomic Energy Division), and from American national laboratories were even more numerous. See Keys, D. A., LAC MG30 B59 Vol. 4.

[77] Owen, L., in Bertin, Leonard, *Atom Harvest: A British View of Atomic Energy* (San Francisco: W. H. Freeman, 1957), p. 109.

[78] 'Choice of site for production pile 1946', NA AB 7/157.

[79] '1946 Organization of Engineering Division Chalk River and Montreal Laboratories—National Research Council: R. E. Newell', NA AB 2/123.

[80] Bertin, Leonard, *Atom Harvest: A British View of Atomic Energy* (San Francisco: W. H. Freeman, 1957), pp. 111–14.

Hinton himself. Harold Tongue (1894–1960), crucial for the design and operation of the Chalk River ZEEP reactor, became head of the Harwell Engineering Division. He had been chief engineer of pre-war heavy industrial chemical and metallurgical research organizations, and spent eleven years in engineering development for the London County Council. Nevertheless, as he later noted, he struggled to obtain adequate experience: 'In atomic energy, the engineering teams at Risley that designed and constructed the industrial establishments at Springfields, Windscale, Capenhurst and Dounreay had, in the most difficult period, no more than a couple of dozen university-trained engineers in them'.[81] Despite ample funding, then, the pools of proficiency were shallow during the early 1950s. In the USA, Canada, and Britain the segregated subject struggled to sustain and extend its borders.

5.4 Controlling information flow

Shortages of technical experts and disagreements about their special know-how were two serious constraints, but a third was the enduring context of security. Disciplines are defined and sustained by texts and teaching. Authors and teachers shape, categorize, and disseminate knowledge to suit both sponsors and perceived audiences. But while the chain reactor was increasingly recognized as the basis for a new technical discipline, its special knowledge was peculiarly curtailed by security. As discussed in Section 4.5, the blanket of atomic secrecy shaped working identities. But it also held back the pressures for the expansion of the field by controlling the flow of information via publications, schools, and universities.

The publication of books and journals was centrally ordained and locally managed for a decade after the war rather in the way that monks protected, selected, and reproduced monastic knowledge. While copious articles and books popularized the aspects of atomic energy that had been revealed in the 1945 Smyth report, none went much further, even for technical specialists.[82] The first journals carrying details on atomic energy had been 'published' for restricted audiences.[83] The first textbooks were heavily censored to protect sensitive engineering details: *Elementary Pile Theory*, published in 1950 from declassified notes of the Clinton Lab, was a mere 71 pages long. As its Foreword emphasized, 'the greatest part of the work is still classified, and probably will remain so for some time'.[84] Yet, as argued subsequently by Edward Shils, controlling the flow of information disadvantaged programme goals.[85]

---

[81] Hinton, Christopher, *Engineers and Engineering* (Oxford: Oxford University Press, 1970), p. 5.
[82] Smyth, Henry D., *Atomic Energy for Military Purposes: The Official Report on the Development of the Atomic Bomb under the Auspices of the United States Government, 1940–1945* (Princeton, NJ: Princeton University Press, 1945). Examples of unclassified technical information include Kramer, Andrew W., 'Atomic energy is here', *Power Plant Engineering*, 49 September 1945: 74–7; Condon, E. U., 'Physics gives us—nuclear engineering', *Westinghouse Engineer*, 5 November 1945: 167–73; 'Atomic piles', *Discovery* 7 (11) (1946): 321–4.
[83] The *Journal of Reactor Science and Technology*. On the spread of specialist periodicals, see Appendix II.
[84] Soodak, Harry and Edward C. Campbell, *Elementary Pile Theory* (New York: John Wiley, 1950), p. vii.
[85] See Shils, Edward, *The Torment of Secrecy: The Background and Consequences of American Security Policies* (Glencoe, Ill.: Free Press, 1956), pp. 176–92.

5.4.1 *American training under the cloak of secrecy*

Research and teaching of applied nuclear topics—albeit of classified material to screened audiences—was similarly restricted. It began on a small scale in the USA in October 1946 when a series of seminars was initiated at the Massachusetts Institute of Technology (MIT). That year, a Laboratory for Nuclear Science and Engineering was also founded there under Jerrold R. Zaccharias, a Professor of Physics, and Edwin R. Gilliland, Professor of Chemical Engineering. They optimistically forecast a new intellectual synthesis:

> Progress in this field requires the coordinated effort of many branches of science and engineering, particularly during the next several years. Gradually, the responsibility will devolve to a new breed of specialists, already dubbed *nuclear engineers*.[86]

Nevertheless, the recipients of its post-war studies were filtered by security: the initial students were principally armed forces officers concerned with the possibilities of nuclear weapons and propulsion. Major General Kenneth Nichols (1907–2000), wartime deputy to General Groves and, at that time, Chief of the Armed Forces Special Weapons Project, took an early public stance on such training and its practitioners:

> the problem presently confronting our universities is to determine what courses should be introduced to train engineers and scientists adequately for the development of atomic energy...Just one course in our engineering schools will not be the answer; what is necessary is the introduction of nuclear engineering courses in several fields – civil, mechanical, electrical, and chemical engineering. Such courses in their respective fields should stress the problems of heat exchange, metallurgy and materials, chemical engineering, gadgeteering and remote control and industrial safety, all handled from the standpoint of how nuclear physics affects the solution of these problems when applied to nuclear engineering.

He noted that 'in attempting to develop proper education courses, secrecy and government control of the atomic energy field may appear to be limiting factors', but concerns about security could be overcome by government sponsorship—via AEC contracts or through temporary employment of professors at the National Laboratories and AEC industrial contractors.[87] American universities were eventually to take up some of these ideas (Section 6.4.3).

More generally, the origins of a profession for American nuclear workers can be traced to the new occupational specialities that emerged within the Manhattan Project. As noted earlier, one such novel speciality was the 'health physicist', a euphemism for those having experience in protection from the biological effects of radiation, still unmentionable during the war.[88] Originating as distinct specialists at the Clinton Laboratories, health physicists began to train technicians in radiation protection methods in 1944. As nuclear activities

---

[86] Goodman, Clark (ed.), *The Science and Engineering of Nuclear Power* Vol. I (Reading, Mass.: Addison-Wesley, 1947), p. i. The volumes incorporated only material declassified by the AEC.

[87] Nichols, Kenneth D., *Nucleonics*, July 1949, reproduced in: *Atomic Energy Indoctrination* (Washington, DC: Dept. of Army, 1950), pp. 91–2.

[88] Proctor, Robert N., '"-Logos," "-ismos," and "-ikos": the political iconicity of denominative suffixes in science (or, phonesthemic tints and taints in the coining of science domain names)', *Isis* 98 (2007): 290–309; Parker, H. M., 'What is a Health Physicist?' presented at *Health Physics Meeting*, Chicago, 1951.

proliferated after the war, armed forces personnel also came to Oak Ridge for training in radiation protection, and the Lab continued this role until 1950.[89]

There were, of course, other levels of training. As early as 1950, the American army had standardized the instruction to be given to different ranks of soldier, bomb commander, and weaponeer involved with atomic warfare as part of a Radiological Defense Plan. This defined new categories of specialist and the training they were to undergo, including:

1. Staff Chemical Officer (Staff Radiological Defense Officer), taking a four-week course, with 72 hours orientation on radiological defence;
2. Unit Gas Officers (Unit Radiological Defense Officers), taking a six-week course, with 50 hours on radiological defence.
3. Radiological Instrument Maintenance Personnel, following a 100-hour course;
4. Radiological Defense Engineers, with training as yet unspecified; and
5. Medical personnel, with the Surgeon General conducting a five-day orientation course in medical aspects of nuclear energy.[90]

Tellingly, this army classification of knowledge and occupational responsibilities was nearly a decade ahead of well-defined niches in industry or academe.

The National Labs also played a central role. The wartime Manhattan Project sites had developed training courses as the numbers of engineers, operators, and other technical staff on the project rose. A more overt role for academics in nuclear technology was also a key selling point for the foundation of the National Laboratories, reconceived as well-resourced regional facilities to serve university research and to teach nuclear expertise. Amidst the post-war uncertainties, three university consortia coalesced. Oak Ridge Associated Universities (ORAU) initially combined fourteen regional universities in the American south; the Associated Universities Inc. (AUI) united the interests of nine north-eastern universities; and, the Participating Institutions Program of the Argonne National Laboratory (ANL) grouped some two dozen institutions centred on Chicago.[91]

As paralleled by the MIT case, Oak Ridge became a nucleus of early education for the developing discipline by fostering post-war government sponsorship and close links with academe. As early as the autumn of 1945, Clinton Laboratories had established a graduate studies partnership with the nearby University of Tennessee. Its educative role was consolidated with the establishment of the separate Clinton Training School in 1946

---

[89] The label was being used at Hanford in 1943; at Argonne, 'health and safety' was the preferred term until about 1948, when 'health physics' replaced it. By 1951, academic training courses of one year duration and ten weeks' site training were available at Vanderbilt University and Rochester University, in conjunction with Oak Ridge and Brookhaven National Labs [Cole, J. E., 'Trip report, Health Physics Meeting, Chicago, January 16–18 1951', memo to file, 22 January 1951, Hagley Series IV Box 44 folder 2].

[90] Dept. of the Army, *Atomic Energy Indoctrination* (Washington, DC: US Army, 1950), Chapter 8, Section III. The equivalent for the British Army was at the Royal Military College of Science, offering an external degree of the University of London in nuclear applications [Wilson, C. G. and D. K. Thomas, 'Teaching nuclear science in the army: The Royal Military College of Science', *Nuclear Energy Engineer* 13 (1959): 443–4].

[91] Ramsey, Norman F., 'Early history of associated universities and Brookhaven National Laboratory', *Brookhaven Lecture Series* 55 (1966): 16pp; Greenbaum, Leonard, *A Special Interest: The Atomic Energy Commission, Argonne National Laboratory, and the Midwestern Universities* (Ann Arbor: University of Michigan Press, 1971).

(known informally as the 'Clinch College of Nuclear Knowledge'). Focusing on nuclear physics and engineering rather than biological consequences of the technology, the Training School was conceived as a small postgraduate program. But directed by solid-state physicist Frederick Seitz (1911–2008) and opened by Eugene Wigner, the School attracted some fifty participants from industry, academe, and the military—particularly the Navy, which had provided funding for their experiments as early as 1939. Its most influential graduate was Captain (later Admiral) Hyman Rickover (1989–1986), who became the champion of a nuclear navy.[92]

In 1947, however, the Monsanto Company, which had managed the post-war laboratory, was replaced by new administration. The Clinton Training School consequently ended, but fourteen south-eastern universities collaborated to form the Oak Ridge Institute of Nuclear Studies (ORINS), which became a government-owned company-operated (GOCO) facility that year overseen by the Atomic Energy Commission (AEC).[93] ORINS disseminated knowledge that was more cleanly divorced from military applications than was reactor design. Its Board of Directors was drawn from senior academics in physics, chemistry, biology, and medicine from eight universities. Faculty at the participating universities had access to the Institute's biological laboratory facilities, and the nascent graduate training programme was focused on final-year PhD students and post-doctoral researchers. Know-how on matters such as radiation protection and the medical applications of radioisotopes was provided to its member universities and was further promoted by travelling lectures by the Institute's scientists. From an initial pair of researchers in 1947, numbers rose and stabilized at some 70 per year by 1950.[94]

Production sites also trained their workers for existing jobs and promotion to new ones. From the late 1940s the Hanford Technology Course taught topics including nuclear physics, instrumentation, chemistry, and handling. A more advanced Pile Technology Course focused on reactor operations.[95]

A separate initiative was the Oak Ridge School of Reactor Technology (ORSORT) founded in 1950. Like the Clinton Training School, ORSORT dealt with classified

---

[92] Hewlett, Richard G. and Francis Duncan, *Nuclear Navy, 1946–1962* (Chicago: University of Chicago Press, 1974).

[93] Board of Directors, ORINS, 'Annual Report of the Board of Directors to the Council of the Oak Ridge Institute of Nuclear Studies Inc', 30 June 1947, DOE ON NV0707737 and Board of Directors, ORINS, 'ORINS Annual Reports Excerpts: Special Training Division 1947–1958', 30 June 1958, DOE ON NV0712404. On the history of the AEC, see Hewlett, Richard G. and Franciscus Duncan, *A History of the United States Atomic Energy Commission* (3 vols) (University Park, Penn.: Pennsylvania State University Press, 1969). On the higher-level AEC measures to educate nuclear specialists, see Poor, R. S., 'The Atomic Energy Commission and nuclear education', presented at *Interrelated Role of Federal Agencies and Universities in Nuclear Education Conference*, Gatlinburg, TN, 1963.

[94] Laboratory, Oak Ridge National, 'ORSORT: Oak Ridge School of Reactor Technology', *Oak Ridge Nuclear Laboratory Review* 25 (3 & 4).

[95] Brown, C. L., 'Hanford Technology Course', personal notebook, 15 October 1948–8 October 1953, DOE DDRS D198027813; Fullmer, G. C., 'Pile Technology Course', personal notebook, 26 January 1950–30 September 1960, DOE DDRS D198027813.

engineering and scientific knowledge that universities could not provide.[96] Organized and directed by Frederick Seitz, its faculty included half a dozen physicists from Wigner's Met Lab theoretical group. With this complement of wartime expertise came a consolidation of the new subject: the first—and influential—text on the theory of reactors, based on ORSORT lectures by four of the Met Lab physicists.[97]

As its prospectus outlined, the curriculum of the one-year School satisfied 'an urgent need for a permanent school to offer specialized education in the field of reactor technology...conceived to supplement the university education and experience of engineers and scientists interested in entering the field of reactor development'. This new domain involved training unavailable elsewhere:

> Many new engineering problems associated with the development and design of reactors are as yet either unsolved or their solution is not incorporated into the general fund of engineering knowledge readily available to those outside the field. Some of these problems are new in the sense that they had not been encountered prior to the reactor; others are old in their basic principles, but are new in the sense of their present intensity. An example of the former is the need to shield against intense radiation of gamma rays, etc. Of the latter type, an example is the thermal stress which arises as a result of the great heat concentrations which are present in a reactor.[98]

As summarized in Table 5.2, the content of the curriculum, by the mid-1950s, was increasingly refined towards reactor design.

Just as importantly, government sponsorship meant that ORSORT had a guaranteed stream of students. Its organizers identified its remit to be the training of 'scientists, health physicists, pilot-plant operators, engineers and administrators who have staffed other vital projects for the AEC'.[99] As Stephen Turner has argued, the creation of a new discipline is bolstered by the creation of dependent clients and a cycle of replacements. Stable internal markets, buttressed by the intellectual exclusivity of reactor technology, promoted the new intellectual field.[100]

Beyond mere technical instruction, the School's curriculum was also intended to combat the recognized problem of institutional isolation in the field of atomic energy:

> It is a common experience for an engineer newly employed by a reactor project to address himself to the problems which are the responsibility and immediate concern of the relatively small group to which he is attached. As a natural result, he may take a long time to attain a perspective from which he can view

---

[96] ORSORT, 'Oak Ridge Operations Information Manual', DOE ON NV0714712. See also Weinberg, Alvin M., *The First Nuclear Era: The Life and Times of a Technological Fixer* (New York: AIP Press, 1994), pp. 52–4.

[97] The 71-page simplified exposition in Soodak, Harry and Edward C. Campbell, *Elementary Pile Theory* (New York: John Wiley, 1950) was superseded by the considerably expanded treatment in Glasstone, Samuel and Milton C. Edlund, *The Elements of Nuclear Reactor Theory* (Princeton, NJ: Van Nostrand, 1952). As the foreword to *Pile Theory* emphasized, 'the greatest part of the work is still classified, and probably will remain so for some time' [p. vii].

[98] 'The Oak Ridge School of Reactor Technology 1953–1954', prospectus, 1 September 1952, Hagley 1957 Series III Box 12 folder 5: ORSORT 1953–57.

[99] Ibid.

[100] Turner, Stephen, 'What are disciplines? And how is interdisciplinarity different?' in: P. Weingart and N. Stehr (eds.), *Practising Interdisciplinarity* (Toronto: University of Toronto Press, 2000), pp. 46–65.

Table 5.2 ORSORT curriculum, 1954–5.

Reactor Analysis (maths—basic theory of nuclear chain reactors)
Reactor Engineering (analysis of system and constituent engineering problems—heat generation, fuel requirements, distribution of fission energy; heat removal for power production; thermal stresses, etc.)
Reactor Technology (reactor shielding, reactor control, and reactor feasibility studies)
    Experimental Reactor Physics (classical atomic nuclear phenomena; types of radiation; ion exchange and solvent extraction techniques, radiation chemistry)
Reactor Materials (properties of materials used, and effects of radiation on them, including metallurgy)
Reactor Chemistry (radiochemistry of elements used, chemical processing of reactor materials; chemical technology relevant)
Reactor Nuclear Physics (stability of nuclides, cross-sections, liquid drop model, data important to reactor designer)
Reactor Design Problems (group theses, led by one student and with a faculty adviser)[101]

---

the whole engineering forefront of reactor development, and his contribution may therefore be limited. Thus, it is most necessary that engineers who will participate directly in reactor research and development first acquire a clear view of the entire field; the purpose of the ORSORT is to provide its student with this view.[102]

But this broad view was still restricted. Unlike ORINS, which considered and occasionally granted access to Canadian and British workers, ORSORT was for exclusive audiences. The profile of the ORSORT classes grew and altered rapidly in the early years: its first year began with a complement of 46 solely government-sponsored students (from the still-young AEC, the military, and government contractors); during 1951, most of its 68 students were from industries becoming involved in reactor design and operation; and the majority of the following year's 81 students were those planning careers as nuclear workers. As for the Clinton Training School of 1946–7, most attendees were college graduates holding Bachelor's degrees in chemistry, engineering, metallurgy, physics, or engineering physics. Applicants were sponsored by the AEC or its contractors, and the school was to be available for some time only to American citizens.[103] Enrolment was stable until the end of the decade because, as one administrator could argue privately, ORSORT had no academic rival:

> While some universities have given occasional special courses, no organization has come in with a program for giving these courses throughout the year. It appears that the group applying at Oak Ridge are attracted by feeling that the most authoritative information is received here and, based on our experience with ORSORT, would not return to the universities for short courses...ORINS has been able to get the

---

[101] 'The Oak Ridge School of Reactor Technology 1954–1955', prospectus, 3 September 1953, Hagley 1957 Series III Box 12 folder 5: ORSORT 1953–57.

[102] 'The Oak Ridge School of Reactor Technology 1953–1954', prospectus, 1 September 1952, Hagley 1957 Series III Box 12 folder 5: ORSORT 1953–57.

[103] News item, 'Training nuclear engineers at Oak Ridge', *Nuclear Engineering* 1 (1956): 115; Atomic Energy Commission, 'Oak Ridge Operations Information Manual, Budget and Reports Division', Dept of Energy Accession No. NV0714712 [on OpenNet as http://www.osti.gov/opennet/detail.jsp?osti_id=16111668], 31 October 1953; Hewlett, Richard G. and Jack M. Holl, *The New World, 1939–1946* (Berkeley: University of California Press, 1969), pp. 184–5. On foreign enrolment, see pp.143.

Fig. 5.3 Students of the evening class of the GE School of Nuclear Engineering, Hanford, 1954.[104] (Courtesy of US Department of Energy, Declassified Document Retrieval System.)

[industrial uses of isotopes] course organized and facilities developed in a manner not available to universities.[105]

So, despite the secondary status ORNL had in reactor development in the AEC ranking, it asserted its pre-eminence in training the next generation of nuclear specialists.

Other National Laboratories played similar, if less prominent, roles, however. One example was the 'School of Nuclear Engineering' sponsored by General Electric at the Hanford site from 1954 (Figure 5.3). And, from a less security-conscious perspective from 1949, Argonne sponsored an Atomic Energy Institute for teachers and administrators in Chicago area schools, with 2400 teachers participating, and had provided special non-credit courses 'in the fields of chemistry, physics and reactor engineering with special emphasis on the unique aspects needed in atomic energy work'.[106] Administrators initially rejected the idea

[104] Department of Energy, 'School of Nuclear Engineering class photo', 1954, DOE DDRS N1D0002164.
[105] Atomic Energy Commission to Telex, 31 October 1953, Dept of Energy, CD 59-5-20/FORM 189. Accession No. NV0702100 [OpenNet http://www.osti.gov/opennet/detail.jsp?osti_id=16289444].
[106] 'Board of Governors' Minutes, ANL', 2 May 1950, UI Box 44.

Fig. 5.4 Kansas State University students at the Argonne International Institute of Nuclear Science and Engineering, 1960.[107] (Courtesy of Argonne National Laboratory.)

of 'formal fellowship programmes akin to ORINS and ORSORT, preferring 'that our temporary staff members learn by direct experience in one of the Laboratory projects'.[108] From 1955, though, Argonne instituted a seven-month 'School of Nuclear Science and Engineering', established by the AEC as 'one of the major projects under the Atoms for Peace program to assist peoples of friendly nations to develop the peaceful uses of atomic energy' (Figure 5.4). The curriculum, restricted to scientists and engineers sponsored by their governments or by US industries, was 'determined by the status of the declassification program at the time the courses are presented'.[109] Brookhaven, too, could report education measures by the late 1950s, including 'a few organized courses and seminars. Chief among these is an annual 10-week summer program, primarily of a laboratory nature and a number of ad hoc courses of a few weeks' duration, chiefly for nuclear engineers'.[110]

[107] UI Box 114, image 139-2409, courtesy ANL. Between 1955 and 1959, the Institute was known as the 'International School of Nuclear Science and Engineering'.

[108] Boyce, J. C., 'To Council of Participating Institutions', memo, 26 January 1951, UI Box 134.

[109] Argonne National Laboratory, *School of Nuclear Science and Engineering* (Lemont, Ill: ANL, 1955), pp. 3, 9.

[110] Brookhaven National Laboratory, 'Brookhaven 1960–82 and its Associated Universities', UI Box 35.

While this hub-and-spokes model of disciplinary education proved effective, especially concerning knowledge that remained largely classified, the AEC further dispersed nuclear engineering skills by sponsoring graduate fellowships from 1950 at Rochester and Vanderbilt Universities in the American north-east.

In 1957, swamped by a growing demand for trained personnel, ORSORT itself linked with a half-dozen institutions to offer a curriculum based for six months at ORSORT and six months at a university;[111] in 1960 it admitted its first non-American students, and in 1965, having graduated some 976 students, it finally closed as more independent university programmes became established. The difficult transition from classified to open dissemination is discussed at greater length in Section 6.4.

5.4.2 *Spreading nuclear knowledge in the UK*

In Britain, formal education also appeared gradually. At Windscale, the Division of Atomic Energy had founded a Nuclear Engineering Society as early as 1948 for internal benefit. But with Christopher Hinton as President and restricted to DAE employees, its object was merely 'the improvement of its members in engineering science, by the reading and discussion of papers, [and] by visits to places of engineering interest', and was seen by Hinton as a means of retaining professional staff in the relatively isolated locale.[112]

As at Oak Ridge, training in the use and applications of isotopes was an early subject of instruction. For engineering matters, the solution of the Division of Atomic Energy and its successor, the UKAEA, was to take responsibility for training its own industrial workers, an activity that continued in various forms thereafter. The introductory courses introduced by Harwell provided the best model at the time. The four-week Harwell Isotope School, begun in 1951, was joined by the twelve-week Reactor School in 1955. Calder Hall instituted an eight-week Operation School to train plant operators; by 1957 it had trained 42, in groups alternating between British and overseas students. John Cockcroft's tentative term for Harwell workers, particularly those trained in the Harwell in-house courses, was 'atomic energy technologists'.[113] Nevertheless, the new Atomic Energy Authority inherited the staffing structures of its predecessor in the Ministry of Supply that had continued through and after the war. The civil service categories for technical workers were well established, and included the Professional class of Scientific Officers (including Assistant, Higher, and Executive grades, identified as Chemists, Physicists, Metallurgists, or Mathematicians) and Engineers (sometimes divided into Chemical, Civil, Electrical, and Mechanical); Industrial grades, incorporating a production hierarchy based on the Ordnance Factory system; and Craft workers.

---

[111] 'The Oak Ridge School of Reactor Technology 1957–1958', prospectus, April 1957, Hagley 1957 Series III Box 12 folder 5: ORSORT 1953–57.

[112] NES, 'Nuclear Engineering Society: Annual General Meetings; notices and nominations, 1958–1978', NA AB 65/459.

[113] Cockcroft, John D., 'Foreword—The Journal of Nuclear Energy', *The Journal of Nuclear Energy* 1 (1954): 1.

In this entrenched system, there was no obvious slot for nuclear workers per se, and no motivation to create a new category.[114]

5.5 Atomic industry for defence: The Savannah River Plant

The British experience had parallels at American industrial sites. In both countries the recognition and status of nuclear specialists was contested, with scientific direction initially dominating the hierarchy. As discussed earlier (Sections 3.2 and 4.2), the distinct flavours of expertise developed at the American sites were also distributed along an industrial–academic axis. The same was true for the American companies that became associated with nuclear energy. Indeed, during the post-war period, when the entire field was unusually malleable, nuclear engineering was shaped by the particular constellation of companies and research laboratories involved. Hinton's remarkable influence in defining reactor expertise, with British experts modelled on ICI culture, was surpassed by Du Pont's success in making invisible nuclear engineers visible in the USA.

According to Abbott's framework, this science–engineering boundary is a common instability in the formation of professional identity. Based on a shared workplace or shared intellectual responsibilities, it is a conflict that can lead to a subordination of one group by another, to the exclusion of the weaker group or, conversely, to the legitimation of new technical experts.[115]

As in the Anglo-Canadian project, some aspects of the wartime nuclear work had been dominated by chemical engineers. Key wartime participants had chemical engineering educations, too, including Lew Kowarski at the Montreal Laboratory, and Eugene Wigner in Chicago.[116] And Du Pont was linked closely to the discipline of chemical engineering; nearly one-third of the authors of the *Chemical Engineer's Handbook* (1934), originated by Du Pont's John Perry, were Du Pont engineers. Indeed, the term *reactor* has a longer lineage in the technical culture of chemical engineering than that of nuclear physics, and had first been used by Du Pont engineers and picked up by the Atomic Energy Commission thereafter. Key Du Pont personnel at Hanford had backgrounds in organic chemistry, high-pressure technique, and fluid flow, and several in the company's research groups were sought without success to teach university chemical engineering courses.[117]

But applying this expertise to nuclear technologies still stretched the company's identity. As explored in Chapter 3, the Du Pont Company had entered into its Manhattan

---

[114] In later years, the UKAEA moved from civil service grading to a simpler salary structure, but largely retained the post-war technical categories with isolated additions, such as Computing and Electronic Engineer ['Personnel policy—proposed new notes for guidance for promotion in the professional grades', Dounreay 07647/47/PP/1-1].

[115] For example, health physicists can be seen as a group subordinate both to nuclear physicists and to nuclear engineers, as their work is dependent on the products of those fields [Abbott, Andrew D., *The System of Professions: An Essay on the Division of Expert Labor* (Chicago: University of Chicago Press, 1988), pp. 72, 167, 286].

[116] Obituary, *The Times*, 9 January 1995, 19.

[117] Thayer, Harry, *Management of the Hanford Engineer Works in World War II: How the Corps, DuPont and the Metallurgical Laboratory fast tracked the original plutonium works* (Reston, VA: American Society of Civil Engineers Press, 1996), p. 34.

Project responsibilities without enthusiasm. The 'Chicago Process' demanded new knowledge and required a significant fraction of its resources, compensated at cost. Profit-making had not been pursued; the company was still sensitive to its identification in 1936 as a 'merchant of death' profiteering from First World War munitions manufacturing.[118] This wariness may, however, have contributed to success and to its role in shaping nuclear experts. Indeed, as David Hall, a physicist hired for the Hanford start-up (and later the Director of the Reactor Division at Los Alamos) observed, 'Du Pont was careful because it cost them nothing'. Their cost-plus-fixed price contract meant there was no competitive cost-cutting, so what some saw as a waste of over-qualified staff ensured new working relationships between scientists, engineers, and technicians. As Hall recalled, 'my wife and I, both with PhDs, were hired to do really quite menial tasks, jobs that could have been done by people without training. It was over-kill, and it paid off'.[119]

Du Pont required profits, however, and initially had other post-war plans. Its managers carefully weighed up the possibilities of commercial atomic energy. Conversations with John Wheeler, Arthur Compton, Enrico Fermi, and Alvin Weinberg proved seductive for Crawford Greenewalt in envisaging new, more efficient pile technologies. He noted the pros and cons of the company's post-war involvement, 'assuming that the government would allow Du Pont to do general research in the field and will make available uranium, P-9 [i.e. plutonium], design data and general know-how':

Pros

1. This is a new field of great scientific importance in which Du Pont has some specialized knowledge which may lead to commercial or by-product developments.
2. Publications in the field have advertising value and should improve Du Pont – university relationships, perhaps thereby aiding technical employment problems.
3. Facilities for tracer preparation should be useful and perhaps vital in many Du Pont researches – radioactive isotopes of practically any element can be readily prepared and techniques be made available for their use...

Cons

1. Suitable physics personnel will be difficult, if not impossible, to get (No good prospects at Met Project except Fermi).
2. Money can be spent in other types of research either fundamental or applied, with better chance of return.
3. No commercial applications can be foreseen now or even guessed at except perhaps in the field of luminous or sterilizing paints.
4. Chance of recovering money spent very remote.
5. Patents if obtainable may run out before commercial use appears.[120]

---

[118] The Special Committee on Investigation of the Munitions Industry, or Nye Committee (1934–6), of the US Senate documented the profits of the American chemical and munitions industries, and led to the popular conclusion that the country's entry into the war had been predicated on, and engineered by, commercial interests.

[119] Sanger, S. L. and Robert W. Mull, *Hanford and the Bomb: An Oral History of World War II* (Seattle: Living History Press, 1990), p. 128.

[120] Greenewalt Diary, Vol. II, p. 254. 27 July 1944.

None of those possibilities included plutonium manufacture. Despite enthusiasm for the novelty of the new domain, Du Pont decided that other commercial fields—notably the marketing of nylon and other synthetic fibres and films—offered greater post-war profits.

This perhaps explains why Crawford Greenewalt's analysis hints that he had not conceived nuclear workers as a new kind of specialist at all ('organization might comprise a director (preferably a physicist), 1 theoretical and 5 experimental physicists and 4 chemists...').[121] In 1946, at the contract's completion, Du Pont had given up the management of Hanford Engineering Works to the General Electric Company.

This change of management carried with it a change of perspective on what piles and its designers and operators were. Labour analyst Melvin Rothbaum has argued that Du Pont, with its working culture of 'almost haphazard, cut-and-try methods' drawn from chemical engineering and industrial chemistry, was replaced by General Electric and its electrical engineering 'systems approach'. Instead of viewing the nuclear reactor, as Du Pont did, as a versatile and malleable collection of plant for producing a chemical product, GE saw it as a product in its own right, akin to the industrial transformers and generators that had made its fortune.[122]

Others, too, were beginning to identify engineering specialisms within reactor design. Looking back from the perspective of 1953, Marvin Fox, Chairman of the Reactor Department at Brookhaven National Laboratory, depicted this as a rapid evolution of industrial experience driven by applications:

> The first reactors, at Oak Ridge and Hanford, were designed and built just to operate, using the meagre background of technical information which existed at that time. The operation of these reactors and the research done with them served to extend both the scientific information about nuclear processes and the technological information about reactor systems. Other reactors were designed and built after the end of the war and they incorporated many improvements. A full decade of development and experience is now behind us, and we have progressed to a point where we think we can build reactors with some competence. Reactor designs of today call for a degree of specialization, and that is why the many kinds of reactors... are receiving the attention of nuclear engineers.[123]

5.5.1 *Industrial paradigm: Du Pont's Atomic Energy Division*

As in the early National Laboratories and training schemes, though, a distinct nuclear engineering identity was slow to coalesce even in American industry. In the years immediately following the war, a clear notion of the scope and content of nuclear engineering had not emerged either at Hanford or other industrial sites, which remained almost completely obscured to the view of the public and, indeed, to contemporary scholars and analysts. The evolution of an industry and associated specialists is best illustrated by the experience of Du Pont with the second major American industrial atomic energy project,

---

[121] Greenewalt Diary, Vol. II, p. 254. 27 July 1944.

[122] Rothbaum, Melvin, *The Government of the Oil, Chemical and Atomic Workers Union* (New York: John Wiley, 1962), pp. 45–58, 88–90, 109–20; quotations p. 109. See also Schatz, Ronald W., *The Electrical Workers: A History of Labor At General Electric and Westinghouse, 1923–1960* (Urbana: University of Illinois Press, 1983).

[123] Fox, Marvin, 'Fundamentals of reactor design', presented at *Atomic Energy in Industry, 2nd Annual Conference*, Montauk, NY, 1953.

at Savannah River, South Carolina. Together, Du Pont's Hanford Engineer Works and the Savannah River Project were the largest American reactor installations for a quarter-century. And—designed for production, not research—the sites were seminal in creating a class of reactor designers, nuclear process workers, and working environments based on the Du Pont model.

In late 1948, Crawford Greenewalt—now president of Du Pont—was approached by the AEC to undertake a fundamental review of all chemical activities bearing on plutonium production.[124] Unlike its Manhattan Project duties, this task appeared to fit comfortably within Du Pont's organizational remit. The objectives included the examination of uranium recycling, plutonium separation chemistry, and handling of fission products and wastes (but excluding plutonium decontamination), all from the standpoint of chemical engineering. The wartime state-of-the-art had been recognized by all parties privy to the secrets to have 'critical process defects', including wasteful treatment of uranium—simply depositing slightly depleted uranium in underground storage—and production of radioactive by-products along with plutonium, a large proportion of which accumulated in bulky and dangerous form.[125]

As the Du Pont survey was being completed in the summer of 1949, however, the political context shifted under the company's feet. That autumn, with the explosion of the first Russian atomic bomb and the success of Mao Zedong's forces in China, the Truman administration and AEC sought to recast the nuclear energy programme to ensure a larger and diversified military stockpile. The programme provided for an increase in the numbers of atomic weapons through new designs and greater production of materials; a programme for increasing the power of the weapons over those at Hiroshima and Nagasaki, and an improvement in the combat usefulness of the weapons through re-engineering. The arsenal of conventional atomic weapons was to be enlarged, particularly by the hydrogen bomb. New construction and administrative additions to support research and development were planned for Oak Ridge, Los Alamos, and other AEC sites together with the establishment of an entirely new project, the Sandia operation in New Mexico. Part of this reorientation of the AEC programme was the inclusion of a new project to produce the plutonium and tritium required for hydrogen bombs and similar weapons.[126] In early 1950, Du Pont consequently was asked to plan another plutonium facility to increase capacity alongside the Hanford Works inherited by General Electric.

The plant site, near Aiken, South Carolina, close to its Georgia border, was announced in November 1950. The Savannah River Plant (SRP) was selected for its ample water supply, lack of frequent floods and storms, large and relatively unpopulated land area conducive to construction, accessibility for transport, and good labour market. Unlike the Hanford site, however, the geology was not conducive to safe storage of nuclear wastes, a realization that was later to motivate projects for waste processing.[127]

---

[124] 'Press release', 22 December 1948, Hagley 1957 Series I Box 2 folder 9.
[125] Genereaux, R. P., 'Object of survey', memo, 29 December 1948, Hagley 1957 Series I Box 2 folder 9.
[126] Atomic Energy Commission, 'SRP Fact Book', 25 November 1960, Hagley 1957 Series V Box 54 folder 15.
[127] In April 1989, site management was passed from Du Pont to Westinghouse Corporation.

As for its wartime Manhattan Project work, the company undertook the design, construction, and operation project as a cost-plus-fixed-price contract. Unlike the wartime work, however, Du Pont engineers now worked with a wider and deeper range of knowledge about atomic energy, and interacted with a nascent American network of specialists. The project was significant in several respects: it incorporated a wide-ranging survey of the rapidly expanding state-of-the-art, revealing the range of skills then available; it adopted technical solutions at variance with the military projects then underway, thus expanding the American state of the art in new directions; and it constructed the first post-war American model for an industrial nuclear workforce.

While this was seen more as a national duty than a commercial opportunity, Du Pont laid the groundwork for future directions. The company set up an Atomic Energy Survey Committee within its Explosives Department in July 1950. Three weeks later this was transformed into a new unit dubbed the Atomic Energy Division (AED)—again, like the wartime TNX Division, within the company's Explosives Department.[128]

Du Pont's Atomic Energy Division conformed to the company's traditional management structures, but also established posts, administrative niches, and network nodes for the new experience in atomic piles. As R. M. Evans, the Assistant General Manager of AED, explained to the Atomic Energy Commission manager responsible for relations with the new plant, Du Pont would be responsible for every aspect of the project. Its Explosives Department would have prime responsibility for the work, defining its scope and specifying the process requirements; the Engineering Department would occupy the role of architect–engineer and general constructor, and its Design Division would develop the final design and handle procurement; the Construction Division would perform the field construction; and the Explosives Department would be responsible for operating the functioning plant.[129]

Supporting this traditional Du Pont integrated management structure for the Atomic Energy Division, however, were atypical Du Pont personnel. Some had had experience at the wartime Clinton and Hanford sites; others were young physicists and engineers chosen to expand company expertise in the field. Hood Worthington, the Assistant Manager of the Research Division (soon renamed the Technical Division), had been picked by Crawford Greenewalt in 1942 to contribute to the earliest Manhattan Project work. He later served as the Chief Supervisor for the Hanford piles, a role involving both technical expertise and administration.[130] Worthington was to be responsible for reactor development at Savannah River. He was twinned with

---

[128] Brown, H. F., 'Atomic Energy Survey Committee', memo, 11 July 1950, Hagley 1957 Series II Box 6 File—Administrative policy correspondence, general, 1950–1963; Brown, H. F., 'Atomic Energy Division', memo, 2 August 1950, Hagley 1957 Series II Box 6 File—Administrative policy correspondence, general, 1950–1963.

[129] Evans, R. M. to C. A. Nelson, letter, 9 August 1950, Hagley 1957 Series II Box 6 File—Administrative policy correspondence, general, 1950–1963.

[130] Worthington, Hood, 'Pile technology—effect of operation on graphite moderator (Wigner and Szilard effects)—experience to August 1, 1945', memo to file, 13 September 1945, Hagley 1957 Series III Box 58 Folder 4. Alternately described as chemist, chemical engineer, and consulting nuclear engineer between the late 1940s and 1960, Worthington's occupational label changed more than did those of his counterparts in the National laboratories such as Walter Zinn and Alvin Weinberg.

Dr J. Elton Cole, focusing on chemical separation processes and related problems. In addition, the skeleton of a design team was constructed within days of the Division's organization: Dr V. R. Thayer on heavy water developments, Dr J. C. Woodhouse on metallurgical problems, and Drs D. F. Babcock and C. W. J. Wende on reactor physics and development.

Dale Babcock, like Worthington, had been involved from the beginning with Du Pont's Manhattan Project work eight years earlier. A physical chemist, he had worked on the start-up operations of the Hanford piles. And C. W. J. Wende, a 'Technical Specialist' at wartime and post-war Hanford, was to become the most active senior pile designer for the Savannah site. Most of those assigned to key posts were transferred from other Du Pont commercial plants and laboratories, but new titles highlighted the new activities: within months, personnel described as 'Atomic Engineering Managers' (more for their Division association than for functional responsibilities) were seeking new employees and arranging AEC contacts.[131]

5.5.2 Exploring American pile networks

Despite this concentration of Du Pont atomic energy experience, most of these senior personnel had lost contact with the growing field since 1946, when the company's responsibility for Hanford had ended. In the intervening four years, an expanding infrastructure had grown: the Atomic Energy Commission and National Laboratories were now established, along with burgeoning projects in the research and development of atomic piles, and the technoscientific staff to support them. To achieve its goals, Du Pont's SRP project required administrators to explore the institutional and intellectual networks of expertise concerned with designing piles. This technical staff would serve as conduits, and were 'expected to maintain contact with other research centers within the AEC framework and to advise the plant technical forces on process improvement and development work'.[132]

The Technical Division of Du Pont's AED consequently began a flurry of activity to liaise with AEC experts, to channel their knowledge into industrial implementation. This was, in some respects, an amplified version of the wartime model. The Met Lab had served first as designers, then trainers, and finally consultants for Du Pont engineers. The shifting relationships had been smoothed by staff serving as technical liaisons (notably Crawford Greenewalt for Du Pont and John Wheeler for the Met Lab, with dozens of personnel from both organizations gaining scientific and engineering experience, respectively). In the same way, the Atomic Energy Division sought the successors of the Met Lab personnel, and they found them principally at the Argonne and Oak Ridge National Laboratories, successors to the University of Chicago Metallurgical Laboratory and Clinton Laboratory.

Hood Worthington and C. W. J. Wende shuttled between Argonne and Wilmington in the autumn of 1950, meeting with Walter Zinn, John Wheeler, and others to discuss design details of the SRP reactors; John Woodhouse visited Brookhaven National Laboratory to consult on the design of fuel elements; and F. S. Chambers went to Oak Ridge to learn

---

[131] For example, Church, G. P. to E. M. Cameron, letter, 13 December 1950, Hagley 1957 Series III Box 12 folder 3; Fulling, R. W. to F. C. VonderLage, letter, 14 May 1954, Hagley 1957 Series III Box 12 folder 6.

[132] Evans, R. M. to C. A. Nelson, letter, 9 August 1950, Hagley 1957 Series II Box 6 File—Administrative policy correspondence, general, 1950–1963.

about plutonium separation.[133] Wende, rapidly assuming responsibility for pile design, felt the need for much more information, petitioning Du Pont and the AEC for:

(1) a 2 to 3 ft shelf of unclassified technical literature useful for preliminary indoctrination of employees prior to clearance;
(2) a critically selected set of classified documents relating to pile physics, pile engineering, metallurgy, separations engineering, and separations chemistry which would cover the basic sciences and pertinent recent developments, and which would serve as a minimum working reference library for people assigned to the Wilmington office;
(3) a need to begin assembling a library and setting up a classified information service which would later be transferred to the plant.[134]

His first few months of work on the project were a scramble for information. Following a visit to the Knolls Atomic Power Laboratory, a post-war General Electric facility focusing on atomic power and breeding applications, he reported how much material was yet hidden:

> GE is not at present conducting a training course for pile physicists. Such a course was given several years ago, and Tonks has a voluminous set of notes on it which he is willing to let us borrow for duplication. A brief survey of this course indicated that it covered much material which is not otherwise available in coherent form.[135]

The question of information transfer was an early and perennial one for Du Pont. Its Atomic Energy Division staff also sought operational training, and drew upon arrangements that were analogous to the wartime relationships. The operating staff of the Hanford site had been trained first at the Clinton Lab and, once it was operating, the Hanford site. Within a month of the AED's organization, plans were mooted for training its operators. It was hoped that some experienced personnel might be obtained from Hanford; indeed, many of them had originally been hired by Du Pont to man the plant during the war, and had stayed on after the assumption of management by General Electric. But GE, like Du Pont a contractor of the AEC, had an equally acute need for operating personnel. Plans thus shifted quickly to having Hanford establish a formal training programme for Savannah River personnel, an idea that appealed to both companies.[136]

From the perspective of the plant operators, though, such decisions were unusually restrictive. Not only did employees have to submit to an unusual degree of site security during this period of the Cold War, but they were also literally cloistered by job conditions: it was decided that Hanford personnel could not apply for new Du Pont posts without seek-

---

[133] Wende, C. W. J., 'Meetings at Argonne, September 13–15 1950', memo to file, 26 September 1950, Hagley 1957 Series IV Box 44 folder 1; Woodhouse, John C., 'Brookhaven fuel elements', memo to file, 21 September 1950, Hagley 1957 Series IV Box 44 folder 1; Chambers, F. S., 'Manpower requirements for separation work at ORNL', memo to file, 19 September 1950, Hagley 1957 Series IV Box 44 folder 1.

[134] Wende, C. W. J., 'Visit to Argonne, September 28–9, 1950', memo to file, 20 October 1950, Hagley 1957 Series IV Box 44 folder 1.

[135] Wende, C. W. J., 'Visit to Schenectady November 7 and 8, 1950', memo to file, 27 November 1950, Hagley 1957 Series IV Box 44 folder 2.

[136] O'Connor, D. F., 'Hanford Personnel and Training Program', memo, 31 August 1950, Hagley 1957 Series II Box 12 folder 3.

ing permission from their present employer, General Electric. The administrators of the national labs also negotiated agreements to prevent the poaching of their senior engineering and scientific staff.[137] Further constraints were presented by the need for Du Pont's subcontractors to implement security measures when working on and storing classified information, requiring isolated offices and locked repositories.

Du Pont managers recorded being 'severely handicapped' by the lack of cleared personnel to handle classified information, and the need for technical personnel to be cleared through the New York Operations Office.[138] Even visitors required AEC approval in secure areas. Correspondence was to be stored in safes, personnel cleared to Q, the highest security level, and identifications recorded by fingerprinting and photographs. Their file would then be turned over to the FBI 'for a comprehensive investigation of the applicant's past…concentrated on showing the applicant's degree of security risk'. The architectural company designing the buildings for Savannah's heavy-water plant, for example, required 175 employees to be cleared to level Q. So time-consuming were these tasks that at one point the AEC had to issue 'Emergency Clearances' for drafting personnel to complete their tasks on schedule.[139]

Nearly all the recruits for the supervisory posts, in their late twenties to mid-thirties, had engineering and often managerial backgrounds. One early cohort of eight trainees included four chemical engineers, two mechanical engineers, one electrical engineer, and one biologist, with experience ranging from supervision of sulphuric acid plants to fibre spinning, and prior responsibilities ranging from maintenance engineering to senior supervision.[140]

Their training for SRP assignments included design liaison work in the Process Section in Du Pont's Wilmington headquarters, and more than half of them spent time at various AEC sites. Sixty-six supervisors received training in Camden, New Jersey, on a mock-up of the reactor tank, hydraulic systems, and reactor loading and unloading systems. This included two components: instruction in reactor physics (control of the reaction, the functions of the various components, and basic supervision), and actual operation of the mock-up. Eight more supervisors received training at ORNL before coming to SRP, instructed on reactor theory and health physics.

The remainder received six weeks' training at the Hanford Works. The Hanford training was customized for operations supervisors, for 'pile operators' and 'separations operators'. Future supervisors, over a period of two months, received instruction in uranium metal quality, 'slug canning' (i.e. uranium fuel jacketing), reactor operations, process problems, and health physics. By contrast, the 'Operation of Pile' trainees were exposed to:

---

[137] Church, G. P. to E. T. Macki, letter, 5 September 1951, Hagley 1957 Series II Box 12 folder 3; Westwick, Peter J., *The National Labs: Science in an American System, 1947–1974* (Cambridge, Mass.: Harvard University Press, 2003), p. 63.

[138] 'Dana engineering and design history—Girdler Corp.' September 1952, Hagley 1957 Series V Box 53 folder 7.

[139] 'Dana engineering and design history—Lummus Co.' September 1952, Hagley 1957 Series V Box 53 folder 8.

[140] McNeight, S. A., 'Hanford cohort of trainees', memo, 6 February 1953, Hagley 1957 Series I Box 2 folder 9.

1. Start-up (check list, preliminary placement of control rods, indication of initial reaction, critical items during power level increase…)
2. Shutdown (normal, emergency, conditions which cause scram, safety circuits)
3. Power level control (control rod operation, safety circuits, reactor control instrumentation)
4. Limitations to power level (slug temperature, graphite temperature, cooling water temperature)
5. Charging/discharging operations
6. Limitations on shutdown time (effect of transients, short and long shutdowns and why)
7. Determining, locating and handling of ruptured slugs
8. Temperature and pressure monitoring system
9. Gas storage, handling and circulation through pile
10. Routine tests of safety devices (safety and control rod tests).

In addition, there were sessions on health hazards, accounting practices for control rods and numerous operational details.[141]

Operators (that is, relatively unskilled technician-level personnel) received considerably less training, and usually from the supervisors themselves via brief on-the-job training. They were also 'subjected to an extensive one-week training program' that included administrative details (meeting area supervisors, description of job progression procedure), technical instruction (an orientation talk on the area and process), discussion of the main equipment, detailed instruction on applicable safety procedures and operation of the assembly, reactor, purification, and disassembly areas'.[142] Most of the operators themselves were recruited locally. Indeed, SRP administrators accumulated a thick 'political correspondence' file relating to recommendations from congressmen and other local politicians urging employment for their constituents.[143]

For Du Pont's design personnel, Hanford training was judged to be largely irrelevant. The first-generation graphite-moderated piles there lacked the distinct features that were characteristic of a heavy-water-moderated, second-generation reactor. Interestingly, design liaison or training at Chalk River, which had, by the early 1950s, extensive experience in design and operation of such reactors, was never seriously considered. As the head of the Technical Division, Lombard Squires, confidently informed his AED General Manager,

> …assistance from Chalk River would be worthwhile and perhaps result in an overall saving of several months to the project. However, we definitely are of the opinion that it is not vital or essential to have this assistance. Hence, if obtaining it is a troublesome or time-consuming procedure, we do not believe it would be justified.

He was equally self-assured about the design team's abilities independent of other production sites, proposing that 'a liaison group of no more than two or three people visit Hanford

---

[141] 'Training of SRP Personnel at Hanford', 1952, Hagley 1957 Series I Box 2 folder 9.

[142] Du Pont, 'Savannah River Plant History—All Areas—August 1950 through June 1954', bound report, 1954, Hagley 1957 Series V Box 51.

[143] 'Admin files—political correspondence', prospectus, Hagley 1957 Series III Box 12 folder 7.

at irregular intervals, as required'.[144] This technical confidence, based on a 'not invented here' mind-set, played a role in defining Du Pont nuclear expertise[145].

On the other hand, an early intention of Savannah River Plant administrators was that a half dozen or more Du Pont designers would be educated by the most formal route then available: at ORSORT, the Oak Ridge School of Reactor Technology. But as its Director cautioned a prospective student, this was a demanding course:

> The wide scope of the school's curriculum and the quality of the courses require sustained and intensive effort by the student. Competition among the students is keen and generally stimulating; therefore, the advantage of prior knowledge is emphasized. In particular, a working knowledge of mathematics through the solution of boundary value problems is essential. The school curriculum provides a short refresher course in mathematics, but you are urged to devote time to this subject during the summer should you feel the need for review prior to enrolment in ORSORT. Factual knowledge and a familiarity with elementary concepts in atomic and nuclear physics are also essential...[146]

Few Du Pont staff, it seemed, had the requisite skills, particularly in mathematics. Only Harry J. Kamack, a Research Project Manager in Chemical Engineering, was admitted to ORSORT from Du Pont's first cohort of seven candidates.[147] Of the 44 government-sponsored students in that first class of 80, he noted that 29 were from 20 private companies and 15 from six government organizations, including the Bureau of Ships, Naval Reactors Branch, Naval Research Laboratory, US Air Force, US Army, and Tennessee Valley Authority. The largest industrial groups included five students from Westinghouse, three from GE, and three from the Electric Boat Division of General Dynamics Corp. The industry students were mostly engineers—mechanical engineers being the largest single group—with a few chemists and a fairly large group of physicists.[148]

As Kamack reported to his superiors, the one-year course covered the topics essential for industrial reactor design:

> The <u>Analysis</u> course is the heart of the school and is concerned with the theory of nuclear reactors... <u>Technology</u> is a set of lectures by visiting lecturers covering particular engineering problems which they have experienced in the design of piles, mainly those at Hanford. The <u>Materials</u> course is concerned with those properties of materials which influence their use in construction of piles. The <u>Laboratory</u> course is one afternoon a week and consists of a series of experiments such as the shielding properties of materials,

---

[144] Squires, Lombard to B. H. Mackey, letter, 9 August 1950, Hagley 1957 Series II Box 6 folder—Administrative policy correspondence, general, 1950–1963. Squires had been head of separation-process organization at Hanford's Technical Division during the war.

[145] Katz, Ralph and Thomas J. Allan, 'Investigating the Not Invented Here (NIH) syndrome: A look at the performance, tenure, and communication patterns of 50 R & D Project Groups', *R&D Management* 12 (2007): 7–20.

[146] VonderLage, F. C. to H. J. Kamack, letter, 21 May 1953, Hagley 1957 Series III Box 12 folder 4 ORSORT 1953–54.

[147] Kamack (1918–2009) had joined Du Pont in 1942 and was assigned to its TNX design group in 1943, serving at the Met Lab, Clinton Lab, and then Hanford. After his ORSORT training, he was to play a senior role in design of the SRP reactors.

[148] Kamack, Harry J., 'Report on year of training at Oak Ridge School of Reactor Technology—1953–1954', memo, 8 September 1954, Hagley 1957 Series III Box 12 folder 4 ORSORT 1953–54.

measurement of back-scattering, etc.... During June, July and August, the class is broken up into small sections each of which designs a reactor... The students also have access to the classified reports from the various Atomic Energy Commission sites and have the opportunity to learn something of the various research and development projects in progress at Oak Ridge... My opinion, based on my conversations at Oak Ridge, is that after taking the course, I will have approximately the same amount of knowledge with respect to the design of nuclear piles as does a recent university graduate in chemical engineering with respect to the design of, say, fractionating columns.[149]

Even so, he was aware of the individualistic slants that were presented on this novel domain, with 'many of the lecturers [having] decided opinions about what is the best type of reactor, or reactor material, etc. for the future, and these opinions [varying] considerably from one man to another'. ORSORT was populated with enthusiasts who proselytized actively: 'the faculty... are most strongly interested in reactors from the standpoint of power production... It is impossible to attend the school without developing a strong interest in the future of nuclear power'.[150]

5.5.3 *Continuing friction in reactor design*

The evolution of American expertise in reactor design offers further insights. Ongoing disagreements between the successors to the wartime Met Lab scientists and Du Pont engineers reveal the evolving responsibilities and status of the specialists who were beginning to call themselves nuclear engineers.

Early in Du Pont's contract to survey plutonium production, it had been decided that any new reactors would be moderated by heavy water rather than the graphite used in the wartime Clinton and Hanford piles. Heavy water initially had been the preferred medium for the Met Lab team, but was summarily rejected owing to the poor prospects for heavy water availability. Post-war perspectives, based on recent experiences, were quite different. The Hanford piles had shown how capricious the properties of graphite could be as an engineering material. The stockpile of heavy water was now better, and new facilities could be built.[151] Moreover, the manager of AEC operations for the Savannah River project, Curtis Nelson, had been a colonel in the Manhattan Engineer District; after serving at Hanford, he had become the AEC liaison officer at the Chalk River site, becoming familiar with the heavy-water moderated technology of the NRX reactor.[152] His lobbying, merging with Du Pont's own engineering analyses, yielded plans for a set of heavy-water moderated production reactors.

The company's engineers had accepted the design and construction of large heavy-water production plants at the Savannah River Project as a relatively routine task in industrial

---

[149] Kamack, Harry J., 'ORSORT curriculum', memo, 10 June 1953, Hagley 1957 Series III Box 12 folder 4 ORSORT 1953–54.

[150] Kamack, Harry J., 'Report on first term of ORSORT year', memo, 20 January 1954, Hagley 1957 Series III Box 12 folder 4 ORSORT 1953–54.

[151] The heavy-water plant at Trail, British Columbia—a by-product of a large electrolytic hydrogen plant—produced about 6 tons/year by 1950, having generated 60 tons altogether, and of which 30 tons remained [memo, 1953, Hagley 1957 Series V Box 53 Folder 10—Dana History, Startup through December 1952].

[152] Joseph, J. Walter and Cy J. Banick, 'The genesis of the Savannah River site key decisions, 1950', presented at *50 Years of Excellence in Science and Engineering at the Savannah River Site: Proceedings of the Symposium*, 17 May 2000, Aiken, SC.

chemical engineering. The challenge of heavy-water reactors was another matter, however. The technology was beyond their experience, and the close reliance of the Savannah River Project designers on Argonne experts proved difficult and, in some respects, mirrored the wartime experiences between Du Pont and the Met Lab. Indeed, some of the same personnel were involved, but the differences can be attributed more to institutional cultures than to individuals. The episodes reveal the distinct perspectives operating at the national labs and production facilities and, by extension, the differentiation of what could be called 'nuclear applied science' from 'nuclear engineering'.

C. W. J. Wende, the wartime Hanford engineer now responsible for the preliminary design of the SRP reactors, was conscious of the scientific uncertainties underlying some of the necessary engineering choices. Four months after beginning his round of visits to the national labs, Wende reported to Hood Worthington that he had learned of an unexplained transient phenomenon in Argonne's heavy-water research reactor. Worthington, in turn, wrote to Walter Zinn for further information, recalling the problems with the first Hanford reactor, and noting their concern about the consequences for unexplained science in an industrial design:

> We remember that what I might call the 'micro behavior' of the Clinton pile would almost certainly have disclosed the presence and importance of the xenon transient if it had been thoroughly studied. For this reason I should like to mention here the phenomenon described by Dr Arnett...[153]

Zinn, perhaps sensing a destabilization of scientific authority akin to the wartime Met Lab troubles, telexed a rapid but dismissive response that the 'BEHAVIOR [...] PROBABLY IS UNDERSTOOD BUT PERHAPS NOT QUANTITATIVELY IN ANY CASE IT IS NOT BEHAVIOR WHICH WE WOULD EXPECT FROM THE EQUIPMENT YOU HAVE UNDER CONSIDERATION'.[154] Wende, junior in status to both Zinn and Worthington, doggedly pursued the issue with Eugene Wigner (who had moved to Oak Ridge after the war) and John Wheeler, both veterans of the wartime skirmishes and, with Zinn, the most senior of American reactor experts. At his next trip to Argonne, Wende underlined that Du Pont engineers 'were very much interested in having analysis of this effect continued', describing the experience with xenon poisoning 'as a case where failure to analyze a small effect in the Clinton pile had resulted in a nearly catastrophic failure to predict a large effect in the Hanford piles'. And in a separate discussion of experimental tests, Wende again admonished Zinn with historical precedent, recalling the inadequate exploration of the Wigner and Szilárd effects in graphite:

> the measurements so far were barely enough to begin to show us where some of the problems are; and that, rather than to terminate such experiments, it was our feeling that they should be continued and probably expanded into full-scale critical experiments. The history of exponential experiments in graphite lattices was cited as an error of omission which should not happen again: namely, that such experiments were dropped by the Metallurgical Laboratory in 1944, and that much important information was not discovered about these systems until the group at Hanford undertook such work in 1949–50.[155]

---

[153] Worthington, Hood to W. H. Zinn, letter, 21 December 1950, Hagley 1957 Series IV Box 54 folder 2.
[154] Zinn, Walter to H. Worthington, telex, 21 December 1950, Hagley 1957 Series IV Box 54 folder 2.
[155] Wende, C. W. J., 'Visit to Argonne, January 3 and 4, 1951', memo to file, 12 January 1951, Hagley 1957 Series IV Box 44 folder 2.

Engineering experiment and analysis, he suggested, had trumped science.

The same visit generated yet another technical confrontation when Zinn asked why Wende had been 'beating the drum' for graphite piles rather than a heavy-water version. Wende argued that that

> if the Commission's primary interest is in getting plutonium economically and with certainty and in minimum time, and at a site more defensible than Hanford, then it makes sense to build graphite piles – or at least one graphite pile out of presently available stocks. While the heavy water piles have the potential advantages of greater flexibility and greater excess neutron production, these advantages are potential and not demonstrated; and Argonne's work over the past six months has certainly demonstrated that major areas of ignorance exist which will not be cleared up until many months after the first heavy water pile is started up.[156]

The episodes reveal attitudes that had been seeded during the war. Wende, an experienced but relatively low-ranking engineer in a technologically conservative company, was prepared to challenge the most senior pile scientists in the world. His criticism did not concern scientific detail, but the manner in which Argonne personnel related ongoing exploratory research to engineering decision-making. Their science—involving the interplay of subtle physical phenomena with large-scale materials—was engineering in orientation but not in diligence. The Argonne Laboratory's brand of applied science, claimed Wende, did not give a viable match with Du Pont's drive for industrial reliability.

Du Pont thus found itself at odds with developing practice in American atomic energy. The company was repeating its wartime role as integral designer, builder, and operator of plutonium plants; it had no corporate intent or technical experience in nuclear power; and its specialists were not ideally configured to mesh with the embryonic training of nuclear engineers at ORSORT.

The new SRP facility was also untraditionally Du Pont in culture. By the end of the decade, the Savannah River Plant had a technical profile quite unlike other Du Pont operations. While a typical 29% of operations work force consisted of managerial, engineering, scientific, or other professional personnel, their mix and qualifications were unusual. Some 6% of them were biological, medical, or physical scientists, and 12% engineers. The Works Technical Section—combining engineers, chemists, and physicists with responsibility for technical, scientific, and safety issues—employed most of this new breed. Of the engineers and scientists, 8% had doctorates, 83% had Bachelors or Masters degrees. The Atomic Energy Division was thus reliant on an operational team for which labelled engineers made up scarcely 3% of the work force.[157]

These episodes suggest a relatively ponderous adaptation of Du Pont to the new field of nuclear engineering—one which, as Melvin Rothbaum argued, was more readily taken up by General Electric. I would argue that this was not merely a matter of differing engineering cultures, but also of corporate reluctance and inertia. In 1956, the AEC asked Du Pont to apply the information it had gleaned from SRP to the design of heavy-water based power reactors (broadly similar in principle to Canada's evolving CANDU designs). The slow

---

[156] Ibid.

[157] Atomic Energy Commission, 'SRP Fact Book ', 25 November 1960, Hagley 1957 Series V Box 54 folder 15.

progress of the Atomic Energy Division's engineers required the AEC to reduce the pressure, shifting from a goal of an operating reactor by 1962 to open-ended design studies. Eventually, these, too, were abandoned, and with them, Du Pont's potential influence on nuclear power engineering.[158]

Du Pont's Savannah River Plant was emblematic of the shifting post-war attitudes about atomic energy specialists in the United States. The 'atomic scientists' so visibly representing the post-war field belied the reality of atomic energy: American engineers were claiming a position beside—not subordinate to—their scientific colleagues. Expertise was to be distributed. Production would be reliant on specialist workers trained within the AEC network according to the emerging traditions embedded at Argonne, Oak Ridge—and Du Pont. The skill-sets of those workers were still malleable, though, and extended scarcely beyond compartmentalization into 'reactor technologies' and 'separation technologies'. Nevertheless, these nascent American specialisms were not tied strongly to existing professions such as mechanical and chemical engineering—a conceptualization quite different from the British perspective.

By the mid-1950s, then—a decade after the production of the first atomic bomb—nuclear engineering and its specialists were part of a shadowy field. Focused on the rapid exploration and development of the chain reactor, they negotiated their status alongside scientists. The working contexts of Zinn's Argonne, Hinton's Industrial Division, and Du Pont's Savannah River generated distinct occupational opportunities and identities. And, uniquely for an aspiring discipline, they had been defined by their work for a single client: their respective governments. Their developing skills remained insulated by secrecy. Their education took place almost exclusively at, or in collaboration with, government laboratories rather than at universities. This segregation hampered integration and dialogue with other scientific and engineering disciplines, limited their professional aspirations, and constrained careers. Thanks to state sponsorship, nuclear engineers were truly a breed apart.

---

[158] 'History of the Savannah River Laboratory Volume III—Power Reactor and Fuel Technology", June 1984, Hagley 1957 Series V Box 54 folder 12.

PART C

Emergence

'It has been suggested in some quarters that "nuclear engineering" is a basic subject which should take its place alongside civil, mechanical, electrical and chemical engineering. This view, however, is not shared by any engineers of standing who are actively engaged in the development of nuclear power.'
John Menzies Kay, Imperial College, 1957[1]

'A rapid expansion of educational opportunities in nuclear engineering in colleges and universities has taken place in the last decade... Nuclear Engineering is defined as that branch of engineering directly concerned with the release, control and utilization of all types of energy from nuclear sources.'
American Nuclear Society, 1962[2]

---

[1] Kay, John Menzies, Imperial College course proposal, 4 January 1957, NA AB 19/84.
[2] American Nuclear Society, American Society for Engineering Education, 'Report on objective criteria in nuclear engineering education', QU Sargent fonds, Series III Box 4.

6

A STATE-MANAGED PROFESSION

Segregation in national laboratories, like a boarding school for a privileged child, protected and shaped nuclear specialists. Their coming of age, though, was marked inevitably by public events and hinted at the further loosening of control by their parent governments. The adolescence for this emerging profession was signalled by higher education and the beginnings of new relationships.

As Abbott has argued, the establishment of a new profession relies on developing its jurisdiction.[1] This competition for, and expansion into, new intellectual territory involved distinct phases for nuclear specialists, during which they staked and consolidated their domain. As we have seen, secrecy cosseted the new experts, allowing time to foster their special body of expertise within the national laboratories. With a significant decline of secrecy in the mid-1950s, however, came new occupational niches. Nuclear specialists were to design and populate nuclear power plants, moving increasingly from government labs into industry and academe.

6.1 Triggered release: Declassifying nuclear knowledge

The first two discrete phases of nuclear engineering—from 1940 to 1945, and then from 1945 to 1955—vaunted the products of nuclear workers but kept the specialists themselves hidden from public view. Expertise in nuclear engineering circulated only slowly and locally, and under the close direction of governmental organizations. As an editorialist for *Nuclear Power* noted in 1958, 'from the earliest postwar period until only about two years ago, knowledge and experience of atomic energy in all its aspects was largely a Government monopoly'.[2]

As discussed earlier (Section 4.5), the preceding post-war decade had been marked by security fears, legislation, and secretive activities at government-funded sites. But as the editorial hints, the environment changed dramatically during the mid-1950s. The transformation to a more open milieu was equally the consequence of discrete political acts.

The opening act was a December 1953 speech to the United Nations by President Eisenhower. 'Atoms for Peace' was the outcome of his government's efforts to publicize to Americans the 'age of peril' created by the loss of the monopoly on nuclear weapons and the

---

[1] Abbott, Andrew D., *The System of Professions: An Essay on the Division of Expert Labor* (Chicago: University of Chicago Press, 1988), Chapter 1.

[2] 'Editorial: the nuclear industry', *Nuclear Power—The Journal of British Nuclear Engineering* 3 (1958): 150. On the similarly 'unprecedented situation in bringing nuclear reactor technology into the marketplace' in the USA, see Hewlett, Richard G. and Jack M. Holl, *Atoms for Peace and War, 1953–1961* (Berkeley: University of California Press, 1969), Chapter 7.

growing momentum of the Cold War. It also sought to address the increasingly unworkable security regime for nuclear technologies. The initiative had been preceded a few months earlier by the Administration's 'Project Candor', a public relations initiative 'to make clear the dangers that confront us, the power of the enemy, the difficulty of reducing that power, and the probable duration of the conflict', and that would stress 'information on Soviet capabilities, and on the rapid growth of the Soviet economy'.[3] Operation Candor had also coincided with growing concerns about the direction the USSR would take following Stalin's death that March. Eisenhower's speech to the United Nations, reconceived in the interim as a 'revelation of atomic power', consequently had a reassuring undertone stressing the power of American nuclear weapons and the nuclear 'secrets' then held by only a handful of countries.[4]

Despite its title, his speech was vague on positive suggestions. It proposed merely that some fraction of nations' fissionable materials be contributed to an international body overseen by the UN, and that this resource be

> allocated to serve the peaceful pursuits of mankind. Experts would be mobilized to apply atomic energy to the needs of agriculture, medicine and other peaceful activities. A special purpose would be to provide abundant electrical energy in the power-starved areas of the world.[5]

These sentences near the end of the 25-minute speech were meant 'not to wind up on a note of destruction, but of hope'. On the other hand, as a briefing paper for an ABC television interview coached administration representatives, 'what it was NOT' was a 'plan for international control of atomic energy', nor a 'plan for exchange of atomic secrets with Russia or anybody.' As the briefing paper prompted, 'on things that require give and take', the USA 'had been unable to work even with Britain and Canada', but 'cooperation between the Americans and British on the non-military aspects of atomic power has increased tremendously'.[6]

Over the following months, the Eisenhower administration noted that the UN speech had put the Soviet Union on the defensive, and decided to follow up the propaganda with examples of 'the social improvement which can be expected to follow from the peaceful application of nuclear energy'. Such examples were nevertheless indistinct. The primary recommendations of a draft plan were to open up the Oak Ridge Isotopes School (ORINS) to 'instructors and students from all of the free world'; to 'intensify public reporting of unclassified current accomplishments', to stage special events for media coverage', and to develop 'orientation tours' of nuclear facilities, particularly power development and radioisotopes, for 'industrial, labor, religious, women's and general civic leadership'. Lower on the list were more substantive proposals: to promote international radioisotope sales (as Canada was already doing); to stimulate discussions 'in the labor and management field' regarding

---

[3] Eisenhower administration, 'Project "Candor"', 22 July 1953, Eisenhower Presidential Library, http://www.eisenhower.utexas.edu/dl/Atoms_For_Peace/Binder17.pdf.

[4] Whitman, Ann to M. McCrum, letter, 27 January 1956, Eisenhower Presidential Library, http://www.eisenhower.utexas.edu/dl/Atoms_For_Peace/Binder19.pdf.

[5] Eisenhower, Dwight D., 'Atoms for Peace', *United Nations General Assembly*, 470th Plenary Meeting 8 December 1953.

[6] Eisenhower administration, 'General outline for Agronsky program', 16 December 1953, Eisenhower Presidential Library, http://www.eisenhower.utexas.edu/dl/Atoms_For_Peace/Binder14.pdf.

the non-military application of atomic energy; and, to promote international nuclear science, perhaps by financially supporting laboratories overseas, and to promote the need to endow new chairs, revision, modernization and improved distribution of text books and technical magazines..., a review of the policies of American scientific institutions in the above fields...and analysis of a large international program of scientific popularization'.[7] 'Atoms for Peace', then, was first and foremost a media-wise re-education of the American public; secondarily, it opened possibilities for American-led dissemination of not-so-secret atomic secrets. The most optimistic of commentators suggested a democratization of access and benefits. One popular writer said of the subsequent bill to liberalize the Atomic Energy Act, 'when it is enacted the future of atomic power will truly belong to the people'.[8]

Cynical or not, Eisenhower's initiative prompted international responses. The Atomic Energy Commission and American Institute of Chemical Engineers sponsored an International Congress on Nuclear Engineering at Ann Arbor, Michigan, in June 1954. For the occasion, it declassified some 75 engineering subjects, and attracted participants from the pro-American countries of Western Europe, India, Argentina, and Mexico.[9] But the following summer—precisely a decade after the use of atomic bombs had ended the Second World War, and nearly fifteen years after the first studies to create them—nuclear specialists and their knowledge became suddenly more visible. The event that signalled this new flow of information was the international conference on the Peaceful Uses of Atomic Energy held in Geneva under the auspices of the United Nations that summer.[10] The conference echoed the announcement, twenty months earlier, of President Dwight Eisenhower's 'Atoms for Peace' programme, but on behalf of some 70 participating countries. An international voice for dialogue was also promoted by the International Atomic Energy Agency (IAEA), proposed by the United Nations in September 1954 and established in July 1957.

In the USA, Eisenhower's political gesture had been followed by legislation, notably the Atomic Energy Act of 1954, which ended the government monopoly of atomic energy processes, production, and materials. In effect, it opened the possibility of a civilian nuclear industry in the United States. But where 'Atoms for Peace' had emphasized the military stalemate caused by fragile secrecy, the Geneva conference foregrounded scientific cooperation.

Alvin Weinberg recalls that nuclear workers, along with governments and power companies, were swept up in an outpouring of information that underplayed the known problems:

---

[7] Operations Coordinating Board, 'A program to exploit the A-bank proposals in the President's UN speech of December 8, 1953', 4 February 1954, Eisenhower Presidential Library, http://www.eisenhower.utexas.edu/dl/Atoms_For_Peace/Binder11.pdf.

[8] Woodbury, David O., *Atoms for Peace* (New York: Dodd, Mead & Co., 1955), p. 21.

[9] 'Progress in declassification', *Bulletin of the Atomic Scientists* 10 (1954): 143.

[10] Even so, the smoke of classification was slow to dissipate. Only 18 months after the Geneva Conference did a new tripartite declassification guide issued in Washington, London, and Ottawa promise to allow free access to all phases of nuclear power, 'from ore recovery and fabrication of fuel elements to the design and operation of plants for the chemical recycling of spent fuel elements from civilian reactors'. As a result, 'the last remaining vestiges of secrecy at Calder Hall are to go, and information will be freely available on fuel element fabrication and reprocessing.' [Editorial, 'Secrecy wraps lifted', *Nuclear Power—The Journal of British Nuclear Engineering* 2 (1957): 5].

...the conference was presented with a vision of a future energy-hungry world for which nuclear power was a panacea. All the major elements of a nuclear world were discussed: uranium resources, nuclear properties of the fissile isotopes, power reactors, research reactors, waste disposal, chemical reprocessing, metallurgy, applications of isotopes, radiobiology... but as one goes through the 15 volumes, one finds hardly any suspicion that the problems mentioned by Fermi in 1944 – immense radioactivity, and the possibility of the clandestine diversion of plutonium – might prove to be intractable. The tone of Geneva I was euphoria! And every one of us was caught up in this enthusiasm: our supreme technical fix, inexhaustible energy from uranium, would set the world free![11]

The Geneva conference, though, represented more than a commemoration and political act; at the human level, as suggested by Weinberg's comment above, it witnessed a collective release of tension by nuclear specialists and their genuine and uncynical hope for a technocentric future. Soviet public expressions were similar in tone and content, with one translated Russian text including chapters 'Limitless Possibilities', 'The Power of Our Age', and 'Miraculous Progeny'.[12] Encouraging them further was the hope of more direct contacts with their international counterparts. Indeed, historian David Holloway ends his historical account of Soviet nuclear developments with the feting in the west of Igor Vasil'evich Kurchatov (1903–60), the scientific director of the Soviet nuclear project—shaking the hand of Walter Zinn, his American counterpart in reactor technology at Argonne, and 'breaking the spell' of secrecy for John Cockcroft at Harwell—interactions made possible by the new openness.[13]

The conference also marked and promoted the first serious attempts to create a new industry. The following year, the Calder Hall power station, the first significant and widely publicized civilian application of nuclear power, was completed next to the Windscale piles.[14] The secretive activity of bomb production was now twinned with an important civilian application. Nomenclature underwent a transition, too: the wide-ranging *atomic energy* projects were being recast as more focused *nuclear power* projects.

[11] Weinberg, Alvin M., *The First Nuclear Era: The Life and Times of a Technological Fixer* (New York: AIP Press, 1994), p. 170.

[12] Semenovsky, P., *Conquering the Atom: A Story about Atomic Engineering and the Uses of Atomic Energy for Peaceful Purposes* (Moscow: Foreign Languages Publishing House, 1956).

[13] Holloway, David, *Stalin and the Bomb: The Soviet Union and Atomic Energy, 1939–1956* (New Haven, Conn.: Yale University Press, 1994), pp. 361–2.

[14] Kurchatov had led a Soviet team to build the country's first cyclotron in 1939, and in 1943 had founded the first Soviet institute for studying atomic energy, subsequently overseeing development of the nation's first reactor (1946), plutonium bomb (1949), hydrogen bomb (1953), nuclear power reactor (1954), and submarine, icebreaker, and civilian ship power plants (1959). In 1956, forecasting mammoth nuclear power stations and vehicles, his Atomic Energy Institute spawned a new research institute at Obninsk dedicated to reactor applications [Josephson, Paul R., *Red Atom: Russia's Nuclear Power Program from Stalin to Today* (New York: W. H. Freeman, 1999), p. 20]. Thus the British Calder Hall, operational over a year before the first American nuclear power station at Shippingport, was not the first time US nuclear engineers had been surpassed by another country. The Board of Governors at Argonne National Laboratory complained in 1949 that 'the best research pile is in Canada, and the second best one in England' ['Minutes of Board Meeting, ANL', 7 March 1949, UI Box 19]. An Argonne director later reported philosophically that 'The US is obtaining valuable data from the "first generation" of nuclear power reactors, which should ensure leadership into the "second generation" phase. Our primary goal in operating the reactors we now have is information, not kilowatts... The immediate future is not likely to see any large-scale process of reactor construction' [Harrer, Joseph R., 'Briefing on atomic science for Missouri newsmen', press release, Argonne National Laboratory, 14 February 1958, UI Box 10].

As this suggests, post-war enthusiasms for all things atomic were redirected and pared down despite active promotion in the months after Atoms for Peace. Radioisotopes proved important in defining the new field of medical physics[15] and became a small, if profitable, by-product of reactors in several countries (particularly in Canada, where AECL focused its early profit-making aspirations in a Commercial Products Division to market them), and somewhat later in the USA.[16]

Agricultural benefits had also been vaunted as a short-term goal. But the impact of radiation on agriculture was, in fact, nothing new. The biological effects of radiation had been investigated from the turn of the century using the newly available radioactive sources such as polonium and radium. Intentional genetic mutations of crops and insects had first been studied during the 1920s using X-rays. This research, made more relevant by the wartime developments in atomic energy, had gained H. J. Müller (1890–1967) the Nobel Prize in 1946. Subsequent gamma-radiation sources such as cobalt-60, however, promised a routine extension to plant breeding and insect eradication. Like nuclear engineering itself, nuclear agriculture promised to scale up the science of the previous half-century.[17]

Yet such potential benefits were more quietly pursued as popular culture developed new associations. With the government-promoted publicity about new applications came a new public consciousness and fears. Portrayals of species mutated by irradiation thrilled American cinema-goers of the late 1950s.[18] Similarly, in the face of public resistance, the irradiation of foods to extend their shelf-life proved difficult to commercialize.[19] Such concerns were undoubtedly underlined by the publication of biomedical studies citing genetic alterations by radiation, and by contemporary protests of hydrogen bomb tests.[20] And schemes for peaceful civil engineering applications of atomic bombs such as the excavating of hills, harbours, and river courses—the goal of an American initiative, Project PLOWSHARE—were similarly sidelined, in part because of their continuing association with weapons of mass destruction.[21]

---

[15] Kraft, Alison, 'Between medicine and industry: medical physics and the rise of the radioisotope, 1945–1965', *Contemporary British History* 20 (2006): 1–35.

[16] Creager, Angela N. H., 'The industrialisation of radioisotopes by the US Atomic Energy Commission', in: K. Grandin and N. Wormbs (eds.), *The Science-Industry Nexus: History, Policy, Implications* (New York: Watson Publishing, 2004), pp. 141–67.

[17] Wilson, A. S., 'Agriculture in an atomic age', report, 8 January 1960, DOE DDRS DA03300049.

[18] Typical titles include *Them!* (1954, arguably the paradigm of the genre), *The Day the World Ended* (1955), *Creature with the Atom Brain* (1955), *The Amazing Colossal Man* (1957), *Beginning of the End* (1957), *The Incredible Shrinking Man* (1957), *Attack of the 50 Foot Woman* (1958), *The H-Man* (1958), *Attack of the Alligator People* (1959), *Attack of the Giant Leeches* (1959), *The Monster That Challenged the World* (1959) and *The Beast of Yucca Flats* (1961).

[19] Buchanan, Nicholas, 'The atomic meal: the cold war and irradiated foods, 1945–1963', *History and Technology* 21 (2005): 221–49; Spiller, James, 'Radiant cuisine: the commercial fate of food irradiation in the United States', *Technology and Culture* 45 (2004): 740–63.

[20] National Academy of Sciences, 'The Biological Effects of Atomic Radiations', National Research Council, 1956; British Medical Council, 'The Hazards to Men of Nuclear and Allied Radiations', HMSO, 1956. On H-bomb protest, see Section 7.1.2.

[21] See, for example, Cosgrove, Denis, 'Introduction: Project Plowshare', *Cultural Geographies* 5 (1998): 263–6 and Frenkel, Stephen, 'A hot idea? Planning a nuclear canal in Panama', *Cultural Geographies* 5 (1998): 303–9.

In their stead, the future of electrical power production by nuclear technology assumed prominence in the new glare of publicity. In the USA, the fostering of a civilian power industry raised problems for the government. The requirements of national defence still jostled with peacetime economics. Power companies were uneasy about government involvement in supplying uranium, inspecting plants, and taking possession of expended fuel rods for plutonium recovery.

The first British power reactors were more clearly government products. Windscale and Calder Hall, built side by side but five years apart in Cumbria in the English north-west, represented opposite faces of secrecy. The purpose of Windscale had been vaunted publicly as plutonium production for nuclear weapons, but Calder Hall, still optimized for plutonium production, was publicized as a power-generating station. The close weaving of military and civil requirements did not begin to unravel until a nuclear industry took shape. A British nuclear industry, in turn, could not emerge until the parallel lines of plutonium and power production were forced to diverge by changes in government policy. Without explicit political acts, nuclear engineering as an art remained hidden and constrained.

6.2 Freedom to publish

Yet some information did begin to flow more freely. The thawing of security concerns and public reorientation towards peaceful applications accelerated the emergence of a handful of journals dedicated to nuclear engineering and its specialists. In the USA, the technical magazines *Atomic Power* and *Atomic Engineering* had appeared immediately after the war, but consolidated quickly into *Nucleonics* magazine, subtitled 'Techniques and Applications of Nuclear Science and Engineering', published by McGraw-Hill from 1946. By contrast, the technological aspects of the field had been heavily censored: the *Journal of Reactor Science and Technology*, for example, was distributed quarterly by Oak Ridge National Laboratory as a classified publication from the early 1950s. A more overt expression of the social conscience of nuclear workers was the *Atomic Engineer and Scientist*, published by the Oak Ridge Engineers and Scientists. This was superseded by the *Bulletin of the Atomic Scientists* from December 1945 and the formation of a parent organization, the Educational Foundation for Nuclear Science. In the UK, sharing this new mixture of technical, social, and policy concerns, the Atomic Scientists Association launched *Atomic Scientists' News* in 1947, which was transformed into *Atomic Scientists' Journal* in 1953, and in turn subsumed by *New Scientist* in 1956.[22]

Not surprisingly, then, a direct response to the increasingly declassified environment initiated by President Eisenhower and Geneva was the appearance of new technical journals. This dissemination of knowledge after a decade of secrecy, argued John Cockcroft, was crucial to the advancement of the subject and to its international expansion. Introducing the first issue of *The Journal of Nuclear Energy* in mid-1954, he complained of the backlog of specialized knowledge that required dissemination:

> An embarrassingly large number of possible nuclear power plants is calling out for investigation, all of them enticing projects which require much development effort and time before their economic prospects can be assessed. It is hoped that the available effort of the world will be dispersed amongst the more promising of these projects and not used to duplicate effort unduly until clearly economic types of reactors

---

[22] See Appendix II.

have emerged…We shall thereby promote that interchange of ideas which is the basis of scientific and technological development in peace.²³

American sources emphasized a similar point. Seeking to develop new markets and readership, an editorial in *Nucleonics* magazine even offered to expedite the supply of available information:

> Although unclassified information is widely available, it can frequently be difficult to locate, particularly for the uninitiated. Because of this, the editors of *Nucleonics* repeat an offer that they made in 1951 – to assist any reader, especially those in foreign countries, in getting non-secret information in the atomic energy field. We shall be particularly happy to help those just starting to work in atomic energy by advising on availability and sources of information.²⁴

In early 1956 the first issue of *Nuclear Engineering* appeared.²⁵ Like *The Journal of Nuclear Energy*, it was a product of Pergamon Press, headed by the British entrepreneur Robert Maxwell. As the first journal of its kind, it devoted editorial space in its first issues to explaining its timing, purpose, and intended readership. The editors voiced attitudes based on over a decade of British nuclear experience, its industrial dimensions and, most significantly, its specialists. They identified both progress and restraint:

> [A]t Geneva last August […] it became obvious that in spite of intense security restrictions developments in the three main countries had moved along similar lines. The divergences that had occurred were forced by local conditions rather than by fundamental conceptions. Similar conditions will prevail in a single country unless an adequate exchange of views and information is encouraged.²⁶

Continued segregation would inhibit, not protect, nuclear expertise. John Cockcroft noted more pointedly the consequences for the teaching provided at Harwell from 1954:

> At that time the courses held at the school were secret and only British subjects could attend them. However, in August 1955, an international conference on the peaceful uses of atomic energy was held in Geneva under the auspices of the United Nations. At this Conference, a very large amount of information on reactor technology was publicly released for the first time; indeed, the amount of 'declassification' which took place at Geneva was so great that it became possible to hold completely unclassified courses at the Harwell school, and meant, of course, that foreign students could attend. Accordingly, the first international course started in September 1955, when roughly one half of all the students attending were from overseas.²⁷

---

²³ Cockcroft, John D., 'Foreword—The Journal of Nuclear Energy', *The Journal of Nuclear Energy* 1 (1954): 1.

²⁴ 'Editorial', *Nucleonics* 9 (1954): 30.

²⁵ As noted above, the first comparable American journal was *Nuclear Science and Engineering*, published by the American Nuclear Society from November 1956.

²⁶ Editorial, 'Another industrial revolution?' *Nuclear Engineering* 1 (1956): 1.

²⁷ Cockcroft, John, 'The Harwell Reactor School', *Nuclear Engineering* 1 (1956): 10–11; 'Britain's premier nuclear training centre: Harwell Reactor School', *Nuclear Energy Engineer* 13 (1959): 246–8. By comparison, that year the American AEC sponsored some 900 students via Oak Ridge School of Reactor Technology (c150); the Argonne National Lab Summer Institute (60); the Radiation Laboratory of the University of Southern California (c225); Brookhaven National Lab Summer Courses (c200); 'on-the-job' courses for engineers in university and industry (c200); and, fellowships for university study of radiological physics, industrial hygiene, and medicine (55) [Perrot, Donald, 'Problems of the supply of scientists and engineers', UKAEA memorandum, 16 February 1956, NA AB 16/1971].

Cockcroft's industrial counterpart, Sir Christopher Hinton, made the same point in the inaugural issue of the *Journal of the British Nuclear Energy Conference*, noting that while the British industry 'had worked behind a curtain of secrecy' and 'had spent its first eight years almost entirely in meeting military commands', 'nuclear engineering had now emerged from that stage…as an important industry in its own right'. He argued that civilian nuclear power would give more freedom and opportunities for employment to its workers.[28] Statistics seemed to confirm the rise: a government report that year found that the 103 chemical engineers at the UKAEA and other organizations in the British nuclear industry filled a variety of new roles, suggesting the emergence of a new variant of the discipline.[29] And appropriately for the country that seemed to be leading in commercial nuclear power production, the government-fostered nuclear energy industry was beginning to outstrip the plastics, dyestuffs, and food industries as occupations for such engineers.[30]

By its second year of publication, *Nuclear Engineering* could publish a three-page list of 'Unclassified Documents' available from the UKAEA.[31] The *Journal of Nuclear Energy* went one step further, and negotiated (via Maxwell's Russian connections) to publish the translated papers of *Atomnaya Energiya* from 1956 as a section in every issue entitled *The Soviet Journal of Atomic Energy*; by 1957 the Soviet papers made up two-thirds of the *Journal*.

The new journals were cautious at first in their articles and editorializing, but became gradually more exuberant. *Nuclear Engineering* and *Nuclear Power—the Journal of British Nuclear Engineering* (both launched in May 1956) even published centre-fold diagrams—later examples in colour and with acetate overlays—of reactor designs, rather like British boys' periodicals of the era (Figure 6.1). Not *Playboy*, to be sure, but equally liberated in its revelation after the years of secrecy surrounding their respective subjects. Nuclear engineering had been more covert than illicit sex, and revelations were more punishable by severe legal strictures. *Nuclear Power*, in particular, sought to bring the hidden field into the engineering fold. It included introductory series of articles on reactor science and health physics, published advertisements for jobs and vacancies, and began running an anonymous commentary column.

This new openness was remarkably free from boosterism of the infant industry and its new professionals, but still tinged with international politics. For their part, the editors of *Atomnaya Energiya* observed that the 'incubation period' was over, and that 'as a result of the considerable experience available in the Soviet Union on the design of nuclear power plant, it has been possible in the sixth five-year plan to increase substantially the effort

---

[28] Hinton, Christopher, 'Inaugural address', *Journal of the British Nuclear Energy Society* 1 (1955): 1–2. On the linking of military agendas to institutional aims, see Forman, Paul and José M. Sánchez Ron, *National Military Establishments and the Advancement of Science and Technology: Studies in 20th Century History* (Dordrecht: Kluwer Academic, 1996).

[29] HMSO, *Scientific and Engineering Manpower in Great Britain* (London: HMSO, 1956); Ross, K. B., 'The chemical engineer in industry', *The Chemical Engineer* (October 1960): A38-A9.

[30] 'The supply and distribution of chemical engineers in Great Britain', *The Chemical Engineer* (October 1958): A41-A5, and Brennan, J. B., 'Chemical engineering manpower in the chemical industry', *The Chemical Engineer*, A44-A45, April 1961.

[31] News, 'Unclassified UKAEA documents', *Nuclear Engineering* 2 (1957): 30–2.

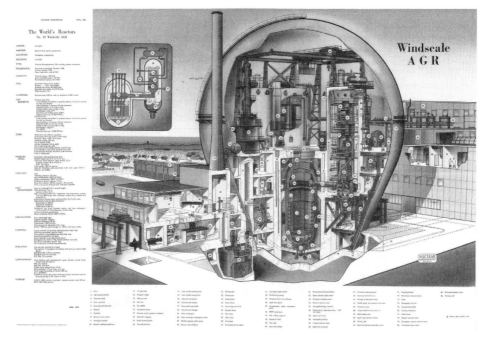

Fig. 6.1 Reactor technology as centrefold, 1961.[32] (Reprinted with permission © Nuclear Engineering International 2011. Other reactor diagrams available on www.neimagazine.com/wallcharts.)

directed into peaceful channels' and called upon 'all scientists and engineers to press forward with the peaceful development of atomic energy for the prosperity and happiness of mankind, and to contend against those military applications of it which hold out such a terrible threat to the human race'.[33] In a more pragmatic vein, *Nuclear Engineering*'s opening editorial stated:

> At no time in the technical development of man has he been so determined to herald the birth of a revolution as today. The discovery of nuclear fission has been described in many extravagant terms – as the dawn of a new era, the beginning of the millennium and so on – but we must leave history to ascribe these high-flown phrases. In the meantime it is better to examine the significance of the 'Atomic Age' in its relation to its immediate industrial and sociological problems.
>
> Its chief impact is in the prospect of a source of illimitable power or, more accurately, an illimitable source of heat – at a cost. Too often vague prophetic allusions have been made to the physical form that atomic energy may take, but fundamentally this energy is available as heat – and heat only. Uranium can, therefore, be described as a fuel in complete accordance with established terminology.[34]

---

[32] 'The world's reactors No. 32: Windscale AGR', *Nuclear Engineering* 6 (1961): 62–3.
[33] Foreword, 'Soviet Journal of Atomic Energy', *Soviet Journal of Atomic Energy* 1 (1956): 75.
[34] Editorial, 'Another industrial revolution?' *Nuclear Engineering* 1 (1956): 1.

6.3 Professional fallout[35]

The decline of secrecy lowered the boundaries for technical identity. As suggested by early journal editorials, Eisenhower's 'Atoms for Peace' initiative opened new ground for its specialists—both in public awareness and in self-recognition. In 1954 the largest American cluster of nuclear workers, including members of the wartime groups, organized in space provided at the Oak Ridge Institute of Nuclear Studies.[36] Its orientation and goals were contested over several months: should it be an Institute, Association, or Society, focusing on nuclear technology, engineering, or science? In the spirit of Atoms for Peace, however, participants agreed that it would focus on peaceful applications of nuclear knowledge.

That October, after rejecting names such as the 'Association of Nuclear Engineers' and 'Society of Nuclear Scientists and Engineers' that vaunted the specialists themselves, the organization adopted the more inclusive 'American Nuclear Society' (ANS). Eight months later a general meeting elected its officers and by late 1956 the ANS launched its first journal, *Nuclear Science and Engineering*, to rival the British publications.

The melding of professional representation, government funding, and the interests of national laboratories is apparent in the choice of Walter Zinn, Director of Argonne National Laboratory and responsible for most reactor research in the USA, as the first President of ANS.[37] In the same vein, ANS involvement in defining university curricula in 1960 (Section 6.4.4, below) and in collaborating with the AEC in 1968 in publishing a monograph series illustrates the concentration of authority and its origins in the wartime Manhattan Project.

In Britain, the emerging profession proved to be equally influenced by pre-existing organizations. The new openness from the mid-1950s created new aspirations for nuclear specialists but also new conflicts. As Abbott argues, groups of new specialists compete with existing professions: 'tasks are created, abolished, or reshaped by external forces, with consequent jostling and readjustment within the system of professions'.[38] The new jurisdictions of university departments of nuclear engineering and nuclear power stations could be filled with labelled nuclear experts bearing new disciplinary labels certified by new professions; alternatively, they could be inhabited by members of existing professions that absorbed that proficiency and made it their own.

---

[35] Aspects of this section are discussed more extensively in the account of the expansion of chemical engineering given in Divall, Colin and Sean F. Johnston, *Scaling Up: The Institution of Chemical Engineers and the Rise of a New Profession* (Dordrecht: Kluwer Academic, 2000), pp. 163–74.

[36] The American Nuclear Society moved to offices in downtown Chicago in 1958 and subsequently remained in Illinois, close to the centre of research and development activities that had begun at the Met Lab during the war.

[37] Subsequent early presidents consolidated these entwined relationships between government, national laboratories, and the profession, and included Leland Haworth (1904–79, third ANS President 1957–8) and Director of Brookhaven National Lab; Alvin Weinberg (1915–2006, fifth ANS President 1959–60) and Director of ORNL; W. B. Lewis (1908–1987, seventh ANS President 1961–2) and Vice President of AECL; and Norman Hilberry (1899–1966, eleventh ANS President 1965–6) and former Director of ANL.

[38] Abbott, Andrew D., *The System of Professions: An Essay on the Division of Expert Labor* (Chicago: University of Chicago Press, 1988), p. 33.

The assessment of the editor of *Nuclear Engineering*, above, can be interpreted in these terms, and hinted at consequences for a profession of nuclear workers, at least in Britain. His view that uranium was merely 'a fuel in complete accordance with established terminology'[39] was mirrored by that of Christopher Hinton, who argued that the 'correct perspective' was that nuclear power 'should not be regarded as a completely new technology, but rather as an extension and development of existing technologies'.[40] Similarly, industrial physicist Sir Claude Gibb cast the new field as a peacetime project requiring a continuation of wartime cooperation and zeal:

> The mechanical problems associated with pile design are for the engineers to solve. They will need the help from the metallurgists in their interpretation of the work and thoughts of the physicists and chemists. It calls for a combined operation – almost an 'Overlord'. We did not allow the possibilities of failure on "D" day to prevent the attempt – the stakes were too high. Careful planning, experiments, a general background of experience in some only of the problems, and then great courage. So it must be with atomic power generation. The stakes are high.[41]

The Director of the AEI Research Laboratory at Aldermaston echoed these views, arguing that 'The nuclear power industry is, in the main, an amalgam of existing industries'.[42] Nor was this merely the industrialists' perspective. At 'the threshold of the Atomic Age', asked the Association of Scientific Workers, where were 'the additional engineers, chemists, metallurgists, etc needed to establish the new nuclear reactor industry' to come from? The solution, they argued, lay in redeployment of technical manpower.[43]

Nevertheless, some argued for the need to shift technical identities, too. Nuclear energy demanded that engineers working in conventional plant adapt to 'entirely new techniques which will require almost complete original thinking'.[44] Given the shortage of adequately trained workers and sparse knowledge and experience outside the Atomic Energy Authority, *Nuclear Engineering* suggested using 'the accumulated knowledge and experience of conventional branches of engineering, adapted in varying degree to meet a new application'. It recommended that on-the-job training programmes be instituted that would add these new specialized skills:

> there is a vital need for a body of engineers in industry who, whilst having a sound knowledge of normal engineering fundamentals and experience, have in addition sufficient understanding of the processes of nuclear energy to equip them to lead design and development work in the field, both for whole power stations and for parts of a nuclear reactor.... Doubtless, within a short time enough experience will have

---

[39] Editorial, 'Another industrial revolution?' *Nuclear Engineering* 1 (1956): 1.

[40] Hinton, Christopher, 'Inaugural address', *Journal of the British Nuclear Energy Conference* 1 (1956): 1–2, quotation p.1. This fit with UKAEA recognition of existing labour unions described in Section 7.1.2.

[41] Gibb, Claude, 'Industry and atomic power', *Atomic Scientists' News* 2 (1952): 98–103, quotation p. 101.

[42] Salmon, A., 'Technologists for the nuclear power industry', *Atomic Scientists' Journal* 5 (1956): 280–5; quotation p. 280.

[43] Association of Scientific Workers, *Peaceful Uses of Atomic Energy* (London: Labour Research Department, 1955), p.15. The problems of building and maintaining large engineering teams was also noted by the Managing Director of English Electric, one of the contractors for the new industry, in Nelson, George, 'British nuclear engineering: the next ten years', *New Scientist*, 27 October 1960: 1115–17.

[44] Editorial, 'Another industrial revolution?' *Nuclear Engineering* 1 (1956): 1.

accumulated within each organization to permit most of the staff to acquire some of the knowledge necessary for their work in the time-honoured way, by learning while working, but the initial problem remains – the dissemination of knowledge to start this process.[45]

But what expertise was required for this specialized breed of worker? An editorialist in the journal *Engineering* saw the need for fluid categories:

> The design of nuclear power plant falls into none of the neat compartments of engineering that we have come to accept. It is not covered wholly by civil, mechanical, electrical or even chemical engineering. It is, in fact, spread across all of them.[46]

Three years later, Sir Leonard Owen, Managing Director of the UKAEA Industrial Group, was willing to consider that 'the basic approach necessitated by the special requirements of the nuclear branch of engineering, demanding rethinking from first principles, may well fertilize the older branches of the profession'.[47]

As in the USA, several engineering professions were interested in protecting or extending their special competences. Sir John Cockcroft, for example, suggested when addressing the Institution of Locomotive Engineers that train-borne nuclear reactors were unlikely, but that a nuclear power station could supply a fully electrified rail network.[48]

Cockcroft himself promoted a division of labour via a science-led, technician-managed profession. In an article on Harwell's work, he described how 'physicists, chemists and metallurgists are carrying out a vast research programme which will enable the engineer to build better power reactors'.[49] Others in the emerging industry allocated praise more equitably: a *Nuclear Power* editorial marked the opening of Calder Hall as 'the triumphant answer of a dedicated team of scientists, engineers and craftsmen'.[50] And the Head of the UKAEA Reactor Division, J. V. Dunworth, arguing that reactors were orders of magnitude more complex than conventional power facilities, rejected 'the frequently expressed view that nuclear engineering is just another branch of ordinary engineering'.[51]

John Cockcroft was concerned that the formation of the American Nuclear Society in 1954, coupled with UN proposals for an international atomic energy agency, meant that

---

[45] Editorial, 'Training the nuclear engineer', *Nuclear Engineering* 1 (1956): 5.

[46] Editorial, *The Engineer* 200 (1955): 569.

[47] 'A message from Sir Leonard Owen, CBE', *Nuclear Energy Engineer* 12 (1958): 33.

[48] News, 'Locomotion', *Nuclear Engineering* 1 (1956): 3. The feasibility of a nuclear-powered train had, in fact, been investigated in a graduate course in Nuclear Technology in the Physics Department of the University of Utah in 1953. The students estimated the construction costs as double those of a diesel locomotive, and considered, but ignored, development costs, maintenance, and the 'elaborate central facility' required for refuelling [Woodbury, David O., *Atoms for Peace* (New York: Dodd, Mead & Co., 1955), pp. 122–4].

[49] Cockcroft, John D., 'Scientific problems in the development of nuclear power', *Nuclear Power—The Journal of British Nuclear Engineering* 1 (1956): 200.

[50] A pay raise for manual workers negotiated between the UKAEA and ten unions in 1956 granted about £10 to craftsmen and £8 to labourers per week [Dunworth, John V., 'Pay rise for 15,000', *Nuclear Power—The Journal of British Nuclear Engineering* 1 (1956): 48].

[51] Dunworth, J. V., 'It's not just ordinary engineering', *Nuclear Power—The Journal of British Nuclear Engineering* 3 (1958): 36.

'the UK would be placed in an unfavourable position unless it had some corresponding body not under the direct control of our own Atomic Energy Authority'. He argued that existing professional institutions were too fragmented to include 'any appreciable fraction of our scientists' who, in fact, 'would not be eligible'. Instead, he proposed forming a new society.[52]

But despite Dunworth's and Cockcroft's perspectives, the centre-of-mass in Britain was nearer Hinton's position: nuclear engineering could be accommodated within existing skills and without a new label. Cockcroft's suggestion of a new society was unwelcome; most of his contemporaries favoured a joint effort by three or more existing groups. Individual professional societies had been scouting territory, too. The Institution of Chemical Engineers, which had long struggled for status in the workplace and among the technical professions, attempted early on to extend its jurisdiction to the new domain of nuclear engineering, with a council member suggesting 'establishing a representative committee on development of atomic energy' as early as November 1945.[53] The prominent chemist Sir Harold Hartley pressed further that this represented new opportunities for existing engineers: 'with the lure of nuclear physics and electronics, classical physicists were a dying race', he claimed, and 'chemical engineers have to do the research'.[54]

John Cockcroft's industrial equivalent, Christopher Hinton, found himself more directly embroiled with professional aspirations. Like Hartley, he saw nuclear engineering, fortified with already-available chemical engineers, as key to the post-war British economy. Hartley had argued for strong parallels between nuclear and chemical engineering, and urged Hinton to make the most of this pool of expertise. On a political level, Hartley interpreted both the nuclear and chemical industries as a new opportunity to assert national prestige against foreign competition. Post-war Europe had been swamped by a wave of imported American technology for the petrochemical industry, and Hartley wrote to one manufacturer:

> What worries me is the cool assumption that the USA is going to export packaged power reactors about the world just as they have done in the case of oil refineries. Chemical engineering at long last is going ahead in Britain and I hope we shall soon be in a position to break the American monopoly in the latter case and that we shan't allow them to get a monopoly so far as power reactors are concerned.[55]

By the mid-1950s, the British Chemical Plant Manufacturers' Association (BCPMA) had recognized the commercial potential and was eager to embrace the new technology, becoming a founder member of the Nuclear Energy Trade Associations Conference. And institutions, too, paid heed to the new potential for employment. In 1955, four engineering

---

[52] Cockcroft, John to unlisted recipients, memo, 22 November 1954, NA AB 19/84. Significantly, the first president of the American Nuclear Society was Walter Zinn, Director of Argonne National Laboratory—Cockcroft's closest equivalent.

[53] Letter, Alliott to Council, 6 November 1945, ICE Gayfere Box III/5; CM 21 November 1945. A Nuclear Processing Group was founded only in 1978; the American Institute of Chemical Engineers, by contrast, formed a Nuclear Engineering Group as early as 1954.

[54] Hartley, Harold, *Chemistry and Industry* (25 April 1953): 404–5; Hartley, Harold, 'The place of chemical engineering in modern industry', *School Science Review* (March 1954): 199–202.

[55] Letter from Hartley to S. R. Armsdon of the British firm CJB, 20 January 1955, CC Hartley papers Box 275.

institutions—Civils, Mechanicals, Electricals, Chemicals—and the Institute of Physics (which then represented primarily British physicists in industry rather than the academically based physicists catered for by the Physical Society) collaborated to form a joint body for the advancement of nuclear technology: the British Nuclear Energy Conference (BNEC). In accord with prevailing British opinion, they agreed that the new specialism was 'not to be regarded as a completely new technology, but rather as an extension and development of existing technologies'.[56] This seemingly stable collaboration was nevertheless short-lived; by 1961 the organization's representation was broadened by the addition of the Physical Society, Iron and Steel Institute, Institute of Metals, and Institute of Fuel. In effect, the organization promoted a pan-engineering grouping that weakened professional aspirations but worked more overtly in the mutual interests of industry, government, and established technical workers. Although no such analogy was drawn at the time, the move rehearsed the flirtation with corporatism that had enthused British organizations immediately after the First World War.

Unsurprisingly, Christopher Hinton became a key player in the BNEC, serving as its first Chairman.[57] He personified this union of disciplines, being a Fellow of the Royal Society and member of the Institutions of Civil, Mechanical, and Chemical Engineers, although he joined the Chemicals only in 1954, probably motivated by his chronic problem of manning the overstretched UKAEA Industrial Division and by Harold Hartley's mentoring.[58] Nuclear workers in the British nuclear industry were thus seeded by the pre-existing occupations and working practices of chemical and mechanical engineers.

Against this institutional and professional momentum, though, others sought to occupy seemingly vacant professional territory. David Rowse (1925–), an architect with no background in the subject, founded *Nuclear Power* after reading about President Eisenhower's Atoms for Peace initiative in the *Financial Times* and attending the Geneva Conference, hiring physics graduate William Davidson as editor.[59] Following the opening of Calder Hall, another trade publisher, John B. Pinkerton (1903–67), adapted his periodical titles to what he perceived as the growing market for a new variety of engineering. A former managing director of a firm of consulting mechanical engineers and a technical journalist, his *Combustion and Boilerhouse Engineering* (1953–6) became the unlikely *Combustion, Boiler House and Nuclear Review* and then *Nuclear Energy Engineer* in 1958. The latter publication set out to survey the teaching and research work in the field, and Pinkerton soon used the journal to float the idea of an 'Institution of Nuclear Energy Engineers'.[60] Its readership consisted

---

[56] Hinton, Christopher, 'Inaugural address', *Journal of the British Nuclear Energy Conference* 1 (1956): 1–2.

[57] Initially, Cockcroft, Ross, and White of Harwell were also members of the governing board [NA AB 19/83]. Hinton had earlier been the figurehead president of the Nuclear Engineering Society, Windscale, the first nuclear workers' group to be sanctioned by the Division of Atomic Energy of the Ministry of Supply.

[58] NA 19/40 Letter Harley to Hinton, 8 December 1954; IME Hinton papers A.170, J.24; CC Hartley Box 144.

[59] Rowse, David to S. F. Johnston, interview, 5 July 2011, telephone, SFJ collection. *Nuclear Power* was later sold to *Nuclear Engineering*.

[60] 'Nuclear engineers in the making', *Nuclear Energy Engineer* 12 (1958): 334; Pinkerton, J. B., 'The Institution of Nuclear Energy Engineers', *Nuclear Energy Engineer* 12 (1958): 49; Pinkerton, J. B., 'The Institution of Nuclear Engineers', *Nuclear Energy Engineer* 12 (1958): 197; 'The journal that spawned an institution', *Journal of the Institution of Nuclear Engineers* 20 (1979): 140–2.

primarily of industrial workers, not academically trained engineers. Nevertheless, Pinkerton quickly gained an expression of interest from over 500 engineers and scientists in fourteen countries, although he was initially undecided about offering examinations to confer professional qualifications, or how it would interact with other bodies.[61] Adopting the role of Honorary Secretary, he selected engineers from a variety of fields working on the industrial application of nuclear energy, inviting them to form a Council. With activities supported largely from his own pocket, Pinkerton gradually passed control of the Institution's affairs to its Council, and was elected its President in 1967, shortly before his untimely death. From these unusual beginnings, the Institution of Nuclear Engineers (INucE) was founded.[62]

Even so, the gradual acceptance of a new engineering field did not extend to recognized specialists. After a mere two issues, *Nuclear Energy Engineer* had mutated into *Nuclear Energy*, again submerging the workers behind the cloak of their specialism. These periodicals and their successors devoted a few pages per issue to institutional issues, but seldom any space to more fundamental professional aspects.[63]

The formation of the INucE—seeking to form a learned society and professional body along the lines of existing British engineering institutions, but unaffiliated with them—provoked energetic responses to quell the threatened invasion. A letter from the Institution of Civil Engineers noted their disquiet, and suggested that each of the members of the BNEC might publish a statement in their journals to discourage members from participating in the affairs of the INucE. After checking with its solicitors, the IChemE echoed this boundary-setting. 'The council sees no reason for the formation of new societies in nuclear energy', ran the statement, because 'the new industry did not bring into being a new kind of engineer, but was a new challenge in the application of existing branches of engineering'.[64] But this, of course, was the very essence of contention in all countries pursuing atomic energy. The Secretary of the INucE countered that

---

[61] During the early years of the Institution the requirements for corporate membership were, in common with earlier engineering institutions, set at the level of Higher National Certificate (HNC) or Higher National Diploma (HND). As more engineers began to receive education with a nuclear orientation and at a degree level, and more opportunities for training in the nuclear field were provided, the Institutional requirements rose. This was particularly so after 1973 when the Institution, like other British professional engineering bodies, raised its entrance standards ['Institution Requirements for Membership', *Journal of the Institution of Nuclear Engineers* 19 (1978): 46–7]. The early international interest (with nearly equal numbers of applicants from the UK and USA) was notable, although not sustained. The (unfulfilled) hopes for international branches of the INucE contrasted with the unmentioned American Nuclear Society founded five years earlier, which seemingly had little interest in nuclear specialists beyond its borders. By 1961 the INucE boasted 1100 members, but the early international links had evaporated.

[62] 'Keeping up with the times [editorial]', *Combustion, Boiler House and Nuclear Review* 12 (1958): 61; 'The birth of the Institution', *Nuclear Energy Engineer* 13 (1959): 579–802; Pinkerton, J. B., 'The Institution of Nuclear Engineers: Officers and Council', *Nuclear Energy Engineer* 13 (1959): 576–7; Callow, J. H., 'A personal tribute to the memory of J. B. Pinkerton', *Journal of the Institution of Nuclear Engineers* 19 (1978): 48.

[63] A British journalistic voice for nuclear specialists proved elusive. After beginning their professional existence with the journal *Nuclear Energy*, The Institution of Nuclear Engineers eventually published the *Journal of the Institution of Nuclear Engineers* from 1971, followed by *Nuclear Engineer* from 1980.

[64] ICE Council Minutes 16 July 1958; 'British Nuclear Energy Conference', *The Chemical Engineer* (October 1958): A47. The same point was made by G. R. Hall, *op. cit.*

we are a professional body, and our work is confined to the spread of knowledge in the field in which we serve no matter how diversified the field becomes; we are in no sense a splinter organization. We are a new creation in a new and expanding field and we are tied to none.[65]

It is ironic the IChemE had voiced a similar argument to support its own formation some four decades earlier. Immediately before the INucE challenge, the young institution had joined the ranks of the 'big three' (Civils, Mechanicals, and Electricals) in receiving a Royal Charter—a status never attained by the INucE.[66]

A year after the formation of the INucE in 1959, the older industries consolidated their opposition by founding the British Nuclear Energy Society (BNES) to replace the British Nuclear Energy Conference. The new organization de-emphasized the disciplinary slant but underscored job competence. The BNES would be open to *qualified* engineers from the constituent societies and to others 'actively engaged' in the technical aspects of nuclear energy and ancillary subjects. And, stealing the thunder of the broadly conceived INucE, it vaunted the virtues of interdisciplinarity:

> One of the most important roles of the society is to provide a channel for the exchange of experience and the cross fertilization of ideas between all engaged in the field of nuclear energy... our members should include not only physicists and engineers, but biologists, medical men, agriculturalists, lawyers and accountants.[67]

The new BNES found a home at the headquarters of the Institution of Civil Engineers but failed to limit the influence of the INucE. The breadth of interests and initially lax standards of entry promoted the expansion of the INucE through the 1960s, attracting a disparate collection of members initially of lower formal attainment than their counterparts in other institutions.[68] The two organizations competed over the decades that followed, obscuring the identity of British nuclear specialists.[69]

6.4 Altering cross-sections: The first university-educated generation

As the mixed fortunes of the early ANS and INucE suggest, staking professional identity was less urgent than establishing a disciplinary presence: nuclear engineers needed to colonize

---

[65] 'From the Secretary', *Nuclear Energy* (February 1966): 31.

[66] In 1977, the INucE was accepted as an Affiliate Member of the Council of Engineering Institutions, a status secondary to the established Engineering professions. Nevertheless, this granted its suitably qualified members to be recognized as Chartered Engineers, an important distinction for British practitioners. Membership classes subsequently included Fellows, Members, Associate Members, and Engineering Technicians, with the first two grades having chartered status.

[67] Owen, Leonard, *Journal of the British Nuclear Energy Society* 1 (1962): 7. The new Society had received some 1700 applications by late 1961.

[68] By 1968, 59% of the corporate (Associate and Full) Members of the INucE held bachelors' degrees or above, and 61% were corporate members of other engineering institutions, but its council, observing that 'in the early days of any foundation it must assemble its main roots from the best talent available at the time', decided that 'future applicants must expect to be set a stiffer test of their experience and knowledge. This is scientific progress.' ['From the Secretary', *Nuclear Energy* (November/December 1967): 153 and May/June 1968: 85].

[69] See Section 8.2.3.

universities to train the next generation and to secure jobs in the new industry. Stephen Turner has argued that the formation of a discipline is essentially the starting of a cartel. It relies on a market for the students who are produced, and disciplinary boundaries that provide privileged access to that market: 'as long as there are benefits for cartelization, there will be disciplines'.[70] That market, for the UK, USA, and Canada, was the nascent nuclear power industry. The corresponding cartel comprised the educational institutions setting out to launch specialized courses. But the benefits for universities, employers, and the trained experts themselves were differently constructed in each country.

6.4.1 Controlling colleges in the UK

The seemingly low entry standards for the INucE were the consequence of a general lack of formal accreditation for nuclear specialists who had, after all, only patchy training available of highly variable depth. The British nuclear programme was constantly in need of professional personnel and, despite proposals to raise salaries and transfer workers from military projects, pressure was growing both internally and externally for training programmes.[71]

In 1954, months after the formation of the UK Atomic Energy Authority, the new organization was approached by the Association of Consulting Engineers and the British Electrical Authority to fulfil its remit to 'distribute information relating to and educate and train persons in matters connected with atomic energy or radioactive substances'.[72] Internal reports further highlighted the gulf between the demand—for design engineers, metallurgists, mechanical and chemical engineers, physicists, mathematicians, and experimental officers—and the 'national shortage, apparent since the end of the war'.[73] The new UKAEA continued the training arrangements it had inherited from the Ministry of Supply, covering Craft, Student, and Graduate apprentices, pre-university trainees with degrees or diplomas in engineering, and Internal and External Vocational Training. In addition, Harwell mounted the Reactor School for more advanced training.

The curriculum of the three-month, 150-hour Harwell Reactor School included background in conventional disciplines and more: reactor theory (nuclear physics and chemistry including critical size, poisoning by fission products, control rod calculations and variation in reactivity with time, various ways of breeding fissile materials); engineering (heat transfer, temperature distributions in reactors, thermal stresses, pumping power, and kinetic behaviour of reactors); chemical engineering and metallurgy (manufacturing moderators

---

[70] Turner, Stephen, 'What are disciplines? And how is interdisciplinarity different?' in: P. Weingart and N. Stehr (eds.), *Practising Interdisciplinarity* (Toronto: University of Toronto Press, 2000), pp 46–65, p. 64.

[71] '1955 Shortage of staff: group proposals', NA AB 16/1770; '1957 Personnel and administration: transfer of scientists and engineers from defence to civil work', NA AB 16/2083. As early as 1950, Lord Cherwell, Churchill's chief scientific advisor during the War, had criticized in the House of Lords the relatively low salaries paid to Harwell staff and their consequent loss to industry. On the other hand, he had also urged stronger security measures to prevent espionage ['Britain's atomic deficiencies', *Discovery* 12 (8) (1951): 235–6].

[72] '1954 Industrial applications: arrangements for training British Electricity Authority and consulting engineers', NA AB 16/1303.

[73] '1955–1956 Report on the recruitment of scientists and engineers by the engineering industry', NA AB 16/1772.

and fuel elements, and extracting bred fissile material (e.g. plutonium) from them); and more advanced topics such as electromagnetic pumps for liquid metal coolants, various types of power generating reactors, and cost estimates.

The School was suitable for men already employed in industry, but Christopher Hinton wanted something different from the model of 'some of the American Universities and Colleges of Technology...running training course for "nuclear engineers".' In just the way that he defined the professional remit of the British Nuclear Energy Conference to exclude labelled nuclear engineers, his earliest educational ideas envisaged an extension of university-taught chemical engineering to embrace nuclear expertise:

> In the chemical engineering schools the application has been to the engineering of chemical plants. Might it not be better to broaden our ideas in this field and to say that the application is not to chemical plants, but rather to process engineering? It then seems to me that we have a fourth major engineering field, and in this field it would be reasonable to include a course of lectures on nuclear engineering.[74]

The journal *Nuclear Engineering* argued further that future educational facilities should exist not only at the atomic energy establishments, but also at British schools, technical colleges, and universities. In its first year, the journal promoted postgraduate education via an annual scholarship for students having a prior degree in engineering, physics, chemistry, or metallurgy, the disciplines that it argued were the basis of nuclear engineering.[75]

A trickle of further education courses began to appear in 1956, mainly in the form of short introductions.[76] Borough Polytechnic in London offered evening classes on nuclear particle techniques, nuclear power, nucleonic circuitry, and reactor instrumentation.[77] Acton Technical College publicized a course to bring experienced engineers up-to-date with the progress in fundamentals of reactor technology. Birkenhead Technical College offered a 16-week Atomic Energy Course attended by 'electrical, mechanical and consulting engineers from shipping companies, the electrical supply industry and other organizations'.[78] Battersea Polytechnic, Leeds College of Technology, and the North West Kent College of Technology also offered a smattering of courses.[79] And, looking to the future, the journal *Nuclear Engineering* noted that at a recent *Schoolboy's Own Exhibition*, the *Daily Telegraph* had sponsored an exhibit

---

[74] Hinton, Christopher to H. Hartley, letter 4 August 1954, NA AB 19/84.

[75] *Postgraduate* in the UK context refers to higher education after the first degree, e.g. to MSc and PhD level education; the corresponding North American term is *graduate school*.

[76] Further education in the UK refers to post-secondary studies at technical colleges or polytechnic institutes, akin to *continuing education* at *community colleges* in North America. This training leads to specialist qualifications such as the Higher National Certificate (HNC) and Higher National Diploma (HND), roughly comparable in complexity to first and second year university attainment, respectively.

[77] News item, 'London evening classes in nuclear physics at Borough Polytechnic', *Nuclear Power—The Journal of British Nuclear Engineering* 1 (1956): 142.

[78] News item, 'Acton Technical College', *Nuclear Engineering* 1 (1956): 95; News item, 'Atomic Energy Course at Birkenhead Technical College', *Nuclear Engineering* 1 (1956): 109. See also Yeo, Frances E. M., *Nuclear Engineering Education in Britain*, MSc thesis, University of Manchester (1997).

[79] See, for example, 'Regular output of engineers for nuclear work—Battersea College of Technology', *Nuclear Energy Engineer* 13 (1959): 136–7. The article notes that there are 'no nuclear engineers yet...Though many of the successful students take up work in the nuclear field, none of them leave the college as nuclear energy engineers—they are chemical, electrical, civil or control engineers' [p.137].

theme of 'Atomic Energy as a Career'. 'Chemical and nuclear engineering are too closely interwoven', argued the editor, 'for us to lose any opportunity of enlisting the bright boys'.[80] The sudden proliferation of disparate courses encouraged the Ministry of Education to limit and approve validated courses. The Ministry wanted Technical School courses to take over the Harwell Reactor School's introductory six weeks so that it could focus on more advanced topics, and judged that three further education courses would be adequate for England's manpower requirements, leaving Harwell's support of the Borough programme awkwardly out on a limb.[81]

Harwell itself was pulled by opposing forces during this period, with the need to promote itself as 'a kind of university where the staff, having obtained knowledge and information, were free to go to other universities and instruct people', to support the UK nuclear power programme, and even by distinct groups within the Atomic Energy Research Establishment who sought diversification outside the field of nuclear energy.[82] Another difficulty was the lack of examinations attached to the Harwell course, meaning that no academically recognized diplomas or certificates could be awarded.[83]

The new further-education courses, initially entrepreneurial forays into what the college directors hoped would be a burgeoning market, were thus reined in and replaced by Ministry-mandated and Harwell-inspired offerings. The most ambitious scheme was launched in 1956 to train technicians 'for atomic power plant work' at Thurso, twelve kilometres from the Dounreay fast breeder reactor then under construction at the northern tip of the Scottish mainland (Figure 6.2).[84]

Once operational, Dounreay instigated an extensive training programme for its craft workers and, in 1964, extended this to scientific workers. This was not the case for further education near all British reactors: the Chapelcross facility in southern Scotland, for example, established no link with local education authorities. The Ministry of Education and Central Electricity Authority agreed that 'once a power station was established the number of men employed in its operation would be quite small, and it would make little difference to the local demand for technical education'.[85]

---

[80] Editorial, 'Jobs for the boys', *Nuclear Engineering* 2 (1957): 85. And, hoping to familiarize children with nuclear power even before adolescence, engineer Chris Schirmer published the illustrated book *Tom Atom and His Magic Spheres* in 1957.

[81] 'Board of Education and successors: Technical Branch and Further Education Branch: Registered Files (T Series)', NA ED 46/1062. The Harwell Reactor School was headed by Derrik J. Littler, who had been part of the British contingent at Los Alamos during the war.

[82] Spence, R., 'Twenty-one years at Harwell', *Nature* 314 (1967): 343–8.

[83] 'Britain's premier nuclear training centre: Harwell Reactor School', *Nuclear Energy Engineer* 13 (1959): 246–8.

[84] News item, 'School for Dounreay', *Nuclear Power* 1 (1956): 47. A much earlier precedent had been set by the Springfields uranium metal factory, which had founded a 'training school for boys and girls of fifteen years and over' as factory workers, and encouraged courses in fundamental science which would equip them for 'more responsible duties of the assistant class or for promotion to the experimental officer class' ['Young chemists needed for atomic energy factory', *Discovery*, 8 (6) (1947): 192].

[85] 'Central Electricity Authority Requirements of engineers with training in nuclear science', interview memorandum, Mr Bellamy, 9 May 1957, NA ED 46/1062.

Fig. 6.2   Craft apprentices at Dounreay, 1956. (Courtesy of Dounreay archives, UKAEA.)

Nor were all nuclear workers in Britain included by such training. The British Navy set up its own Department of Nuclear Science and Technology at the Royal Naval College, Greenwich, in 1959 to train officers responsible for operating nuclear submarines (the first of which, *Dreadnought*, was then being built) and to provide an advanced course for technical officers and civilian engineers who would work on nuclear propulsion in the Admiralty's Design departments. The original six-month navy course for operators ('Marine Engineering Officer (Nuclear)', or MEO(N)) covered nuclear physics, chemistry, and metallurgy, shielding, heat transfer, control systems, and instruction on the planned submarine reactor details. The *Dreadnought* officers then received a period of practical training on sea-going submarines of the US Navy. The one-year course for designers included detailed study of a mobile reactor system, along with lectures from UKAEA personnel and visits to their sites, concerning safety, radiation damage, fuel element design, steam power systems, and radiochemistry.[86] This break-away from academic training was a result of the American release of its PWR-1 submarine reactor design to the British Navy (a design subsequently developed further at other British establishments, notably at the naval reactor research facility set up next to—but isolated from—the civilian establishment at Dounreay).

6.4.2   *Shaping British curricula*

As prospects for a nuclear power industry grew, the shortage of trained personnel offered a chance for nuclear experts to diversify.[87] At the same time, however, it gave opportunities to existing disciplines to merge this fledgling niche knowledge with their own.

---

[86] Edwards, J., 'Nuclear training facilities at the Royal Naval College, Greenwich', *Nature* 200 (1960): 545–7.

[87] Rising demand frequently produces differentiation of professions, often by expanding lower-tier expertise necessary for routine tasks [Abbott, Andrew D., *The System of Professions: An Essay on the Division of Expert Labor* (Chicago: University of Chicago Press, 1988), 77].

From 1953 to 1955 Hinton lobbied the Universities of Manchester and Cambridge for nuclear energy or engineering courses suitable for current and anticipated Authority staff. From 1954, Manchester offered a Post-Graduate Nuclear Engineering Course, well sited for the concentration of nuclear sites in north-west England. Even supporters of the plan reported that nuclear engineering courses were impracticable given the already heavily loaded syllabus. The discipline could not be stretched to incorporate the new expertise.[88]

Frustrated by their lack of response, Hinton was pragmatic about the disciplinary home for such teaching, complaining that Prof T. R. C. Fox (1912–1962), in the Cambridge Chemical Engineering Department and member of the Atomic Energy Council, was cool to the proposal for a UKAEA-funded research reactor at Cambridge, unlike his counterpart in Mechanical Engineering: 'Although I always thought that chemical engineering was the proper framework for the new nuclear subjects, if Baker is prepared to take action on the mechanical engineering side I think he should be encouraged'. Similarly, Hinton judged that Manchester's weaker staffing in the 'heavily biased…chemistry side' of its College of Technology would make it easier to 'graft a course in nuclear engineering' onto the Mechanical Engineering offerings.[89] By contrast, Heriot Watt University was considering a course on Power Generation by Nuclear Reactors within its Electrical Engineering Department. And a survey of Imperial College departments in early 1957 indicated that several of them taught a smattering of nuclear subjects. The department of physics, for example, provided 20 hours of instruction to its students on nuclear physics, 15 on nuclear theory, about 50 hours in chemistry on radioactive tracers and advanced radiation chemistry, four or five from metallurgy, and an introductory course of 12 hours from mechanical engineering. All were lecture-based courses to undergraduate students.[90]

Influenced by such local university contexts, the emerging discipline thus developed along the parallel lines of expertise in chemical separation and transformation of materials (through chemical engineering) and power-generation, thermodynamics, control mechanisms, and shielding (through mechanical and, occasionally, electrical engineering). Hinton consequently hoped to shape the nature of its university teaching by situating nuclear engineering as a postgraduate or on-the-job adaptation of conventional engineering skills—rehearsing, in fact, the very experience of his cohort of Industrial Division engineers.

Hinton was averse to a labelled discipline. Approached to recommend an appointment for a new chair of Nuclear Engineering at Queen Mary College, London, Hinton replied that it would be 'difficult to find a man with experience in this field' and suggested instead that they pick a conventional academic engineer who could be further trained at Harwell for six months.[91] He argued to Prof Owen Saunders of the Mechanical Engineering Department of Imperial College, 'The last thing we want to do is to treat Nuclear Power as a separate technology' and, a few days later, 'I am still nervous about the proposal to set up a Chair of Nuclear Engineering. It seems to me that it is quite wrong to regard this as a separate branch of engineering technology. I think that it would be better to think of it as

---

[88] NA AB 19/84.
[89] Hinton, Christopher to J. M. Kay, letter 4 January 1955 and 22 December 1954, NA AB 19/84.
[90] 'Nuclear Science Instruction—UGC enquiry 1957 (File No. 571)', letters, 1957, ICA GB 0098 KNP/3/1.
[91] Hinton, Christopher to T. P. Creed, letter 23 November 1955, NA AB 19/84.

an additional field in which students can practise the application of the basic engineering principles which ought to form the backbone of their University course'.[92] He reiterated the same message a year later to the newly appointed professor John Menzies Kay (1920–1995), who had taught chemical engineering at Cambridge from 1948 and had then been Chief Technical Engineer at Risley from 1952.[93] Not surprisingly, Kay correspondingly proposed an Authority-inspired and BNEC-sanctioned curriculum:

> It has been suggested in some quarters that 'nuclear engineering' is a basic subject which should take its place alongside civil, mechanical, electrical and chemical engineering. This view, however, is not shared by any engineers of standing who are actively engaged in the development of nuclear power...The design and development of nuclear power stations and of the associated chemical processing plants involves a very wide range of engineering skill. It has been found necessary within the Atomic Energy Authority and in the engineering industry to build up large teams of mechanical, civil, electrical and chemical engineers to handle these projects. These engineers, while working in the field of nuclear power, and acquiring special knowledge that might be described under the term 'nuclear engineering', in fact remain primarily as mechanical, civil, electrical or chemical engineers respectively...This point of view also leads naturally to the present proposal for a postgraduate course in nuclear power at Imperial College which would be open to men who have already graduated in one of the recognized basic branches of engineering.[94]

As an alternative to a curriculum focused on the novelties of nuclear engineering, Hinton proposed exercises in basic theory, accompanied by detailed models of the Calder Hall nuclear plant to be donated by the UKAEA and, eventually, subcritical or research reactor facilities at universities.[95]

The first brochure for the Imperial College postgraduate diploma course was more explicit about the institutional pressures shaping the curriculum:

> The arrangements for the course will be based on the belief that nuclear power is not a distinct technology, but a new field of application for the existing basic branches of engineering and applied science. This view is supported by the leading figures both in the Atomic Energy Authority and in the engineering industry, who have been consulted in planning the course. It also reflects the policy of the Institutions of Civil, Mechanical, Electrical and Chemical Engineers and the Institution of Physics in setting up the

---

[92] Hinton, Christopher to O. A. Saunders, letters 5 and 9 January 1956, NA AB 7/40. Hinton's desires were satisfied: a former Harwell scientist, Walter Murgatroyd (1921–), was appointed professor at Queen Mary College and dubbed the subject matter covered by the course 'applied nucleonics', likening its relationship with nuclear engineering to the part played by aerodynamics and electromagnetics in aeronautical and electrical engineering, i.e. as a specialist body of knowledge that extended the existing engineering disciplines. On the later curriculum, see 'Teaching nuclear engineering in the mile end road', *Nuclear Energy Engineer* 12 (1958): 335–6.

[93] On the Imperial College department under Kay's successor, see Hall, G. R., 'Nuclear technology in a university environment', presented at *Imperial College of Science and Technology inaugural lecture*, 1964.

[94] Kay, John Menzies, 'Imperial College course proposal', 4 January 1957, NA AB 19/84.

[95] Hinton, Christopher to J. M. Kay, letter 9 January 1957, NA AB 19/84. A subcritical assembly is an arrangement of fissile material insufficient to generate a sustained chain reaction ($k < 1$), but which can produce sustainable fission via an external source of neutrons. Because they cannot reach criticality, such assemblies are relatively safe and often used in teaching contexts to demonstrate the multiplication of fission or geometric factors.

British Nuclear Energy Conference on a collaborative basis to cover the interests of engineers and applied scientists working in the atomic energy field.[96]

Even so, Kay's text hinted at inconsistency:

> Engineering is an art and the design engineer must be able to keep the conflicting demands of the various specialized sciences in balance. It may even be hoped that one of the special merits of nuclear power as an academic subject will be to exercise an integrating effect at a time of increasing specialization in the pure sciences.[97]

Was nuclear power, then, a straightforward extension of existing disciplines, or the seed of a new one? Interestingly, this paragraph disappeared from subsequent versions of the brochure, and the views of the existing engineering institutions were no longer foregrounded.[98]

Basing curricula on guesses for the potential market as well as experts such as Hinton meant that the complement and level of university courses were rather varied.[99] The increasingly declassified environment of the late 1950s witnessed growing university involvement. King's College, Newcastle, predicting the rise of marine applications, mooted a nuclear engineering course oriented towards the naval industry.[100] The University of Glasgow instituted a more general course in 1956, with three-quarters of the lectures by the Natural Philosophy (i.e. Physics) department, and the remainder of lectures and practical sessions provided by Engineering. Its students—undergraduates and extra-mural students from local industry—received forty hours of lectures and twenty hours of practical work that covered scientific topics that included elementary nuclear physics, neutron thermalization, and reactor physics, alongside the more practically oriented 'problems' of irradiation, damage to reactor materials, health physics, and radiation shielding. The engineering component of the course dealt primarily with heat transfer and power generation.[101] Birmingham University began a one-year MSc course in reactor physics and technology in its Physics Department in 1956, with entry to students holding a degree in pure or applied science, normally physics, metallurgy, mechanical engineering, or chemical engineering. Beginning in 1957, Queen Mary College, London, and Nottingham University offered six-week familiarization courses in nuclear engineering for graduate-level students having no prior knowledge of nuclear physics or nuclear engineering, and with Queen Mary offering a nuclear engineering option in its BSc in Electrical Engineering. Cambridge offered a course on reactor theory. And following Kay's appointment, Imperial College, London, began

---

[96] 'Imperial College of Science and Technology (University of London) Postgraduate Course in Nuclear Power', brochure, February 1957, ICA GB 0098 KNP/3/1.

[97] 'Imperial College of Science and Technology (University of London) Postgraduate Course in Nuclear Power', brochure, February 1957, ICA GB 0098 KNP/3/1.

[98] The 1957/8 and 1959/9 brochures made no mention of 'integration' of engineering disciplines, but highlighted the practical skills of plant design and construction. ['nuclear power leaflets', brochure, 1958, ICA GB 0098 KNP/2].

[99] 'University: nuclear engineering 1954–1957 Train/1', NA AB 19/84.

[100] Minutes, meeting of the first Advisory Panel in Chemical Engineering, King's College, Newcastle, 20 February 1957. See also Walker, John, 'Training nuclear engineers', *New Scientist*, 18 July 1957, 23–5; 'Training nuclear engineers at Southampton University', *Nuclear Energy Engineer* 13 (1959): 305–6.

[101] News item, 'Course of nuclear engineering at the University of Glasgow', *Nature* 177 (1956): 508–9.

post-graduate courses in Nuclear Power (via the Department of Mechanical Engineering) and Nuclear Technology (via their new Reader in Nuclear Technology, Geoffrey R. Hall, in the Department of Chemical Engineering) in 1958. Within one university, then, nuclear specialists were trained with distinct disciplinary slants.

Hands-on training facilities were instituted more gradually at British universities. In 1959, the universities of Scotland and Northern Ireland discussed a joint facility, and in 1963, with funding provided by the Department for Scientific and Industrial Research (DSIR), the Scottish Research Reactor Centre was opened near Glasgow. Initially it provided one- to two-week courses on reactor physics, nuclear engineering, health physics, radiochemistry, and radioisotopes for students of the universities.[102] Two other university consortia—Manchester–Liverpool and the University of London—were funded at the same time. The Liverpool and Manchester Universities Research Reactor at Risley, Cheshire, operated during 1964–1991; The University of London Reactor Centre, located at Silwood Park, Berkshire, opened shortly after. All three reactors originally generated up to 100 kW of power.[103] A fourth training/research reactor, operated by Queen Mary College, London, 1964–1982, was the only other university teaching facility in the UK. The British Navy operated JASON, a 10 kW reactor at Greenwich (1962–1996), for training Navy and related personnel.

The view percolating from the UK nuclear institutions and academe influenced students, too. Most identified themselves with their undergraduate speciality. Tony Goddard (1937–), for example, later Professor of Nuclear Science at Imperial College, saw himself as a budding applied physicist. His contemporary, Mike Williams (1935–), later a Professor himself at Queen Mary College, also identified with physics:

> It was quite an eye-opener for me, being introduced to engineering stuff. Doing mechanics, which we'd done from a physics point of view, suddenly you'd find these weird recipes for dimensionless numbers and things, and ask 'what is this fudge factor here'? It was quite a revelation...I found myself, somehow, as sort of in-between. I got an appreciation of engineering, and although what I've done since has been theoretical physics, in a sense, applied to reactors, I'd been influenced by engineering concepts.[104]

Both studied at the University of London, Goddard for a postgraduate Diploma in Nuclear Power at Imperial, and Williams for a Diploma in Nuclear Engineering at Queen Mary. Both found themselves confirmed in their leanings by their subsequent posts. Williams spent a fellowship year in Brookhaven, USA, and then took up a readership at Queen Mary

---

[102] Wilson, Henry W., 'The Scottish Research Reactor Centre', *Nature* 205 (1965): 10–14; 'The Scottish Research Reactor Centre, East Kilbride, Glasgow, Scotland', *Journal of Radioanalytical Chemistry* 6 (1970): 273–83.

[103] 'Universities of Manchester and Liverpool Joint Research Reactor', *Nature* 203 (1964): 348; Newton, G. W. A., 'Nuclear and radiochemistry training in the UK ', *Journal of Radioanalytical and Nuclear Chemistry* 171 (1993): 45–56. The University of London Reactor Centre was later renamed the Imperial College Reactor Centre and in 1992 (perhaps reflecting public concerns about nuclear facilities) the Centre for Analytical Research in the Environment (CARE). This rebadging is similar to that undergone by the Willow Run Laboratories, a contract military research centre of the University of Michigan which, following student protests of the late 1960s, was reborn in 1972 as the independent Environmental Research Institute of Michigan (ERIM).

[104] Williams, Michael M. R. to S. F. Johnston, interview, 2 July 2008, London, England, SFJ collection.

College, and Goddard spent three years alongside physicists at the UKAEA reactor research facility at Winfrith followed by a lectureship in the physics department at Birmingham.[105]

Such examples among the first university-trained nuclear workers hint at the inconsistent and sometimes interdisciplinary definitions of the new subject within British universities. The provision of formal training created its own tensions within the UKAEA. The original generation of nuclear workers recruited by Christopher Hinton—pioneers in the developing art but without academic training—increasingly found themselves sidelined or even demoted in the organization's hierarchy as a newly trained but inexperienced group joined their ranks. As one of them, relegated to a study of workplace contamination, complained, he was reduced to polishing brass doorknobs.[106]

6.4.3 *Finding a niche at American universities*

The outlook was different in the USA. There was a strong continuity from wartime sites via national laboratories to higher education institutions, supported by industry pressures. In the face of predictions of an annual national demand for some '2,000 engineers and scientists—mostly engineers', power company executives in the southern states lobbied for support of college and university training programmes.[107]

But the evolution of the subject and its profession were firmly grounded in the experiences of the wartime and cloistered periods up to the mid-1950s. Key administrators in American nuclear engineering until the 1970s often had backgrounds bereft of university training in the subject. A typical example is Martin Shaw, who served at Argonne National Laboratory as Director of the Division of Reactor Development and Technology from 1964 until the 1970s. Like Walter Zinn, the first Director of ANL, Shaw had notched up an amazing list of accomplishments as his career grew with the subject itself. He was a career government engineer. Before the Second World War, Shaw had worked in engineering works for the Tennessee Valley Authority, Roosevelt's government-funded power, irrigation, and flood-control project. He served with the Navy during the war, and obtained degrees in mechanical engineering after it, supplemented by completion of ORSORT courses. From the late 1940s, Shaw had worked in the Naval Reactors Program under Vice-Admiral Rickover. As Director of the Advanced Design Division for the Navy, he was responsible from conception to completion for plant and fluid systems work for the first United States central-station nuclear power plant at Shippingport and for systems and arrangements aspects and the conception and initial design of all submarine plants until 1956. During the late 1950s, as Project Manager for Surface Ships Projects,

---

[105] Williams, Michael M. R. to S. F. Johnston, email, 31 July 07, SFJ collection; Goddard, Tony to S. F. Johnston, interview, 2 July 2008, London, England, SFJ collection.

[106] Millar, R. N., 'Future power programme in Britain', *Nuclear Engineering* 1 (1956): 304. For a more recent summary, see Skelton, B., 'The education & training of nuclear engineers in the UK', *Nuclear Engineer* 42 (2001): 175–8.

[107] Folger, J. K. and M. L. Meeks, 'Educated manpower: key to nuclear development', in: R. Sugg Jr (ed.), *Nuclear Energy in the South* (Baton Rouge: Louisiana University Press, 1957), pp 105–38, quotation p. 116. See also Anderson, E. E., F. C. Vonderlage and R. T. Overman, 'Education and training in nuclear science and engineering in the United States of America ', presented at the *International Conference on the Peaceful Uses of Atomic Energy*, Geneva, 1955.

Shaw was responsible for design and development of the nuclear power plants for the nuclear aircraft carrier *Enterprise* and the nuclear cruiser *Long Beach*, as well as for all other surface ship propulsion plant projects in their conceptual and design stages. During the early 1960s, Shaw played a senior technical role in the Navy and was responsible for policy and management of research, development, test, and evaluation for Navy and Marine Corps, for liaison for engineering, scientific, programme, and planning offices within the Department of Defense. And, after transfer to the AEC at Argonne, Shaw carried impressive responsibilities: development and improvement of nuclear reactors and isotopic systems for civilian and assigned military applications; development of nuclear technology for it; nuclear safety of reactors and systems; and fostering civilian applications of nuclear power.[108] Career biographies such as Shaw's illustrate the close alliance between government-sponsored research, military development, and civilian nuclear power, as well as the relatively weak role of universities in this even a quarter-century after the first atomic bomb.

Only in the late 1950s did American universities begin to usurp the role of ORSORT in training nuclear workers. The initial university responses were hampered by security, lack of personnel, agreed standards, and precedents. North Carolina State College (NCSC) was the first to institute a (classified) Nuclear Engineering programme, beginning in 1950 in its School of Engineering under Clifford K. Beck. For more than a decade, its academic staff consisted of Beck, who had been recruited from Oak Ridge to head the NCSC Physics Department; Arthur C. Menius, Jr (1916–96), a physicist with experience of proximity fuse design; Raymond L. Murray (1920–), who had worked at Berkeley and Oak Ridge on isotope separation during the Manhattan Project, and obtained a doctorate in physics at NCSC in 1950; and Arthur Waltner and Newton Underwood (1906–), who had undertaken their own training in physics before the war.[109]

Similarly, in 1951 MIT had established a School of Nuclear Engineering in its Chemical Engineering Department and appointed its first professor of the subject, Manson Benedict (1907–1996).[110] Benedict had been responsible for developing gaseous diffusion processes during the Manhattan Project, and was joined by Thomas H. Pigford (1923–2010), who had headed the post-war Engineering Practice School at Oak Ridge Laboratories. By 1953 an MSc degree was offered, and the first PhD was awarded in 1958 as part of the newly established Nuclear Engineering Department.[111] Like NCSC, the programme had strong echoes of ORSORT, though. Its students were mainly US Air Force officers; its syllabus was ring-fenced by security classification; and the MIT department's staff had backgrounds heavily weighted towards physics rather than engineering.

More university departments grew from engineering foundations following the Atoms for Peace initiative. The University of Arizona, for example, began a programme in 1959

---

[108] 'Biography, Milton Shaw', file, 31 January 1969, UI Box 13.

[109] Beck, Clifford K., 'Undergraduate nuclear engineering curriculum at North Caroline State College', *Nucleonics* 8 (1951): 54–9.

[110] Benedict was to serve as the eighth president of the American Nuclear Society, 1962–3.

[111] MIT, 'History of the MIT Department of Nuclear Science and Engineering', http://web.mit.edu/nse/overview/history.html, accessed 20 March 2006.

under Lynn Weaver (1930–). Weaver had not worked on the Manhattan project or in the post-war national laboratories. Instead, with two degrees in electrical engineering under his belt, he gained experience with aircraft companies working on radar, guided missile control, and, from 1953 at Convair Corporation, on the programme to design a nuclear-powered bomber, where he began to study reactor control problems. Weaver took one course from a graduate of ORSORT, and absorbed more from the handful of texts then becoming available.[112] At the age of 29, then, Weaver had the privilege of directing one of the first degree programmes as well as a university-based teaching reactor.[113] Gradually he gathered a half-dozen teaching staff, most of whom had worked on the General Atomics project for the SNAP reactor (System for Nuclear Auxiliary Power) intended for space flight applications. They taught a growing number of students—'not a whole lot, but sufficient to have a department'.[114] At Arizona then, like several other new departments, the first generation of academic staff had relatively brief industrial experience—some on the Manhattan Project itself—and engineering degrees in other subjects supplemented by short courses that could be traced back to ORSORT.

The national laboratories, along with their associated university consortia, played a conscious role in defining educational standards for the new subject and shifting teaching responsibility to academic institutions. In 1955, Argonne National Laboratory sponsored an 'inter-institution committee for considering a cooperative effort for advanced nuclear engineering education' with eight of its participating universities. Norman Hilberry, the Deputy Director of ANL, argued that 'engineers trained in one of the conventional disciplines need a greater appreciation of the nuclear problems of the other engineering disciplines than would come from their cursory associations during their undergraduate years', and suggested an increase in Argonne's teaching facilities.[115] The University of Illinois representative more pointedly surveyed the situation and suggested the frustrations felt by university departments urging the transfer of nuclear engineering to universities. His argument, coherently voiced and widely shared among his colleagues, is worth quoting in detail:

---

[112] The principal texts then available were Lapp, Ralph Eugene and Howard Lucius Andrews, *Nuclear Radiation Physics* (New York: Prentice-Hall, 1948); Glasstone, Samuel and Milton C. Edlund, *The Elements of Nuclear Reactor Theory* (Princeton, NJ: Van Nostrand, 1952); Stephenson, Richard, *Introduction to Nuclear Engineering* (New York: McGraw-Hill, 1954); Glasstone, Samuel, *Principles of Nuclear Reactor Engineering* (Princeton, NJ: Van Nostrand, 1955); and, Murray, Raymond L., *Introduction to Nuclear Engineering* (New York: Prentice-Hall, 1955).

[113] The TRIGA reactor ('Training, Research, Isotopes, General Atomics') was conceived by Freeman Dyson and Edward Teller from 1956 as a design unlikely to suffer thermal runaway, and so suitable for teaching purposes. From 1958 some 35 were installed around the USA and an equal number in other countries. A Canadian equivalent was SLOWPOKE, conceived by John Hilborn at AECL as a low-cost source of neutrons and eventually installed in six universities [Buyers, William J. L., 'Neutron and other stories from Chalk River', *Physics in Canada* 56 (2000): 145–51]. The seeding of nuclear engineering around the world owed much to the export of such teaching reactors.

[114] Weaver, Lynn to S. F. Johnston, telephone interview, 3 March 2007, SFJ collection.

[115] Argonne National Laboratory, 'Inter-institution committee for considering a cooperative effort for advanced nuclear engineering education', minutes, 20 December 1955, UI Box 100.

> The past decade has seen the development of controlled nuclear power engines. From its inception during World War II, the burden of this development rested upon essentially a relatively small number of scientists. There had been no nuclear engineering prior to that war, and the requirements and complexity of nuclear power techniques, criteria of purity of materials, etc., and the sheer expense of development had had no counterpart in any other field of engineering. The people that pioneered nuclear technology were of various backgrounds and developed their proficiency by on-the-job training and experience. The success of the Nautilus submarine and projected applications of nuclear powered plants create a demand in industry for design engineers, construction supervisors, operating and maintenance engineers that are specifically trained for this field and fully aware of its stringent requirements. The latent dangers in the devices and the problem of waste disposal leave no room for capricious expansion. Rather, the expected growth into unthought-of fields demands highly trained, imaginative people…It is the function of universities to provide a 'door' into this field that will stimulate potentially competent people to enter this new technology and to train them…The prerogatives of education rest with the universities. To keep them there, the universities must fill the gap in nuclear engineering as soon as possible.

The extended use of nuclear power was, he suggested, of strategic importance and would require that the federal government and Atomic Energy Commission employ a growing number of qualified nuclear engineers. Even so,

> The establishment, in different universities, of courses and curricula in nuclear engineering comparable in scope and quality with those in other branches of engineering will be difficult owing to the lack of adequately trained staff and the high capital and operating costs of the desired physical facilities. Government laboratories and the rapidly expanding nuclear power industry quickly absorb the few persons receiving on-the-job training or formal education in those schools offering anything resembling appropriate courses.
>
> Presently, there is a great shortage of nuclear engineers. While one or two universities might be able to attract an adequate and competent instructional staff that preferred the academic life, certainly the prospects are not good for many of them to do so, and they would either have to wait before instituting nuclear engineering curricula or cope with an inferior staff'.[116]

The transition from restricted education by the national laboratories to a university environment was fraught. The 'great shortage' of trained staff could not support the rapid expansion, leading the University of Cincinnati in 1959 to advertise desperately for 'a staff scientist who is preferably an Experimental Physicist, Electrical Engineer, Nuclear Engineer or someone with a PhD in any of the other engineering areas. Experience in the area of reactor operation is desirable, but not necessary…'.[117] Some early nuclear engineering programmes were forced to define criteria of expertise in relation to the handful of basic texts, with one advertisement calling for 'a good working knowledge of reactor analysis at the Glasstone and Edlund level'.[118]

6.4.4 Contesting curricula in the USA

Beyond these pragmatic problems of costs and resourcing, however, were the intellectual foundations of the discipline itself. The Argonne education committee noted:

---

[116] Longacre, Andrew, 'Proposal for a cooperative facility for education in nuclear engineering', memo, November 1955, UI Box 100.

[117] Roberson, John H. to W. F. Stubbins, letter, 28 September 1959, UI Box 83.

[118] 'Iowa State University Dept of Nuclear Engineering advertisement', 27 September 1960, UI Box 83. The Glasstone and Edlund text had been the primary resource for ORSORT courses at least as early as 1953 [Kamack, Harry J., 'Report on first term of ORSORT year', memo, 20 January 1954, Hagley 1957 Series III Box 12 folder 4 ORSORT 1953–54].

In most of the accepted academic fields, there is at least tacit agreement fostered by text books and professional societies, etc., as to what constitutes adequate preparation of the students in the several fields of engineering. At this early stage, no such formulation exists for nuclear engineering.[119]

The committee decided that each institution should, with whatever government support might be found, provide new courses and facilities necessary for a core programme in nuclear engineering on their own campuses. In return, Argonne would provide more elaborate facilities and serve as a 'finishing' school for graduate students and for short courses, conferences, teacher training, cooperation with medicine, and so on.

Three years later, after the formation of the Associated Midwest Universities (AMU) that collaborated at Argonne, it established a Nuclear Engineering Education Committee (NEEC). Its objectives were (1) to advise on the formulation of policies on university cooperation with ANL; (2) to advise and consult on educational programmes originated by AEC and/or ANL; and (3) to advise on special educational programmes and activities between AMU schools and ANL. Its vision of nuclear engineering was nevertheless contested then and later. A University of Michigan professor complained to a colleague in 1966:

When these [NEEC] meetings started 7 years ago, Nuclear Engineering, as a well defined discipline of Engineering, was in its infancy, and in the minds of many it was vague and quite misunderstood. The early meetings were – I believe – intended and did serve to clarify the picture and to stimulate the interest of many in this young and challenging field. However, in spite of the splendid effort on the part of the planning committees in selecting topics and speakers these meetings seem to always convey the notion that Nuclear Engineering is essentially reactor engineering. This notion continues to prevail and has perhaps been intensified due to the immense activity of Argonne in this field…it is only a part and a part which has perhaps reached the saturation point as evidenced by the decline of these activities even at Argonne. In other words it is time for diversification…In this regard, the Middlewest seems to be a 'one-field' region (much like Detroit being a one-product city); namely reactor engineering. I cannot help but observe the serious lack of educational and research activities in this region in many fields which rightfully belong under the heading of nuclear science and engineering'.[120]

Argonne's dominance in reactor know-how threatened those with broader definitions of nuclear engineering. As work continued to expand at the national laboratories, military services, and industry during the 1950s, the AEC consulted with the Nuclear Committee of the American Society for Engineering Education (ASEE) to establish programmes to procure appropriate equipment, publish texts, and develop faculty. From 1956 it published a *Nuclear Energy Education Newsletter* to help build a network of educators. Even so, the result was piecemeal coverage, both geographically and intellectually.[121]

The first generation of courses at American universities in nuclear engineering illustrates the problems of professional and disciplinary identity. The national labs at Oak Ridge,

---

[119] Longacre, Andrew, 'Proposal for a cooperative facility for education in nuclear engineering', memo, November 1955, UI Box 100.

[120] Kammash, T. to C. E. Dryden, letter, 22 February 1966, UI Box 100.

[121] Disputes about curricula were not new. A half-century earlier, American educators had wrangled over the appropriate mix of practical skills and theoretical science appropriate to engineering courses, noting a gradual 'academic drift'. See Harwood, Jonathan, 'Engineering education between science and practice: rethinking the historiography ', *History and Technology* 22 (2006): 53–79 and Grayson, Lawrence P., *The Making Of An Engineer: An Illustrated History of Engineering Education in the United States and Canada* (New York: John Wiley, 1993).

Argonne, and Brookhaven had taken responsibility in 1946 for providing facilities for regional universities. For a decade, this meant arrangements for university academics and their students to visit the Labs to conduct research or to serve a form of academic apprenticeship. Only during the late 1950s—when commercial power reactors were first coming online—did universities begin to mount educational programmes in nuclear engineering. Academics were faced with the new problem of defining adequate curricula and the characteristics of their graduates.

Notre Dame University, in Indiana, opted for a graduate programme 'with the emphasis on fundamental concepts rather than on technology. The program of instruction was originally conceived as one in reactor physics and in instrumentation, built on a sound background in atomic and nuclear physics and on some chemical physics'.[122] Other institutions stressed engineering design, but all struggled to obtain adequate laboratories and competent teachers.

The difficulties in defining an adequate curriculum are illustrated by the typical case of Wayne State University in Michigan. Its Associate Professor of Engineering Mechanics felt 'an obligation to provide training and education in the rapidly expanding field of nuclear science and engineering' imposed by 'the hundreds of millions of dollars allocated by the AEC to the objective of promoting education in this field'. But his enthusiasm was tempered; he argued that 'Nuclear Engineering is really little more than a group of specialized applications of basic engineering principles, supplemented and modified by the laws of nuclear physics'. He cautioned,

> The process of educating students in Nuclear Engineering is, however, neither a simple task nor one to be undertaken without careful consideration, forethought, and planning. Nuclear Engineering is still extremely young as branches (or sub-branches) of engineering go; it is still changing with a rapidity almost unprecedented in the evolutionary history of technology (even so-called 'fundamental constants' may vary from day to day); opinions of both professional educators and practitioners in the field still diverge widely over what constitutes a proper nuclear engineering education; in brief, it is apparent that any curriculum or program established for this discipline must be flexible in both conception and administration.
>
> There are a number of problems associated with an attempt to educate nuclear engineers. For example, the choice of textbooks from which to teach the largely nuclear phases is very limited, and the majority of the texts published to date have been written by nuclear scientists rather than professional educators. As a result, books in this field are characterized (from the teacher's viewpoint) by poor arrangement of topics, lack of continuity and integration of approach to the subject, completely inadequate problems for student solution, both in quantity and in nature, and by inconsistency of numerical data. Not only do the numerical constants vary markedly from book to book and within particular books, depending on the sources used by the authors, but much of the data in any book more than a year or two old are obsolete and in error – sometimes by as much as an order of magnitude. It is also noticeable from the textbooks that, unlike the situation in many other engineering fields, very few standard analytical methods and design techniques have been developed.
>
> In addition to the shortage of satisfactory textbooks there is a scarcity of qualified teachers. The latter factor is somewhat aggravated by the former, since, in general, the poorer the textbooks, the more capable must be the teacher. Because so few universities have very elaborate or even complete nuclear facilities, it is often necessary to send teachers to an AEC laboratory for at least some first-hand experience with nuclear reactors and equipment. With nuclear engineers currently in shorter supply than most other

---

[122] Newman, M. K. to J. H. Roberson, letter, 17 February 1960, UI Box 85.

varieties, it is to be expected that competition from industry and government laboratories will continue to make qualified teachers of nuclear engineering difficult to obtain.[123]

Such concerns were commonly voiced by academics and university administrators at the end of the decade. Tellingly, the Director of the Associated Midwest Universities, closely linked to the Argonne Laboratory, responded to one potential student of the subject, 'I am somewhat reluctant to offer any guidance with respect to the comparability of the nuclear engineering programs because it is a difficult subject and I am not too well versed'.[124] A contemporary sociological survey of American nuclear training noted that courses ranging from 16 to 133 hours of instruction were being provided by employers and adult-education programmes, although 'not all management officials in industrial plants affected by nuclear technology are convinced that training workers in the fundamentals of nuclear theory is necessary or desirable'.[125]

From 1960, the American Nuclear Society (ANS) and the ASEE, noting that 'current programs differ significantly from one another because of varying educational philosophies among institutions as well as differences in calibre, education, and experience of faculties', undertook a joint study with the ASEE to define 'the attributes of formal education in nuclear engineering', and with it the means of accrediting nuclear engineers.[126] Their report—unparalleled in Britain or Canada—was seminal in defining the American discipline and profession.

The ANS/ASEE committee specified nuclear engineering as 'that branch of engineering directly concerned with the release, control and utilization of all types of energy from nuclear sources':

> Nuclear engineering includes the design and development of systems, such as fission or fusion reactors, for the controlled release of nuclear energy and the applications of radiation. It is recognized that nuclear engineering is closely associated not only with nuclear and solid state physics but also with other branches of engineering such as chemical, electrical, and mechanical.
>
> The central area of interest in nuclear engineering involves the solution of those problems in which the unique nature of nuclear energy presents the major challenge. Typical nuclear engineering problems arise, for example, from the high effective temperatures which occur at the instant of energy release, the intense radiation of photons or particles which accompany the release process, the factors involved in maintaining the energy release on a continuous, controlled basis, the necessity for protection of personnel from radiation, including the safe handling of reactor materials, irradiated materials and radioisotopes, and the use of radioactive materials.[127]

Nuclear engineers required skills in reactor analysis and design; analysis of radiation effects; shielding design; utilization of radiation, processing and control of radioactive materials, and nuclear energy systems design. Neutron expertise—or the freshly coined 'neutronics'—was foremost. The list of essential content made a strong case for demarcating a discipline (Table 6.1).

---

[123] Perry, C. C., 'Nuclear engineering at Wayne State University', memo, 6 April 1959, UI Box 22.
[124] Roberson, John H. to E. L. Multhaup, letter, 11 April 1960, UI Box 83.
[125] Vollmer, Howard M. and Donald L. Mills, 'Nuclear technology and the professionalization of labor', *American Journal of Sociology* 67 (1962): 690–6, quotation p. 692.
[126] American Nuclear Society, American Society for Engineering Education, 'Report on objective criteria in nuclear engineering education', QU Sargent fonds, Series III Box 4.
[127] Ibid.

Table 6.1 Content of education in nuclear reactor engineering recommended by the American Nuclear Society and American Society for Engineering Education.[128]

1. Neutronics and energy production
   Fuels
   Moderators
   Coolants
   Methods of control (rods, soluble poisons, voids)
   Power distribution in the core and power flattening
   Long-term reactivity effects

2. Heat flow and fluid flow
   Laminar, turbulent, two-phase flow
   Dimensional analysis, heat transfer coefficients
   Conduction and convection, including boiling heat transfer
   Heating of radioactive materials
   Hot channel factors, hot spot factors
   Thermal and flow transients

3. Applied mechanics
   Structures
   Elastic and plastic behaviour of tubes, plates, shells, beams, etc.
   Thermal stresses
   Vibrational behaviour

4. Control theory and systems
   Instrumentation
   Feedback loops

5. Design
   First-order consistent design
   Successive iterations
   Integration with non-nuclear heat transfer and power conversion system
   Economic evaluation; effects of fuel cycles, heat transfer, thermal cycles
   Optimization procedures; numerical methods of analysis
   Description of typical and advanced power systems

6. Advanced concepts
   Fusion reactions
   Properties of plasma
   Plasma containment
   Magnetohydrodynamics

The report argued that these novel competences relied on existing disciplines, and cited the importance of mathematics, the physical sciences, the engineering sciences, and professional engineering in the subject. While calling for a range of education and training at universities, technical institutes, industry, and the government laboratories—essentially the model then being implemented in Britain—the committee deemed postgraduate education to be the most suitable for the complex subject. The prerequisites would be a bachelor's

[128] Ibid.

degree in engineering, including nuclear physics, advanced mathematics and competence in the analysis, design, and synthesis of engineering systems.

As a degree of uniformity spread around American university curricula, higher education adapted to teach workers newly employed in nuclear-related industries and, even more importantly, university educators themselves. The typical mode of delivery was with one- to two-month summer courses, funded jointly by the AEC and the universities. During 1964, for example, prospective students could choose from a 'basic' summer institute at NCSC, or advanced courses on Reactor Kinetics Theory and Experimentation at Iowa State University, Engineering Applications of Radiation and Nuclear Techniques at Kansas State, Neutron Transport Theory and Applications at the University of California at Berkeley, Intermediate Reactor Theory at the University of California at Los Angeles, Reactor Engineering at the University of Michigan, or Fast Reactor Physics at Idaho State.[129]

As a discipline, American nuclear engineering thus saw a two-decade transition from direct government sponsorship to increasing industrial and academic definition.[130]

6.4.5 Developing a demand in Canada

Canadian training appeared later than in either the USA or UK, and indeed had features in common with the second wave of countries to pursue atomic energy, particularly in Europe, following the first Geneva Conference and into the early 1960s.[131]

Only in 1954 did Chalk River administrators, still focused on the NRX research reactor and its pending successor, the NRU (National Research Universal), commit themselves to studying nuclear power generation. David Keys, the scientific advisor to the President of AECL, observed that a feasibility project involving Canadian power companies (notably the Ontario Hydro Electric Power Commission) had begun, and

> In view of the fact that both the Americans and British are proceeding with the construction of plants to produce reasonable quantities of electrical energy from nuclear fission, it is important that Canada should also be considering such possibilities, since our scientists and engineers have made a very successful contributions to nuclear pile operations.[132]

W. B. Lewis became the champion of nuclear power in the Canadian programme. In early 1962, Canada's first pilot nuclear power station, the NPD (Nuclear Power Demonstration) reactor, went critical in Rolphton, Ontario, just 19 km up-river from Chalk River itself.

---

[129] American Nuclear Society, 'Summer institutes on nuclear energy for 1964', QU Sargent fonds, Series III Box 4.

[130] On more recent consolidation, see Martin, W. R., 'Undergraduate education in nuclear engineering in the USA', *Journal of Radioanalytical and Nuclear Chemistry* 171 (1993): 183–92.

[131] For a comparison of teaching programmes and curricula in Austria, France, Germany, Greece, Holland, Italy, Portugal, Spain, Sweden, and Switzerland, see Barca Salom, Francesc X., *Els Inicis de L'Enginyeria Nuclear a Barcelona: La càtedra Ferran Tallada (1955–1962)*, PhD thesis, Departament de Matemàtica Aplicada 1, Universitat Politècnica de Catalunya (2002), Chapter 8. For example, from 1956 in France, l'Institut National des Sciences et Techniques Nucléaires in Saclay and l'Institut Polytechnique de l'Université de Grenoble initiated courses on 'Genie Atomique'. Soviet republics, too, began to train engineers for indigenous nuclear power plants by the early 1960s [Josephson, Paul R., *Red Atom: Russia's Nuclear Power Program from Stalin to Today* (New York: W. H. Freeman, 1999), p.34].

[132] Keys, D. A., 'Monthly report', March 1954, LAC MG30 B59 Vol. 5.

A joint project of AECL, the Hydro Electric Power Commission of Ontario and Canadian General Electric, the NPD was the indicator of another phase change: The Atomic Energy Project and the AECL had evolved into a new government-driven industry with high aspirations and a new type of technical labour.[133]

Nevertheless, Canadian consideration about suitable training began much earlier. In 1948, David Keys, as the chief scientific advisor for the Atomic Energy Project, was asked by the Canadian Navy to suggest appropriate college engineering courses for cadets 'interested in the development of atomic power and research in the atomic bomb', and what mixture of 'Engineering Physics, Electrical Engineering, Chemical Engineering, pure Physics or Chemistry' would be most appropriate. Keys' answer reflected his own background:

> Mathematics Honours, Mathematics and Physics, Honours Chemistry, or Engineering Physics, in that order. They then should proceed to graduate work in either Physics or Chemistry. Actually a man going into this field should have a good solid foundation in Science and although we have Mechanical, Electrical and Chemical Engineers, the research end is performed more by physicists and chemists than by engineers.[134]

By 1951 night school classes had been established at Chalk River, ranging from first-year university classes to the more advanced subjects of calculus, pile theory, and nuclear physics. This was not open training for new careers, though: it was limited to existing Chalk River workers who had passed the usual security procedures.

Formal training in Canada, like the power-generating programme, trailed behind Britain and, in many respects, echoed the British case. During the early post-war years David Keys noted a chronic 'scarcity of available scientists and engineers in every field'; the situation improved for craft workers, but not for the scientific and engineering professionals.[135] Security concerns, physical remoteness of the site, and limited employment prospects further exacerbated the shortage.

As early as 1953, B. W. Sargent at Chalk River fielded enquiries from prospective engineering students, counselling one that 'a graduate nuclear engineer can today practise his profession in Canada' at AECL and the C. D. Howe company (then responsible for building NRU, the next reactor), and that 'within a few years' various power companies would also be involved. He envisaged two classes of employment: nuclear power plant operators working for power utilities, and designers and constructors of power plant.[136]

---

[133] The 200 MW CANDU ('Canadian Deuterium') reactor at Douglas Point, Ontario, on Lake Huron, was the first full-scale Canadian nuclear power station, and became operational in 1968. Canadian expertise was by then more dispersed: the AECL Nuclear Power Plant Division (NPPD, 1958) in Toronto, Ontario, was responsible for NPD and subsequent reactor designs; at the Whiteshell Nuclear Research Establishment (WNRE, 1963), near Winnipeg, Manitoba, new reactor designs were investigated. Chalk River staff transmitted design expertise to the C. D. Howe Company, CGE, and to India, which designed and constructed a series of power stations based on the CANDU design. On CGE, see Cantello, Gerald Wynne, *The Roles Played by The Canadian General Electric Company's Atomic Power Department in Canada's Nuclear Power Program: Work, Organization and Success in APD, 1955–1995*, MA thesis, Frost Centre for Canadian Studies and Native Studies, Trent (2003).

[134] Keys, D. A. to C. C. Cook, letter, 21 January 1948, LAC MG30 B59 Vol. 4.

[135] Keys, D. A., 'Monthly report', August–September 1947, LAC MG30 B59 Vol. 5.

[136] Sargent, B. W. to G. M. Everhart, letter, 4 November 1953, QU Sargent fonds Series III Box 4 file 4.16. Bernice Weldon Sargent (1906–) had worked with G. C. Laurence in Ottawa on his chain reactor experiments from 1941, and was an early member of the Montreal group.

But such certified graduates did not exist and could not be produced. Like Harwell's earlier example in England, AECL at Chalk River employed summer students—some of whom eventually became permanent employees—and began its own training courses for its various professional and craft workers in 1958.[137] A year later, just as Harwell was doing, AECL began to produce recruitment publicity. Aimed at graduates of chemistry, physics, mathematics, and biology, and chemical, mechanical, and electrical engineering—i.e. the pre-existing disciplines—it highlighted the opportunities in novel areas such as 'operation of nuclear reactors', 'biochemistry of nucleic compounds', 'technology of reactor operations', 'disposal of radioactive wastes', and 'statistical studies on mutation rates'. In particular, the brochures noted that 'reactor research and development presents problems for people with post-graduate training in engineering and nuclear physics and in mechanical engineering'.[138]

'Nuclear engineers' went unmentioned, because there was nowhere in Canada to obtain suitable university training; technical college courses appear to have been equally absent. In 1953 an introductory extension course had been offered briefly at McGill University, but five years later, at Queen's University in Kingston, Ontario (some 375 km from Chalk River and 175 km from Ottawa, the two principal AECL sites), B. W. Sargent launched a one-year course in nuclear engineering. Leading to a Diploma in Engineering (Nuclear), the course centred on 'Nuclear Power Reactors, Nuclear Physics, Heat Transfer, Fluid Mechanics, Stress Analysis, Controls, Safety, Metallurgy, and Corrosion'.[139] Students could select five courses from five domains: physics, chemistry, metallurgical engineering, electrical engineering, and mechanical engineering. Nevertheless, industrial support was diffident. Following a visit to W. J. Bennett, the President of AECL, and to C. J. Mackenzie, one of the Queen's organizers admitted 'feeling rather depressed':

> I gathered that he had forgotten that he had been openly enthusiastic about our proposal a year or more ago. He now feels that good Nuclear Engineers are produced by experience in the field rather than by any formal training. Dr Mackenzie felt...that it would be much easier to obtain support for a programme already under way than to launch a new one.[140]

Not all of these difficulties were attributable to nuclear engineering as a subject. Canadian self-confidence was arguably low. Good students had traditionally pursued advanced degrees in other countries. Post-war Canada had 28 universities and, while a number of them offered Master's level degrees, only two (University of Toronto and McGill University) had active research programmes supporting doctoral degrees. As the pre-war President of the University of Saskatchewan had judged, 'the University has no intention of preparing candidates for the Doctor's degree...It would be folly...to add another feeble graduate school to those that encumber the land'.[141] Nevertheless, as nuclear engineering was being mooted as a new

---

[137] Mackenzie, C. J., LAC RG24 Box 5002 File 3310-50/7. In 1960, AECL further pursued the British model of training by instituting the 'Chalk River Reactor School', open to international students and making 'the basic principles of such reactors available to those qualified engineers and scientists who desire to gain practical knowledge in their design and operation'.

[138] AECL, 'The university graduate and Atomic Energy of Canada Limited', 1959, LAC MG30 B59 Vol. 8.

[139] Sargent, B. W., 'file 4.1 Nuclear engineering', QU B. W. Sargent fonds, Series III Box 4.

[140] Conn, H. G. to P. Mackintosh, letter, 2 January 1958, QU Sargent fonds Series III Box 4.

[141] Murray, W., cited in Preston, Mel A. and Helen E. Howard-Lock, 'Emergence of physics graduate work in Canadian universities 1945–1960', *Physics in Canada* 56 (2000): 153–62; quotation p. 155.

subject area during the late 1950s, the idea of educating home-grown specialists seemed more plausible. By 1960, swollen by war veterans and government funding, undergraduate enrolment and programmes had trebled, but graduate studies lagged behind.

Not until 1961 did the Canadian Nuclear Association (an industry-focused organization rather than a professional body) address technical education and training.[142] It noted that a handful of universities offered relevant courses, but that most were too narrow to accommodate the most suitable candidates (namely graduates of engineering physics), forcing them to enrol in either physics *or* engineering departments. McMaster sought to become the major Canadian university for nuclear research and nuclear engineering training, building the first university reactor in the British Commonwealth.[143] Both McMaster and the University of Toronto adapted by appointing professors of nuclear engineering but building course choices around individual students to meet their particular educational lacunae.[144] Even so, there were fewer jobs in the industry than there were graduates. Courses at Queen's and the University of Ottawa had been suspended owing to lack of demand; indeed, the Queen's proponents had been forced to vacate the building that housed their small 'low energy pile' because no students had registered for the nuclear power engineering course. By 1966 an AECL report noted that 'because there are so few university graduate students with the appropriate kind of academic training and orientation, the need for staff for the design and operation of nuclear reactors is being met at present by the hiring of graduates from foreign countries'.[145] The home-grown field was threatened by imported expertise.

Nuclear training, then, distinguished the former allies. Unlike their British counterparts, Canadian educators defined new specialist programmes unproblematically as 'nuclear engineering'. But they struggled to define the content of their curricula and faced perennial questions of viability. At best, education and training programmes surfaced intermittently to satisfy the unpredictable demand for nuclear workers in Ontario and Quebec. By contrast, American nuclear engineers existed both in name and reality by the end of the decade.

---

[142] The CNA was founded to promote the nascent Canadian nuclear industry rather than to represent its expert workers. In 1979, the Canadian Nuclear Society was established as a 'technical society' of the CNA, incorporating independently in 1998. Its learned society functions did not extend to defining or assessing professional qualifications.

[143] The post-war university had grown from a Baptist liberal arts college and focused on nuclear topics under the influence of Harry Thode (1910–1997), who had been associated with the atomic energy project and subsequently specialized in isotopic chemistry.

[144] Canadian Nuclear Association, 'Survey of Nuclear Education and Research in Canadian Universities 1961', QU B. W. Sargent fonds, Series III Box 4. At the University of Toronto, Boris Davison taught reactor physics during the late 1950s. Davison had joined the Montreal Group in 1944 from the University of Birmingham, moved to Harwell in 1947, and back to Canada in 1954 following security concerns in Britain surrounding his Russian background [Buyers, William J. L., 'Neutron and other stories from Chalk River', *Physics in Canada* 56 (2000): 145–51]. Davison's *Neutron Transport Theory* (Oxford: Oxford University Press, 1957) was a definitive text.

[145] AECL-PD-323, QU W. B. Lewis fonds Box 12 file 11. As late as 1975, the situation had not changed materially: only the University of Montreal had established a Nuclear Engineering Department authorized to Master's level, and the University of Toronto and McMaster University offered nuclear engineering options within their engineering degrees.

7

NUCLEAR SPECIALISTS AT WORK

As to be expected for a field-in-the-making, the protagonists of this story have often been hard to visualize. Chapter 3 explored the ambiguous wartime definition of nuclear specialists. The roles of isolation and key administrators at post-war institutions were the subjects of Chapters 4 and 5, while Chapter 6 surveyed the uncertain disciplinary opportunities that followed political initiatives. But on a more pragmatic level, technical identity was defined in the workplace. This chapter focuses on nuclear specialists via their jobs. During the 1950s, workers from a spectrum of disciplines expanded at sites specializing in reactor development (ANL, ORNL, Chalk River, Harwell, Brookhaven, and Dounreay), plutonium production (Hanford, Savannah River, and Windscale), and power generation (Calder Hall, Shippingport, and Chapelcross).

Not until the early 1960s did nuclear engineers begin to emerge as a recognized occupational speciality. While the American ANS/ASEE committee was agreeing the content of nuclear engineering curricula, two American sociologists, Howard Vollmer and Donald Mills, were exploring what they termed the 'rudimentary professionalization of labor' in the American nuclear industry. In a study of workers at the Shippingport Atomic Power Station and another planned for Indian Point, New York, they identified the development of 'specialized techniques supported by a body of general theory', and the beginnings of community recognition.[1]

In the USA, the nascent nuclear industry shaped disparate working identities. As discussed earlier (Section 5.5), Du Pont's Savannah River Plant based an increasingly refined industrial hierarchy around its nuclear workers and emphasized engineering over science for the first time. And in the first generation of commercial nuclear power stations, craft lines might, or might not, be maintained. For example Commonwealth-Edison, responsible

---

[1] Vollmer, Howard M. and Donald L. Mills, 'Nuclear technology and the professionalization of labor', *American Journal of Sociology* 67 (1962): 690–6; quotations pp. 690 and 691. This appears to have been the first example of consideration by sociologists of this internal aspect of nuclear energy. The wife of a Du Pont Hanford engineer had written her PhD dissertation on the social milieu of Manhattan Project scientists—broken down into 'pure' and 'applied' concerning the atomic bomb—but paid scant attention to engineering perspectives [Smith Stahl, Margaret, *Splits and Schisms, Nuclear and Social*, PhD thesis, Wisconsin (1946)]. A post-war article [Ogburn, William Fielding, 'Sociology and the atom', *American Journal of Sociology* 41 (1946): 267–75], breathlessly forecasting the *social effects* of atomic energy and warfare, urged sociologists to undertake studies of social adaptation for future planning. By 1950 *public resistance* was predicted [Whitney, Vincent Heath, 'Resistance to innovation: the case of atomic power', *American Journal of Sociology* 56 (1950): 247–54], and over the following twenty years attention focused on *public reaction* to nuclear technologies [e.g. Duncan, Otis Dudley, 'Sociologists should reconsider nuclear energy', *Social Forces* 57 (1978): 1–22].

for the Dresden Power Station in Illinois, had trouble establishing craft responsibilities between pipefitters, electrical, and instrument craftsmen. The company combined millwrights, pipefitters, painters, and others in the work of one occupation, labelled a Mechanic A or B, and helper. Electrical craftsmen reported to a master mechanic, and instrument craftsmen to instrument engineers. Commonwealth-Edison allowed either 'qualified' or 'licensed' control room operators to operate any of the controls, including reactor operation. Estimates of suitable numbers also varied. The Yankee Atomic Power Station in Vermont was operating with fewer personnel 'due to the increasing difficulty of the test being given to operators'. Their maintenance personnel were receiving no special craft training, and radiation monitoring and instrument technicians were not formally recognized as a craft. By contrast, Shippingport, Illinois, operated according to Navy conventions which visitors from ANL hinted were unusually relaxed.[2]

The Atomic Energy Commission played a direct role in making this professionalization more uniform via its regulatory powers. As illustrated by the Shippingport and Indian Point specialists, reactor operators and workers using radioisotopes, as well as the facilities in which they worked, had to be licensed by the AEC. As a result, training standards were indirectly promoted. In addition, the AEC had responsibility for regulating working conditions and settling labour disputes in the industry, sometimes in conjunction with Army, Navy, or Air Force influence when work was carried out in connection with defence contracts. This rendered nuclear workers more dependent on their national institutions, and less affected by local employers, than were most industrial workers.

With the emergence of a civilian nuclear power industry, the categorization and training of industrial engineers became more potent issues.[3] The Atomic Industrial Forum, a policy body organized in 1952 to promote nuclear energy, made detailed forecasts of technical labour:

> Industry's needs in the order of decreasing magnitude call for [conventional] engineers, chemists, physicists, nuclear engineers, metallurgists, and other. Institutional needs, as would be expected, are somewhat different: the order being physicists, other, chemists, nuclear engineers, metallurgists and non-nuclear engineers. Contrary to what might be expected, industry does not appear to consider the field to be one of such specialty as to require a greater number of nuclear engineers than other types of engineers, chemists and physicists.

The study observed that growth of the industry would require some ten to twenty thousand more personnel by 1965. Should the numbers graduated and post-graduated in nuclear studies be inadequate to meet these needs, conversion training of personnel from other technical fields was hoped to provide a solution.[4]

[2] DeVoss, H. G., D. C. Keck, G. V. R. Smith and E .W. Wilson, 'Report of trip to Dresden Power Station, Shippingport Atomic Power Station, Yankee Atomic Power Station, Savannah River Plant, and Combustion Engineering, Chattanooga Tennessee', letter, 9 January 1961, DOE DDRS D4763348.

[3] By 1952 the AEC was about equally staffed with scientists and engineers [Glennan, T. Keith, 'The engineer in the AEC', *Bulletin of the Atomic Scientists* 8 (1952): 55].

[4] Atomic Industrial Forum, *A Growth Survey of the Atomic Industry, 1955–1965* (New York: Atomic Industrial Forum, 1955), p. 18. The total engineering requirement of 46.2% compared quite closely with the 52.2% make-up of the professional and scientific staff of AEC and its contractors. The 'other' category for institutions included biologists, physicians, and other skilled professionals in the life sciences.

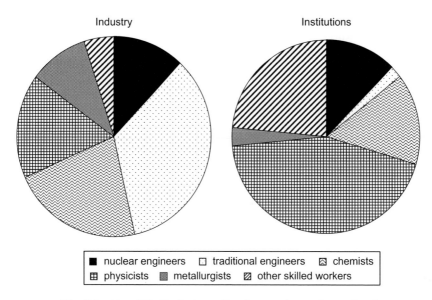

Fig. 7.1 1955 skills distribution attributed to American nuclear workers.[5]

The survey of American personnel (Figure 7.1) revealed variations in the scientific make-up, but remarkably similar proportions of specialists identified as nuclear engineers. However, the 1955 numbers—amounting to scarcely two hundred workers in institutions and a similar number in industry—were predicted to rise to 400 in institutions and 2300 in industry by 1958, as nuclear power expanded. The late 1950s were to see atomic energy activities tilted towards a more frankly engineering slant.

In the UK, the Atomic Energy Authority had similar concerns about nuclear occupations: how were their workers to be recruited for tasks and responsibilities, categorized in the civil service hierarchy, and suitably paid? An even more fundamental question exercising educationalists was: how many professionals would be needed in the coming years? Was the British nuclear programme to be an enduring industry or a post-war government-funded blip in labour statistics?

From 1956, when the Calder Hall power station joined the UK electrical grid, government surveys sought to forecast and plan for the labour needs of the new industry.[6] These predictions challenged the civil servants in the Ministries of Labour and Education, who sought to categorize and count nuclear specialists according to existing categories. All participants had been trained by orthodox engineering courses or apprenticeship; those with experience in the new field were not represented by specific professional institutions or unions. The only occupational and disciplinary pigeon-holes were the pre-war categories.

---

[5] Data plotted from ibid., p. 17.

[6] '1956–1960 Personnel and administration: manpower statistics; Ministry of Labour surveys of qualified scientists and engineers', NA AB 16/2063; '1956–1961 Surveys of apprentice schemes in industry', NA AB 16/343.

As a result, administrative momentum continued to plough a furrow that had been begun before the Second World War.

Attracting a new generation to the field was another matter. By 1959, The UKAEA had adopted a public definition of the nuclear engineer in its promotional literature:

> Fifteen years ago, when Fermi's first pile went critical in Chicago, the nuclear engineer did not exist. When Harwell was set up in 1946 its engineers came from many traditional fields and learned their craft as they went along. In deciding what duty a piece of metal can perform in a reactor, the engineer now has to consider how its atomic structure will change in radiation fields – whether it will become brittle, or will creep more, how many neutrons it will absorb, and how radioactive it will become. Radioactivity calculations and reactor theory are the accepted tools of the trade. It is his job to bring together the theories of the physicist and the professional skill of the engineer so that practical machines and equipment can be designed, built, operated and maintained.[7]

By the following year, the second sentence had been edited to read '...engineers were drawn from many traditional fields to take over this embryo technology from the physicists and to develop it into an applied science...Nuclear energy is such a field in which the engineer faces the challenge of creating a new engineering discipline'.[8] This was considerably more coherent than the ad hoc extension of existing disciplines that had been promoted by Christopher Hinton. For public consumption, at least, British nuclear engineering was now reshaped as an 'applied science' incorporating new physics to create a productive field.

Nevertheless, the UKAEA labour infrastructure embodied Hinton's original organization. According to these criteria, the Authority employed some 2500 technical professionals in 1956, categorized by government pay scales into various disciplines of scientist and engineer (6300 by 1962; see Figure 7.2). The fraction classified as 'scientists' fell from two-thirds of the total in 1956 to one-half six years later.[9] These 'Qualified Scientists and Engineers' (QSEs) were recognized according to the Ministry of Labour by membership in one or another of the recognized institutional bodies.[10]

This labelling of expertise became well established in Britain. Twenty years later the British approach was questioned, but not shaken, in a general review of engineering institutions. Sir Monty Finniston, who as early as 1947 had toured Chalk River, complained that 'to date the nuclear industry has relied to a large extent upon the general stock of QSEs to

---

[7] 'Harwell: Careers in Nuclear Engineering 1959 February' NA AB 17/231.

[8] 'Careers in Nuclear Engineering at Harwell and Winfrith 1960 December' NA AB 17/234.

[9] '1956–1962 Programmes: professional manpower in the atomic energy industry including Advisory Council on Scientific Policy and Ministry of Labour survey.' NA AB 16/1971.

[10] Those labelled 'scientists' were predominantly chemists, physicists, metallurgists, and mathematicians; only four biologists were employed in 1956. The principal engineering disciplines represented were mechanical, electrical, chemical, and civil, respectively; there was no category for nuclear specialists or mention of the Institution of Nuclear Engineers, founded in 1959, as a recognized professional organization. On the other hand, blurring the scientist/engineer divide, the UKAEA tabulated roughly equal numbers of chemists/chemical engineers, physicists, and mechanical engineers, with metallurgists and electrical engineers about one-third as numerous. The QSEs represented 11% and 13% of the UKAEA workforce in 1956 and 1962, respectively [NA AB 16/1971].

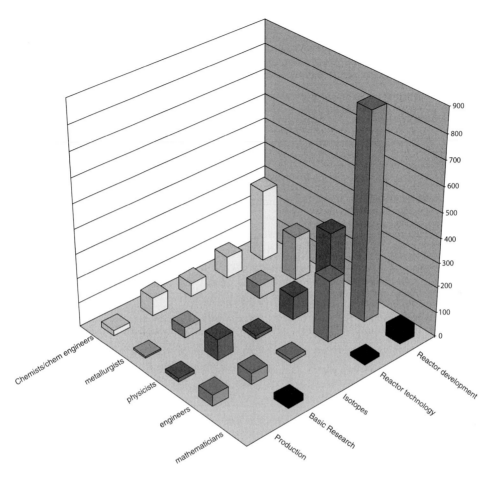

Fig. 7.2 Qualified scientists and engineers (QSEs) in the UKAEA, 1960–61.[11]

provide its engineering proficiencies—i.e. physicists and mechanicals etc have been converted into nuclear engineers according to demand... Indeed, it is only in the past few years that it has been possible to gain a first degree in nuclear engineering at any British university'.[12]

7.1 Nuclear unions: Segregation and identity

As we have seen, the growth of a collective identity for nuclear specialists of all varieties had been opposed by the need for wartime and post-war security. The technologies of nuclear weapons and even atomic energy were to be safeguarded by choosing trustworthy

[11] Data plotted from figures in NA AB 16/1971.
[12] Finniston, Montague, 'Opening Address, Institution of Nuclear Engineers 1977–1978', NA BT 251/198.

scientists, engineers, and skilled workers to contribute to the field. The organization of the post-war research and development facilities followed the wartime model of regional centres working in the national interest. These isolated cloisters bred distinct types of expert and tended to fragment notions of a coherent identity.

For lower-grade specialists such as process workers and technical artisans, labour representation played an important part. The roles of the American, British, and Canadian administrators in defining occupational categories and opportunities for unionization of their nuclear specialists were among the most significant differences between the countries.[13]

Labour laws in each of these countries had recognized the right of workers to unionize and to bargain collectively for improved working conditions and pay. From the 1880s, craft unions had attracted a growing number of workers in industrialized countries. During the First World War, American union membership had increased significantly, in part because of the support for the war effort from the American Federation of Labor (AFL) and by the benign sanction of President Woodrow Wilson's administration, seeking to boost public approval for the war. The interwar period witnessed further labour legislation there, and America's entry to the Second World War was strongly supported by its large labour unions. A twelve-day strike by the United Mine Workers of America in 1943, however, swung public opinion against the labour movement; its most significant consequence was the passing of the Taft–Hartley Act in 1947, which drastically reduced the power of labour unions. From the viewpoint of governments, then, the unionization of labour could effectively map the workforce to national objectives during times of war, but equally could oppose and disrupt those objectives for the more self-interested goals of its members.

7.1.1 *Patriotism and dissuasion in the USA*

Under General Groves, the Manhattan Engineer District had constrained union activities even more than had other wartime secure projects. It placed strict limitations on union activities, establishing special grievance procedures in lieu of public hearings by the National Labor Relations Board (NLRB) and providing its own internal administration of the Fair Labor Standards Act. As Government Owned, Company Operated (GOCO) sites, the major corporations with Manhattan Project contracts followed government practice on labour issues. Construction contractors such as Stone & Webster at Oak Ridge recruited scarce labour via unions in the American Federation of Labour; in return, the AFL required the company to run a closed shop, that is, to employ only union labour.

By contrast, the operations staff presented new problems. Unlike construction workers, the wartime process workers at Du Pont's Hanford piles and plutonium separation plants and at Kellex's Oak Ridge uranium separation plants were privy to classified information,

---

[13] There were equally great differences in labour representation with other countries. In France, for example, unions representing nuclear workers were vocal in orienting them along political and nationalistic lines [Hecht, Gabrielle, *The Radiance of France: Nuclear Power and National Identity after World War II* (Cambridge, Mass.: MIT Press, 1998), Chapter 5].

skills, and equipment to carry out their jobs. Those processes were particularly vulnerable to disruption or damage if, for example, work stoppages or strike actions were taken.[14]

Moreover, most of the operations personnel were not union members, and the new atomic processes did not fit conveniently into existing job categories. And, to further limit possibilities, Du Pont—having full wartime responsibility for Hanford and supplying senior staff to tend the piles—had, in fact, always followed an open shop policy and had never unionized its commercial plants. The same was true of Tennessee Eastman, the contractor for the operators of the electromagnetic plant at Oak Ridge.

What was true for the technical/engineering staff was equally true for the scientists: the universities, acting as contractors for several Manhattan Project sites, were mostly non-union employers. Thus the University of Chicago had an agreement with a CIO union—the State, County and Municipal Workers of America—but did not extend its conditions to its Met Lab or Clinton Lab personnel. In a similar way, the University of California recognized the rights of unions of a regional trade council to set pay scales and conditions for maintenance employees, but the academic and technical staff of the Radiation Laboratory were not affected. Some Berkeley personnel were active in the Federation of Architects, Engineers, Chemists, and Technicians (FAECT).[15] And a number of firms manufacturing materials or equipment for the Manhattan Project were already unionized (including Allis-Chalmers, Chrysler, and Hooke Electrochemical) and so employed union labour on their contracts.

But by the summer of 1944, some unions were pressuring officials of the Manhattan District for bargaining rights over operating personnel. These included the electromagnetic and gaseous diffusion plants at Oak Ridge, and the Hanford plutonium plant. That November, General Groves proposed banning all outside agencies (such as the National Labour Relations Board) and banning all union activity for Project workers. As this did not accord with existing laws, he then suggested allowing union membership only for employees of firms that had signed a secrecy agreement, and which agreed to exclude outside union representatives and to allow inspection and control by the Army, with the Secretary of the Army arbitrating any unresolvable disputes. By December 1944, however, it was agreed that Clinton Laboratories would not unionize during the remainder of the war, and this decision carried through to the other Manhattan Project sites.

Hanford and Los Alamos, in fact, presented few organizational problems for the Manhattan Project administrators. Hanford construction workers, many of them union members, had briefly attempted to organize operating employees but, as the plant history records, 'the Area Engineer promptly intervened, persuading the workers to call off their organizational activities'.[16] Similarly, outside unions requested the War Department for permission to organize Hanford workers but agreed to postpone their effort as long as it would constitute a threat to security.

After the war, particularly with the change of responsibility of site management (Du Pont being replaced by General Electric at Hanford, for example, and Manhattan stepping

---

[14] See Hales, Peter B., *Atomic Spaces: Living on the Manhattan Project* (Urbana: University of Illinois Press, 1997), pp. 169–71.

[15] Strickland, Donald A., *Scientists In Politics: The Atomic Scientists Movement, 1945–46* (Lafayette: Purdue University Studies, 1968), p. 82.

[16] 'Hanford Story, chap 12–14', Hagley 1957 Series V Box 50.

in for the University of Chicago at Clinton Laboratories), there was more manoeuvring room for unionization of atomic workers. At Argonne, the Federation of Architects, Engineers, Chemists, and Technicians attempted to organize workers to improve their working conditions, but despite support from some two hundred employees, the effort was defused by administrators' attention to the grievances.[17] The 1950s saw the peak of unionization in North America. Representation of American and Canadian workers by branches of the same unions had become increasingly common, although the distinct nuclear industries of the two countries were to challenge this.

In this context, then, the notion that competing interests could separate employees from their employer (the government) was a growing concern. Collective representation of workers was doubly disturbing in the post-war period owing to the perceived linkage between calls for workers' rights and socialist or communist influence. For many administrators, unionization represented a loss of control. It threatened to provide a back door for divisive voices, and even for external subversion. At best, the bargaining tool of strikes or other withdrawal of labour would be unpatriotic, limiting production of nuclear materials deemed essential for national defence. At worst, they could inhibit removal of suspect employees and so facilitate espionage.[18] Even more subtly, unions could provide a voice other than the official State interpretation. As explored earlier (Sections 4.5 and 5.4), nuclear workers were sequestered and defined by their governments more than any other profession. Union representation—particularly on the national scale—challenged not just operating costs and working conditions, but the public message about the aims and achievements of atomic energy.

For such reasons, unionization of post-war American workers was initially a cause for concern for their administrators. Having interviewed workers in 1946, agents of the Intelligence Office at the Hanford Engineer Works suggested that there were no grievances requiring collective action but that most employees would find it 'difficult to refuse to join since union representation is now an accepted part of our National Industrial policy'. They judged unions unnecessary, however. The agents reported that the 'benevolent paternalism' prevailing at the site was 'strikingly illustrated by the often repeated query, "What does the government and General Electric wish us to do?"'. The Special Agent responsible consequently suggested that 'a statement [by the War Department or a designated agency] opposing unionization would most certainly end the matter for a decided majority'.[19] In the absence of such a proclamation, though, the American Federation of Labor (AF of L)

---

[17] Strickland, Donald A., *Scientists in Politics: The Atomic Scientists Movement, 1945–46* (Lafayette: Purdue University Studies, 1968), p. 82.

[18] Concerns about unionization were often embedded in the larger-scale governmental distrust surrounding disarmament and espionage, and was affiliated with right-of-centre political leanings.

[19] Daerr, R. L. to Officer in Charge: Intelligence Office, memo, 31 July 1946, Hanford, DOE DDRS D4763417. The Intelligence Branch at Hanford was set up during the war by Du Pont and followed security procedures typical of a secret wartime project, including FBI and credit checks [Mountjoy, P. B., 'Security report on Hanford Engineer Works', memo to District Engineer, Manhattan Engineer District, 4 November 1943, DOE DDRS D4851457]. Transferred to General Electric in 1946, the Branch tightened security after the blanket of wartime secrecy had ended, investigating 'espionage, sabotage, sedition, treason, disaffection, subversion, etc'. It also assumed responsibility for security at the heavy-water plant in Trail, British Columbia [Mountjoy, P. B., 'History of Intelligence and Security Division, HEW', memo to file, 1947, DOE DDRS D4740793].

quietly organized craft workers at the Hanford site—responsible for all American plutonium production—from the summer of 1946. The Vice President of the Chemical Workers' Union argued that, even though wage conditions were currently above some union pay scales, its representation was necessary to prevent the rival CIO union from organizing at the plant.[20] This activity was monitored by the Chief of the Intelligence and Security, and the formation of a union by staff of GE (responsible for managing the site) the following year was communicated up the line to the AEC.[21] In the eyes of the security personnel, unions represented unpatriotic self-interest and a dangerous element of organizational instability. Their Chief reported disappointedly that

> most of the membership [of one of the new unions] do not understand the importance of the Hanford Engineer Works and... are not aware that about a year ago for security reasons President Truman asked unions not to organize, nor that in April, 1946, the Secretary of War requested... the C.I.O. and... A. F. of L. to not organize at Hanford.[22]

Concerned that interruption of strategic work 'would be very serious in terms of the national security', another research manager asked for advance warning of strike action.[23] At the Argonne National Laboratory, the Board of Governors discussed 'the union problem', and forecast that the existing unions might create 'some trouble' when the Atomic Energy Commission Review Panel began dealing with the 'security cases' in 1948.[24]

Hanford, like Oak Ridge and Chalk River, was a relatively isolated site amenable to firm direction by administrators. The 'benevolent paternalism' claimed by security officials was underpinned by the few opportunities for transfer to other Du Pont facilities after the war, and the smooth transfer of responsibilities to the new administration by General Electric. Indeed, the two companies later worked together to restrict inter-site transfers in order to maintain their workforces at the less desirable locations. Local poets recorded the God-like influence wielded by GE manager Mr Milton, at least in economic terms:

> ***HEW Theme Song*** (sung to the tune *Jesus Loves Me*)
> Wilmington's too far away,
> So with the atoms I will stay,
> Richland will be home instead,
> Sand and sage until I'm dead,
> Cause Du Pont left me, yes, Du Pont left me, But G. E. loves me,
> Mr. Milton told me so.
>
> ***9/1/46 Psalm***
> Milton is my shepherd,
> I shall not question,
> He maketh me lie down in Richland

---

[20] Hull, J. W., H. T. Kelley and R. L. Daerr to Officer in Charge: Intelligence Office, memo, 29 July 1946, Hanford, DOE DDRS D4763587.

[21] Daerr, R. L. to Officer in Charge: Intelligence Office, memo, 17 October 1946, Hanford, DOE DDRS D5763387; Mountjoy, P. B. memo to file, 5 June 1947, Hanford, DOE DDRS D4763339.

[22] Mountjoy, P. B. memo to file, 5 June 1947, Hanford, DOE DDRS D4763339.

[23] Suite, C. G. to L. R. Boulware, letter, 5 May 1948, Schenectady, DOE DDRS D3800167.

[24] 'Minutes of Board of Governors' Meeting, ANL', 2 May 1948, UI Box 19.

> He leadeth me beside the Columbia
> He restoreth my job.
> Yes, though I walk through the valley of Wilmington
> Thou art with me.
> Thy insurance and thy housing comfort me,
> Thou prepareth a pension for me in the presence of my creditors,
> Thou anointeth me with soft soap,
> My service runneth over.
> Surely raises and bonuses shall follow me
> All the days of my life
> And I shall dwell in the house of G.E. forever.[25]

At the American national labs, unionization was also seen as a nuisance that could oppose the security investigations of potential and existing staff. At Argonne National Laboratory, for example, the Board of Governors discussed 'the union problem', and forecast that the existing unions might create 'some trouble' when the Atomic Energy Commission Review Panel began dealing with security cases in 1948.[26]

Despite continuing covert surveillance, administrators at Hanford and other production sites cooperated with site unions once they were established. In 1954, General Electric signed a contract with the Hanford Atomic Metals Trades Council, a collection of most of the craft unions on site. The following year, American nuclear workers represented principally by the Chemical Workers' Union combined with those in the oil and chemical industries to form the Oil, Chemical, and Atomic Workers' Union (OCAW).[27] The recognition of this new technical domain was further illustrated in the mid-1950s by OCAW's affiliation with the newly merged AFL-CIO, which also established an Atomic Energy Technical Committee to promote occupational safety.[28]

Thus American nuclear workers' labour representation was distinguished by a clear national slant on professional identity. Relationships between the AEC, its industrial contractors, national labs, and their employees emphasized patriotism and the maintenance of security, and shaped labour policy to ensure continued production on still segregated sites.

---

[25] Hope, Nelson W., *Atomic Town* (New York: Comet, 1954), pp. 107–8.

[26] 'Minutes of Board of Governors' Meeting, ANL', 2 May 1948, UI Box 19.

[27] Representing some 80,000 workers by the end of the century, OCAW remained problematic for successive American administrations. Its then-president was included on Richard Nixon's list of political opponents during the early 1970s, and Karen Silkwood (1946–1974), a chemical technician who gained media attention (and eventually portrayal in the biographical Holywood film, *Silkwood* (1983)) by her investigation of health and safety irregularities at an Oklahoma nuclear plant, was an OCAW member. Membership declined precipitously from the 1980s when oil companies moved to off-shore refining; OCAW merged with the United Paperworkers International Union in 1999 to become the Paper, Allied-industrial, Chemical, and Energy Workers International Union, thereby sacrificing overt connection with the nuclear industry. On labour unions at Oak Ridge, see Olwell, Russell B., *At Work in the Atomic City: A Labor and Social History of Oak Ridge, Tennessee* (Knoxville: University of Tennessee Press, 2004).

[28] The two largest North American labour federations, the American Federation of Labor (AFL) and the Congress of Industrial Organizations (CIO), merged in 1955. Leo Goldman (1910–1982), Secretary of the Atomic Energy Technical Committee 1956–67, was instrumental in the American labour movement in voicing concerns about radiation safety and occupational conditions for nuclear workers.

7.1.2 *British organizations and nuclear subversion*

British labour policy during the same period had a different emphasis. The post-war security regime was an undercurrent that influenced professional identity and labour representation even more overtly than in the USA and focused on the potential danger of professional organizations and unions as harbours of destabilizing political subversion. The UK Atomic Energy Authority, as the near-monopolistic employer and educator of British nuclear workers until the mid-1950s, sought to shape its workforce and its public voice.

The relaxation of nuclear security in 1955 did not dramatically alter security measures for UKAEA workers, whose political affiliations continued to be scrutinized closely. The Atomic Scientists Association, discussed in Section 4.5.2, was to fold in 1957 as a result of its opposition to British H-bomb testing.[29] The collapse was triggered when a subcommittee of the ASA, under Professor Dame Kathleen Lonsdale, a chemist at University College, London, and Professor Sir Joseph Rotblat of Liverpool University, published a report on the medical effects of nuclear fallout.[30] In a radio interview the Conservative Foreign Secretary, Selwyn Lloyd, suggested that 'a great deal of this kind of argument comes from people with strong fellow-travelling tendencies and leanings'. Lonsdale protested at the implication that the ASA was a subversive organization; Cockcroft and the ASA President, H. S. W. Massey of University College, complained to the other Officers at the embarrassment and disturbance caused by MPs and newspapers attributing the findings directly to them, and threatened to resign from the Association. Upon threats of legal action from Rotblat and outraged letters from pacifist groups such as the Quakers, the government issued an apology, while a civil servant noted privately, 'The Atomic Scientists Association, as is normal with associations of scientists, has its share of political innocents (and at least one known Communist sympathizer): but as a whole both it and its Committee on Radiation Hazards are reputable, if somewhat naïve, bodies'.[31] By 1959, however, the pressures dividing public conscience from Authority allegiance proved too much for the ASA: it disbanded, with most of its activities being taken up by the Division for the Social and International Relations of Science of the British Association for the Advancement of Science.[32]

Where the relatively elite atomic scientists were cowed by such clashes, nuclear engineers and technicians were even more effectively controlled. Security concerns continued to isolate British nuclear workers as an occupation and filtered their ranks as an organized

---

[29] On contemporary American concerns, see Kopp, Carolyn 'The origins of the American scientific debate over fallout hazards', *Social Studies of Science* 9 (1979): 403–22.

[30] Joseph Rotblat (1908–2005), working under James Chadwick, was a participant in the Manhattan Project until late 1944, when Germany's defeat appeared certain. He was a long-time advocate of nuclear disarmament and Secretary General of the Pugwash Conferences on Science and World Affairs 1957–73. He was awarded the 1995 Nobel Peace Prize for this work.

[31] 'Apology for impression possibly given by Secretary of State in a BBC broadcast that the Atomic Scientists' Association are "fellow travellers"', NA FO 371/129241.

[32] 'The Atomic Scientists Association Ltd: policy and associated correspondence 1950–1951: 1954–1959', NA AB 27/6; 'Atomic Scientists Association's general correspondence 1946–1951', NA AB 16/52. See also Hodgson, P. E., 'The British Atomic Scientists' Association, 1946–59', *Bulletin of the Atomic Scientists* 15 (1959): 393–4.

workforce. For example, William Griffiths, Labour MP for Manchester, complained in 1960 to the Minister of Science that job advertisements for nuclear workers did not mention security clearance, despite 'the past record of the security services in dealing with persons of "left-wing views"'. The Minister's draft response observed that:

> it was not the practice of the Authority to state in an advertisement whether work at the particular establishment is classified as secret, as they do not wish to reveal where secret work is undertaken... today the chief risk is presented by Communists and other persons who for one reason or another are subject to Communist influence. The great danger, as you know, is that the Communist faith overrides a man's normal loyalties to his country and induces the belief that it is justifiable to hand over secret information to the Communist party or a Communist foreign power.[33]

On a level less apparent to the public, labour unions that were specific to this still-isolated domain also proved unsuccessful. Categorization, shaped by security concerns, restricted the emergence of new occupational labels. The National Union of Atomic Workers (NUAW), founded at Capenhurst to represent non-skilled, non-craft workers in 1958, had branches and over 2000 members at five UKAEA facilities by the following year. It challenged the existing (and Authority recognized) unions, the National Union of Municipal and General Workers and the Transport and General Workers Union, as 'anachronistic' and poorly representative of the skills and environment of its members. The Authority was steadfast in refusing to recognize the upstart group, however, and membership rapidly declined. Its Council decided 'to act on the desire of an overwhelming majority to seek membership of another Union common to our industry'—the Chemical Workers' Union. The NUAW disappeared in acrimony and confused accounts by 1963.[34]

Fifteen years later, security concerns had faded substantially, but a role for unions in consolidating professional identity was no clearer. The Institution of Nuclear Engineers adopted a neutral stance, inviting four unions to make the case for representation. Their arguments followed the conventional lines of job security, negotiation of salaries with appropriate differentials according to skills and training, and the resolution of work grievances. But none of the unions proffered clusters along the lines of disciplinary specialty.[35]

7.1.3 Canada: Occupational labels for all

Canadian activities were distinctly different. Just as the UKAEA had inherited Ministry of Supply practices regarding trade unions, the National Research Council's Chalk River site encouraged its workforce to unionize. By 1947 many of the rate workers (i.e. non-professional tradesmen paid by weekly wage) joined union locals affiliated with the American

---

[33] '1960 Minister's case: security clearance for process workers', NA AB 16/3675.

[34] 'National Union of Atomic Workers 1958–63', NA FS 27/406; '1961 Minister's case: National Union of Atomic Workers, recognition by United Kingdom Atomic Energy Authority.' NA AB 16/3678.

[35] 'Nuclear engineers and the trades unions', *Journal of the Institution of Nuclear Engineers* 20 (1979): 77. The unions representing INucE members included the Association of Scientific, Technical, and Managerial Staffs (ASTMS), the successor to the Association of Scientific Workers; Technical, Administrative, and Supervisory Section of the Amalgamated Union of Engineering Workers (AUEW), the Engineers' and Managers' Association (EMA), and the UK Association of Professional Engineers (UKAPE).

Federation of Labor (AFL). Following chronic complaints about inadequate wages, and following the creation of Atomic Energy of Canada Limited in 1952, administrators made a more concerted effort, and AECL accommodated most of its workers in existing unions of the Canadian Labour Congress.

Unlike the UK, though, the Canadian Atomic Energy Project was not averse to distinguishing its employees with fresh identities. Thus the 'Atomic Research Workers' Union' (1952), 'Association of Atomic Energy Technicians and Draftsmen' (1953), 'Atomic Energy Workers' Union' (1957), and 'Chalk River Nuclear Process Operators' Union' were founded to represent AECL employees. Where British nuclear workers were accommodated as 'general workers', their Canadian counterparts fell into novel and presumably status-bearing categories. The unique activities and specialists associated with the AECL were not in doubt: the Canada Labour Relations Board listed the 'nature of the employers' business' as 'creation of atomic energy'.[36]

Remarkably, though, these seemingly exclusive bodies represented widely disparate skills that were not explicitly tied to the peculiar context of nuclear radiation. They were marked out principally by their circumscribed working locales (initially only at Chalk River, and later with AECL sites at Ottawa and at the Whiteshell Nuclear Research Establishment near Pinawa, Manitoba) but much less by novel job functions. Thus the Atomic Research Workers' Union accepted AECL employees classified as 'bricklayer, painter, stores counterman, labourer, seamstress, laundry operator, process operator and process trainee, maid and animal attendant, excluding foremen, employees of higher rank, salaried personnel, office staff, scientist staff, guards, firefighters and hospital nurses'. The Ottawa Atomic Energy Workers—associated with the AECL Commercial Products Division, which focused on radioisotopes—included 'carpenter, painter, tool and die maker, electronics technician, lead burner, machinist, trades helper, welder, shop boy, labourer, stores counterman, inspector, sheet metal worker, truck driver'—indeed, all those employees who were not numerous enough to be fitted readily into an existing union.[37] AECL, in conjunction with union representatives, was careful not to usurp the territory of existing craft unions. In short, the Canadian labour groupings that were tied most clearly to nuclear craft work had the weakest occupational identity and yet probably the highest prestige to outside eyes (for more on prestige, see Chapter 8).

Such labelling was slower to develop for specialist salaried workers at AECL. The Association of Atomic Energy Technicians and Draftsmen fissioned in the mid-1950s to form a separate craft union for draftsmen. Dissatisfied with their representation by the American Federation of Technical Engineers and impelled by 'a feeling of national pride in the atomic Energy Project and a resultant preference for a Canadian union', the technicians petitioned in 1956 to form a Canadian Association of Nuclear Energy Technicians and Technologists. Their occupational demarcation was clear: they were 'all employed at Chalk River in the following fields of nuclear energy: (1) Biology and Health Physics; (2) Chemistry and

---

[36] 'Atomic Research Workers Union, No. 24291, Applicant—and Atomic Energy of Canada Limited, Respondent', LAC RG145 Vol. 114 File 766:336:52.

[37] 'Ottawa Atomic Energy Workers, Local No. 1541 (CLC), Applicant—and Atomic Energy of Canada Limited, Ottawa, Ont., Respondent (Commercial Products Division)', LAC RG145 Vol. 160 File 766:811:57.

Metallurgy; (3) Physics Research; (4) Reactor Research and Development; (5) Operations Division (Reactors NRX and NRU); and, (6) Engineering'.[38] Through their respective roads to union representation, then, Canada highlighted but did not always clearly characterize its nuclear workers while Britain, on the whole, hid them.

The transition from high security to a more open and visible field, then, was played out differently in each country. Secrecy had been a defining feature of wartime labs and factories; post-war hiring practices had been profoundly shaped by the McMahon Act, McCarthyism, and subsequent legislation; and occupational categories and hierarchies were deeply embedded in the civil service practices of each country. The aftershocks of Eisenhower's Atoms for Peace initiative and the Geneva Conference opened opportunities for nuclear power generation and the growth of related occupations. But security concerns and industrial cultures continued to influence occupational identity: in Britain, the categorization of nuclear specialists continued to be inhibited; in the USA, their identity was shaped by a combination of the AEC, private companies, and the national labs; and in Canada, more cleanly divorced from military applications, nuclear specialists assumed a variety of status-bearing labels.

7.2 Risk and radioactivity

Nuclear workers were also differentiated from those in other fields by the peculiar dangers of radiation.[39] The earliest wartime nuclear workers had been well aware of biological dangers of radiation, but this appreciation extended only gradually to the greater dangers and modes of failure unique to specific types of reactor.[40]

Broadly speaking, the American, British, and Canadian programmes were concerned primarily with local environmental consequences of accidents—principally the danger of explosion or radioactive fallout having consequences within a few miles of the reactor—and secondarily with radiation exposure to its workers. It was expected that inadvertent radiation exposure, like exposure to dangerous chemicals, could be controlled by suitable containment and handling procedures. The earliest policy was a conservative one: the assumption that a reactor, once put into operation, would thereafter become a permanent no-man's land (this assumption also flavouring notions of eventual decommissioning). But

---

[38] 'Canadian Association of Nuclear Energy Technicians and Technologists, Local 1568, CLC, Applicant—and Atomic Energy of Canada Limited, Chalk River, Ont., Respondent (Technicians)', LAC RG145 Vol. 168 File 766:886:58. The identification of biology as a significant discipline also contrasts with British organization of the UKAEA, which employed few biologists. In the later AECL, the principal union representing Canadian nuclear workers (but excluding Chalk River) was the Society of Professional Engineers and Associates. Founded in 1974, the professional society had 1200 members in 2010, some 900 of whom were engineers and scientists.

[39] The role of risk in defining the occupation has been studied for French nuclear engineers, in Hecht, Gabrielle, 'Enacting cultural identity: risk and ritual in the French nuclear workplace', *Journal of Contemporary History* 32 (1997): 483–507.

[40] On the gradual realization of the dangers of radiation, see Caulfield, Catherine, *Multiple Exposures: Chronicles of the Radiation Age* (London: Secker & Warburg, 1989) and Patterson, H. Wade and Ralph H. Thomas, *A History of Accelerator Radiation Protection: Personal and Professional Memoirs* (Ashford, Kent: Nuclear Technology Publishing, 1994).

Fig. 7.3 Hanford workers, February 1954. A contemporary radiologist's report found advances in protective clothing 'disappointing' and 'a tacit admission of the basic failure of engineering in the field'.[41] (Courtesy of US Department of Energy, Declassified Document Retrieval System.)

with the growing complexity of nuclear reactors and their associated instrumentation, manipulation, fuel loading, and monitoring operations, workers became increasingly involved in routine maintenance tasks in potentially high-radiation environments. Fuel reprocessing—particularly plutonium recovery—required even more routine handling of radioactive materials (Figure 7.3). As a result, the occupational skills of nuclear specialists diverged increasingly from those of the professions that had spawned them.

7.2.1 *Hazards of operation*

Not surprisingly, skills in safe nuclear engineering were often informed by predictions or past experience of accidents. Christopher Hinton, in the earliest site planning for a British

---

[41] 'Reactor activities—100 Areas', 1944, DOE DDRS N1D0002151; Parker, M. M., 'Radiation protection in the atomic energy industry: a ten-year review', draft conference presentation for meeting of the Radiological Society of North America, Los Angeles, 7 November 1954, DOE DDRS D198195963; quotations p. 8–9.

production reactor, used the rule-of-thumb advice provided by Ronald Newell as the criterion for the Chalk River location:

> A water cooled Pile may explode with chemical violence and violent dissipation of radioactive substances (i.e. with the force of 300 tonnes of TNT) and for this reason the Canadian site for a much smaller Pile was chosen as 5 miles from a town of 1000 people or 50 miles from a town of 50,000 people. It may be impossible to meet this specification in Great Britain; the conditions can only be approached on a site of the type which may be found in the Western Highlands.

Nevertheless, Newell's estimate had been based on the limited experience of two varieties of reactor: the water-cooled Hanford reactors (about which he had second-hand knowledge, at best), and the recently completed small ZEEP heavy-water research reactor at Chalk River. Hinton was aware of the rather arbitrary nature of these assessments:

> The safety distances which were adopted for the American and Canadian Piles would drive us to the North of Scotland for our site.... if it were decided that it is safe to make considerable reductions in the safety distances for a water cooled Pile then areas other than the North West Highlands are opened up... Urgent consideration should be given to the safety distance specification for a water cooled Pile... If it is not possible to relax the safety distances it may be impossible to find a British site which is suitable for a water cooled Pile.[42]

Hinton recommended Arisaig, in the western Highlands of Scotland, as a suitable sparsely populated site but, following advice from Harwell in June 1947 that an air-cooled design would be superior, the Plant Location Panel was disbanded. Surviving documents do not explain the rapid choice of Windscale—in the much more densely populated English northwest—and the reasons for its purportedly greater safety. Indeed, the summary of a 1948 Technical Committee meeting tersely reports that it 'considered that chance of major catastrophe remote except by enemy action or sabotage. British view that gas-cooled piles were safer than water cooled and that major disaster may affect areas up to 40 miles down-wind' (Figure 7.4).[43]

This concurred with American views. That year, the medical director at Hanford—which had then been operating for four years—noted that disaster plans were restricted to the operating plant and immediate environs, but suggested that the AEC consider the consequences of an accident that would require the evacuation of thousands of residents of the nearby town of Richland.[44] Cloaked from American experiences by the McMahon Act, British accident planning pursued a similarly lackadaisical course.

The Canadian NRX reactor turned out to host the first major nuclear accident: the catastrophic overheating and radioactive breaching of the reactor vessel in December 1952. Chalk River Director C. J. Mackenzie was not deterred by the accident: 'I do not think anyone ever suggested that all attempts to develop aviation would be stopped by the crash of one plane, and I can see nothing in the incident at Chalk River to prevent our getting on with the development of industrial power. In many ways such accidents, although not very pleasant while they are occurring, do provide experience for future designs which could not

---

[42] 'Choice of site for production pile 1946', NA AB 7/157.
[43] 'Production Piles—Air Commodore Rawley's History', NA AB 8/557.
[44] Norwood, W. D., 'Disaster plans', memo, 2 June 1948, DOE DDRS D8417395

Fig. 7.4 Windscale 'general worker' in crash suit, April 1957. Such garb was used six months later by many of the 470 men responsible for the control and clean-up operation following the first widely publicized nuclear accident: A fire and radioactive release from the core of Pile 1.[45] (Reprinted with permission of National Archives (UK).)

be obtained in any other way'.[46] In the same way, early experience with the Windscale piles, despite encountering problems for which there was 'no adequate explanation of what was wrong in the design', did not raise concerns by the operators.[47]

Such incidents did, however, provoke a rare 1953 meeting between British, American, and Canadian engineers focused on reactor safety.[48] At it, Christopher Hinton observed that 'the penalty for failure of a reactor is larger than for any other man made object' and recommended provocatively that, as engineers base designs on experience of failures, 'a series of tests might be planned involving blowing up of reactors'. A Phillips engineer noted that, as the power densities of new reactors were rising, 'the margin of safety is decreasing'. While other participants soothed that their latest safety systems and automatic controls minimized the known risks, Walter Zinn organized experiments at the Idaho Reactor Test Station to test a reactor (BORAX-I) to destruction, demonstrating that the use of boiling water as a moderator and coolant was relatively safe.[49]

---

[45] NA AB 16/340.

[46] Mackenzie, C. J. to C. D. Howe, letter, 15 August 1952, LAC MG30-B122 Vol. 3.

[47] 'Incidents in the life of the Windscale piles, by A. A. Farmer and J. M. Hill, 26 May 1953', NA AB 7/18254.

[48] Lewis, W. R., 'Trip report: UK—US—Canada Reactor Safety Conference', memo, 23 October 1953, DOE DDRS D8490526.

[49] Stacy, Susan M., 'Proving the Principle', http://www.inl.gov/proving-the-principle/, accessed 12 June 2009, p. 260.

In an industrial field still dominated by fundamental scientific questions, the passage of time itself provided some reassurance. At the American Hanford site, 'operated for a period of ten years and accumulating a large number of pile-days without experiencing a pile disaster', an engineer's internal report inferred logically that the probability of another decade of safe operation was 50%.[50] His mechanical judgement overlooked the fact that the original three reactors had risen to eight, and that all were now operating at power levels some ten times higher than during the war.[51] A much more detailed analysis by the Hanford Radiological Department evaluated the most likely scenario: a thermal runaway, in which 'the uranium [will] melt, and the aluminum jackets and process tubes vaporize along with a portion of the fission products', leading to 'a sudden release of radioactive material in the form of hot gases'. The 86-page report estimated the radioactive cloud's dilution in the atmosphere, fallout, lung deposition in workers and residents, water supply contamination, and crop spoilage. Its worst-case estimate suggested several hundred deaths and some six hundred million dollars' damage.

Emphasizing the novelty of this first analysis of what was later dubbed a 'meltdown', the report cautioned that the results should be treated 'with considerable skepticism at the present time because of the unreliable state of our knowledge of the events leading to such a disaster, the dispersion of the material in the atmosphere, and many of the biological data needed to estimate the effects'; even so, the results were 'probably quite conservative'.[52] A colleague, consulting with the Radiation Laboratory at Berkeley, highlighted the more routine but 'peculiar and as yet unresolved dangers attendant to active waste disposal in a metropolitan area'.[53]

But the Windscale Pile 1 accident in 1957 forced greater sensitivity to safety estimates. The UKAEA faced lawsuits from as far away as Spain for the agricultural consequences of radioactive fallout. Following the accident, the Americans provided extensive information about their experiences with the Hanford piles, which until then had been strictly withheld from their British counterparts. Similar experiences of Wigner problems, fuel cartridge ruptures, and other discoveries and malfunctions were discovered on both sides of the Atlantic.[54] The post-mortem enquiries revealed that such an accident at the Windscale piles was inevitable, given the storage of Wigner energy and the lack of adequate design safeguards to make annealing of the piles possible without risking overheating. The accident itself was the source of considerable new knowledge about reactor design for all reactor designers, and

---

[50] Gast, P. F., 'Comment on the probability of a pile disaster', memo, 20 October 1954, DOE DDRS D8372268.

[51] 'The Hanford Atomic Project and Columbia River pollution', memo, 20 December 1957, DOE DDRS D8413586.

[52] Healy, J. W., 'Computations of the environmental effects of a reactor disaster', memo, 14 December 1953, DOE DDRS D198160849. The term *meltdown* was employed in popular accounts by 1955, if not earlier. Arguably some aspects of such scenarios had not been adequately envisaged even at the time of the Chernobyl and Fukushima accidents some 33 and 58 years later, respectively.

[53] 'Report of visits to the Radiation Laboratory, Berkeley, California, Clinton Laboratories, Oak Ridge, Tennessee, and the Argonne National Laboratories, Chicago', memo, 13 July 1956, DOE DDRS D197214145.

[54] Arnold, Lorna, *Windscale 1957: Anatomy of a Nuclear Accident* (London: Macmillan, 1992), p. 95.

one reason that American sharing of information opened up was to gain access to this valuable new information on failure modes.[55] A research programme on graphite properties followed rapidly. Hinton himself noted that Pile 1 would be 'a monument to our ignorance'.[56]

The Windscale accident also hinted at endemic problems for nuclear workers, however. The Authority programme had grown rapidly, and was doggedly challenged by a shortage of workers. The Windscale piles for military plutonium, Calder Hall and Chapelcross as the first UK power reactors, the Dounreay fast reactor to study future design concepts, and Winfrith, a new research site, all siphoned experienced manpower from the small Industrial Group to undertake the duties of operating new plants or managing new staff elsewhere.[57] Sir Leonard Owen noted that 'the [Industrial Group of UKAEA] has always put as the paramount issue the necessity of meeting the Defence Programme. Shortages of men or materials or knowledge were not allowed to jeopardize this. The times given were such that risks had to be accepted'.[58]

The acceptance of such matter-of-fact trade-off between technical goals and safety was difficult to shake off. Describing one recent design, British attendees at a 1956 American conference noted that 'The MIT reactor was described as a...close relation to [Chicago Pile 5, at Argonne National Laboratories] but designed with greater attention to safety due to the built-up location in Cambridge, Mass'.[59]

In his unpublished memoirs, Christopher Hinton justifies the flexibility—and underlying uncertainty of knowledge—in safety estimates in the subsequent selection of Dounreay, at the northern tip of the Scottish mainland, for the first British fast breeder reactor, a design considerably more sophisticated than the Windscale plutonium production and Calder Hall power reactors and that generated much higher temperatures and radiation.[60] The decision was taken by Hinton, James Kendall, his former Filling Factory associate, and John Hill, later Chair of the UKAEA. Supported by Hinton's Production Group, they selected the remote location owing to design uncertainties. His engineering staff estimated that, in the event of an accident, about 99% of the core material would be retained within the containment vessel; when the health physics committee judged this an unacceptably large radiation

---

[55] One contemporary report, which found the British 'gratifyingly cooperative' and suggested that they would achieve 'a substantial lead over the US in the next few years' in some aspects of fuel technology, recommended more cooperation [Bush, S. H. and E. A. Evans, 'Trip to Harwell, November 27–8, and to Risley and Culcheth, November 29 1957', memo, 31 December 1957, DOE DDRS D8298841]. See also Fishlock, David, 'The world's nuclear engineers compare notes', *New Scientist*, 3 September 1964: 551–2.

[56] Arnold, Lorna, *Windscale 1957: Anatomy of a Nuclear Accident* (London: Macmillan, 1992) p. 123.

[57] As new recruit Tony Goddard recalled, 'Winfrith was a fairly raw site in 1960; the physics side was divided into two camps: the graphite reactor outfit who were supporting the design for the Magnox programme, but there was also a little group who had just come from Harwell...I suspect they really did not know what to do with themselves! There were two camps, and a bit scratchy at the interface' [Goddard, Tony to S. F. Johnston, interview, 2 July 2008, London, England, SFJ collection].

[58] Mitchell, D. S. to L. Owen, memorandum, 1 January 1958, NA AB 9/1869.

[59] Codd, J. and R. F. Jackson, 'Notes on the second American Nuclear Society conference, Chicago, 6th–8th June, 1956', NA AB 15/5081.

[60] On the corresponding Soviet breeder reactor programme directed from 1949 by Aleksandr Leipunskii (1903–1972), see Josephson, Paul R., *Red Atom: Russia's Nuclear Power Program from Stalin to Today* (New York: W. H. Freeman, 1999), Chapter 2.

release for the population in the vicinity, the engineers re-estimated the likely containment as 99.9%. Some twenty years after that choice, Hinton argued that 'provided the design of the reactor is done by competent engineers working in an organization which is not complex and in which the ultimate responsibility is clearly defined, I am sure that I should be able to say the same things about it as I said about the first Dounreay Fast Reactor in 1955 and the things that I am saying today about the Prototype Fast Reactor'.[61] In effect, the hardware, the administrative organization, and its expert specialists had to form a reliable technological system. All three components—particularly the specialists—were new and relatively untested in the mid-1950s.

The political and economic pressures to create a rapidly expanding programme both for national prestige and the pursuit of export markets directly affected its technical specialists. Fundamental research and practical implementation tripped over each other in a race simultaneously to discover, apply, and correct. The endemic culture of the British programme short-circuited conventional engineering development. In discussing the Dounreay fast reactors, a researcher noted:

> ideally, each stage in the evolutionary sequence builds upon the experience gained from operation of its predecessor. In practice, the time lag between the conception and commissioning is so long, the rate of technological advance so rapid, and the incentive to establish the fast breeder reactor as the fundamental system for nuclear power exploitation so urgent, that design innovations in advanced units inevitably anticipate operational endorsement...[62]

The engineering complications engendered by the government direction of the nuclear programme in turn had their own political ramifications: inaccurate forecasting of costs and achievements and, even more importantly, lapses of safety that proved just as unpredictable.

7.2.2 Radiation rites

The unmapped uncertainties of early nuclear reactors, and their creation of life-threatening radiation, required and produced a new specialist occupation: the radiation-protection technician. Such specialists were much more numerous than in other industries, and radiation safety was one of the key forms of knowledge required of all nuclear workers. As Crawford Greenewalt of Du Pont had noted in his Manhattan Project diary, these concerns had demarcated the field for Met Lab personnel early on:

> ...doing experimental work now on radiation dose and its effect. Recuperation of dose, if not too large, is rapid. Individuals likely to be exposed are being studied clinically now to establish normal variation before regular exposure to radiation. Individuals differ enormously in the amount of radiation they can stand...Continued observation on exposed individuals is expected to prevent extreme danger.[63]

---

[61] Hinton, Christopher, 'The birth of the breeder', in: J. S. Forrest (ed.), *The Breeder Reactor: Proceedings of a Meeting at the University of Strathclyde 25 March 1977* (Edinburgh: Scottish Academic Press, 1977), pp. 8–13.

[62] Bagley, K.Q. and J. A. Gatley, 'The application of core and fuel performance experience in British reactors to commercial fast reactor design', presented at *International Symposium on Design, construction and operating experience of demonstration liquid metal fast breeder reactors* Bologna, Italy, 1978, p. 2.

[63] Greenewalt, Crawford H., 'Manhattan Project Diary, Vol. I', 1942, Hagley 1889, p. 28.

As early as July 1942, Arthur Compton had established a health group in his Metallurgical Laboratory, the only site then having experience in designing and hoping to operate atomic piles. With participation from X-ray radiologists and Robert S. Stone, a member of Ernest O. Lawrence's cyclotron group at the University of California, the group surveyed available data to set crude exposure limits, information that defined working practices at Oak Ridge and other sites of the Manhattan Project, and the boundaries of a post-war speciality.[64] While much knowledge about biological effects had been obtained from the turn of the century via X-ray sources, and before the war from commercial misuse of radium sources, nuclear reactors and their products yielded incomparably more radioactivity.[65] Reactors were profligate in generating radiation of a variety of types not just from the core, but from materials exposed to it. At Hanford, the plutonium production process in 1944 involved so many unknown radiological hazards that Du Pont insisted that the government establish a $20 million claims fund, to which General Groves readily agreed.[66]

There were distinct problems in assessing and controlling exposure. Each became the basis of an emerging technical specialism. First, as appreciated by radiobiologists, the criteria of danger were only approximately known. The basic nature of the interaction of radiation with living matter became a subject of intense investigation. The operating limits of exposure were based principally on the effects of radium—by then known for a half-century—and on the short-term effects of high radiation dosages from the Japanese bombs. The safe limits still had to be determined from human accidents and animal studies, which were complex and time-consuming. The most obvious danger was presented by the reactor itself. Neutrons emitted in the core to sustain the chain reaction could be lethal if absorbed by a human body, where they acted as a hail of invisible bullets to disrupt the composition of cells.

Radiation and life-forms were incompatible. So-called 'biological shields'—thick walls of concrete, lead, or other dense neutron-absorbing materials—prevented exposure in normal operation, but had to be bypassed or penetrated during repairs. For most nuclear workers, though, the principal dangers related to the chemistry of radioactive species in their environment. There was often a relaxed tolerance for leaks of radioactive species at early power plants.[67]

Even so, the danger was centred on reactors more than on radioactive materials. Uranium, even in its purified form as the radioactive isotope U-235, can be held safely in the

[64] Hewlett, Richard G. and Jack M. Holl, *The New World, 1939–1946* (Berkeley: University of California Press, 1969), p. 206.

[65] On the evolution of legal limits of radiation exposure, see Caulfield, Catherine, *Multiple Exposures: Chronicles of the Radiation Age* (London: Secker & Warburg, 1989).

[66] Groves, Leslie R., *Now It Can Be Told: The Story of the Manhattan Project* (New York: Harper & Row, 1962); Gerber, Michele S., *On the Home Front: The Cold War Legacy of the Hanford Nuclear Site* (Lincoln: University of Nebraska Press, 1997).

[67] Hanford engineers visiting Shippingport, the first American nuclear power plant, noted that there were no filters on the ventilation effluent from the plant (unlike Windscale) and that 'the fact that tritium is present in reasonably significant quantities does not seem to be of much concern to the operating force' [DeVoss, H. G., D. C. Keck, G. V. R. Smith and E .W. Wilson, 'Report of trip to Dresden Power Station, Shippingport Atomic Power Station, Yankee Atomic Power Station, Savannah River Plant, and Combustion Engineering, Chattanooga Tennessee', letter, 9 January 1961, DOE DDRS D4763348, p. 11].

hand. So, too, can plutonium-239 if wearing rubber gloves: its radioactivity is in the form of alpha particles which can scarcely penetrate skin or other solids. However, the dust from these heavy metal elements is incomparably toxic: ingested plutonium or other radioactive isotopes tend to lodge in bodily organs where their radioactivity, over a period of years, is highly likely to cause cancer. Quantities undetectable by conventional chemical tests—and traceable only by their radiation—are of biological significance. The dangers are dramatically increased when either element is available in amounts that approach a critical mass. For plutonium, this amounts to a volume about the size of a grapefruit. Two lumps amounting to this size and brought into close proximity will fission inefficiently, burn, and create lethal dust-laden smoke, as well as generating lethal neutrons and gamma-rays. Because of these factors, the processing of uranium and plutonium—whether for bombs or fuel rods—walks a knife-edge between routine materials processing and criticality crises.

All these concerns were magnified in the context of nuclear reactors (and, to a smaller extent, particle accelerators). Some compounds—wholly unknown before the war—appeared in the form of gases, dust, or liquids produced in separation processes, or by transmutation of elements produced in the reactor. Radiation biochemists were investigating how some isotopes were selectively absorbed by human bone, by the thalamus, or other organs; their location, decay rate, and time of residence in the body could determine the onset of cancers and other complications of concern to radiobiologists. The worrying concentrations varied by many orders of magnitude: while 0.1 part per million (ppm) of lead was a known danger, it was estimated that as little as 10^{-8} ppm of plutonium, or 10^{-11} ppm of certain other fission products, were a concern. The spectrum of such hazards varied from site to site, and depended on the materials employed in reactors and the chemical separation techniques that were used. As a result, workers trained for one operation might be inadequately safeguarded for another. Separation processes were particularly dirty. At Hanford, the extraction of plutonium from fuel rods routinely released large quantities of radioactive iodine and ruthenium gases into the air; the chemical plants and reactors discharged contaminated cooling water and activated products into the Columbia River and local soil; and long-term storage vats for radioactive waste, rapidly but inexpertly assembled to meet production targets, seeped their contents over a period of months and years. In the UK, the Sellafield and Springfield sites were chronic emitters from the early 1950s.[68]

And, disturbingly, hazards were also a matter of bureaucratic construction. A 1963 memo from the Association of Midwest Universities' National Engineering Education Committee (NEEC) complained:

---

[68] Gerber, Michele S., *On the Home Front: The Cold War Legacy of the Hanford Nuclear Site* (Lincoln: University of Nebraska Press, 1997); Warneke, Thorsten, *High-Precision Isotope Ratio Measurements of Uranium and Plutonium in the Environment*, PhD thesis, School of Ocean and Earth Science, University of Southampton (2002). For a summary of the comparable situation in the Soviet Union over the same period, see Egorov, Nikolai N., *The Radiation Legacy of the Soviet Nuclear Complex: An Analytical Overview* (London: Earthscan, 2000).

A common problem cited by representatives of the several universities was the difficulty of negotiation and of maintaining good relationships in matters involving licensing. As one pragmatic participant put it, 'if we notify the AEC about certain incidents, we'll be shut down or, as a minimum, be involved in many months of report preparation; if we keep our mouths shut, we will avoid no end of useless paper work and trouble'. Expressed in more polite terms, the end effect of an extremely legalistic approach to many practical licensing problems is to cover up hazards rather than to promote safety.[69]

Even with growing biomedical knowledge, a practical problem related to methods of quantifying exposure. The wartime American project had relied on the white blood cell count as 'the principal criterion as to whether a person suffered from overexposure to radiation', and the affected individuals 'would be shifted to other jobs or given brief vacations'.[70] Electrometers and film badges were later introduced as physical analogues to record irradiation.

An additional difficulty, as hinted above, concerned the specialists assigned to radiation safety tasks. By the mid-1950s, medical participation had been deemphasized, in part because there had been few incidents requiring direct intervention. Radiation dosimetry became the preserve of radiological physicists, instrument engineers, and analytical chemists. A contemporary report reflected the prevailing view that 'annual physical examinations have their place but probably no more so in a radiation industry than elsewhere' and that

> it was natural that leadership of the radiation protection programs should pass to a group of scientists (principally physicists) versed in the scientific problems involved, and with enough general knowledge of biological and medical procedures to know when and how expert assistance is needed. This led to the golden age of the health-physicist.[71]

The report noted that although during the war 'the qualified leaders were few in number, and came generally from a common background of association with radiotherapy', thanks to formal training in the intervening decade there was now 'an adequate corps of men in the intermediate field of technical attainment' and 'at the lower echelons of control there is a general movement toward the development of a recognized craft of radiation monitoring'. At Hanford, a dozen named subdisciplines had been integrated under the catch-all title of Radiological Sciences. Even so, radiation protection had swung to the incorporation of radiation protection as part of a more general package of health and safety, and even assumed a public relations function. As the author observed, this had professional implications for radiologists: 'the industrial health physicist is limited to a position of moderate stature; somewhere else must be an atmosphere in which leading scientists in radiation protection can develop the field for further advancement'.[72] In short, radiation safety did not have the same cachet as reactor design.

---

[69] Isbin, Herb, 'Licensing and regulation of university research reactors', memo, February 1963, UI Box 100.

[70] Smyth, Henry D., 'A general account of the development of methods of using atomic energy for military purposes under the auspices of the United States Government 1940–1945', 1 July 1945, LAC MG30-E533 Vol. 1, p. VIII–15.

[71] Parker, H. M., 'Radiation protection in the atomic energy industry: a ten-year review', draft conference presentation for meeting of the Radiological Society of North America, Los Angeles, 7 November 1954, DOE DDRS D198195963, p. 2.

[72] Parker, H. M., 'Radiation protection in the atomic energy industry: a ten-year review', draft conference presentation for meeting of the Radiological Society of North America, Los Angeles, 7 November 1954, DOE DDRS D198195963, pp. 11 and 2.

The 1957 Fleck Committee investigating health and safety concerns following the Windscale accident noted that there were few experts in radiological safety within the UKAEA. It recommended training for medical and nuclear safety staff to increase the complement of these specialists. Consequently physicists competent in radiological monitoring (rather than biologists competent in the medical effects of radiation) rose to prominence in the Authority. And just as 'atomic energy' became 'nuclear power', and 'piles' became 'reactors', the terminology evolved here, too: 'radiological safety' mutated into 'health physics' and later 'operational assurance' at UKAEA sites.[73] Thus the dangers of nuclear radiation produced the seed of a new discipline and a more positive label.[74]

Safety was sometimes seen as a nuisance by administrators, but a nuisance that they could define and control. In proposing a training course in radiation safety, the Director of Argonne National Lab wrote to his AEC superior in 1960:

Fig. 7.5 'Irksome duty'. Craftsmen wearing pressurized rubber 'frog' suits for maintenance of the HERMES reactor, 1958, received an extra 7/6d weekly pay.[75] (Reprinted with permission of National Archives (UK).)

[73] Personal communication, Ian Pearson, Dounreay Records Manager, 19 September 2007.

[74] On medical aspects of radiation, see Cowell, S. F. and S. B. Dowd, 'We are what we think we are: Professionalization in nuclear medicine technology', *Journal of Nuclear Medicine Technology* 24 (1996): 336–41.

[75] NA 16/340.

Historically, health and safety regulations have been adopted and enforced by the individual states and municipalities. In time they will include atomic energy. If atomic energy is to grow as predicted, it is clear that competent, well-trained individuals must be available to handle the health and safety problems of nuclear energy at the state and local levels. Excessive cautions and controls, based upon the erroneous concepts of untrained persons, could seriously interrupt the development of atomic energy.

Soundings from state and city public health agencies suggested likely support from them, because pending state legislation related to the control of radioactive materials.[76]

Safety concerns were similarly important in the early UK nuclear programme, but not notably more so than they had been for the chemical industry. Indeed, the post-war Atomic Energy Division inherited procedures and risk assessments from the very different context of wartime chemical factories.

This culture of pragmatic risk assessment was a direct carry-over from the war. Gethin Davey, the Windscale works manager who had managed wartime chemical plants, was cited by one senior colleague as being used to taking too many risks.[77] But described as a 'chemist, physicist, engineer, schoolmaster and local politician, all rolled into one', he was typical of the founding generation of nuclear workers in Britain.[78] As Ernie Lillyman (1925–), an ex-ICI chemist who became a works manager at Dounreay in 1957, observed,

I had operated things akin to [nuclear] reactors, because in my job at ICI I was making a substance called pthalic anhydride, which is made from naphthalene. You pass it over a catalyst at 900 deg centigrade and it splits one of the benzene rings and produces some nasty side products, as well, that get in the way. [The job involved] keeping it running, preventing breakdowns, dealing with breakdowns. Those chemical reactors were liquid cooled – so I had experience with liquid metal [chemical] reactors, mercury cooled with cadmium added.... We were using an explosive mixture: naphthalene vapour. It wasn't particularly pure, and was at a high temperature, and it was an exothermic reaction – so if you got a little bit of impurity it might go on fire, and if it was big it would go bang...There was no automatic shut-off. We could have suffered from mercury poisoning, but we never did.[79]

For such specialists, radiation was a danger comparable to other chemical production plants, and could be handled with equal attention to safety measures (Figure 7.5).

Early Atomic Energy Division documents say relatively little about the dangers of routine exposure to radiation, but there were subtle consequences nevertheless for the organization. Gender differentiated British from American nuclear workers. American practice allowed women to work in process buildings and on monitoring work in the pile building, both of which presented a moderate radiation hazard. On the other hand, the fraction of women in post-war American factories was rapidly declining as post-war culture emphasized the social stability and benefits of being a full-time housewife.

Christopher Hinton appears to have drawn on chemical industry practice here, introducing both equitable and cautious policies concerning exposure. From his first planning of nuclear

---

[76] Hilberry, Norman to M. B. Powers, letter, 24 February 1960, UI Box 85.
[77] Schonland, Basil to J. Cockcroft, letter, 8 January 1958, NA AB 6/1971.
[78] Bertin, Leonard, *Atom Harvest: A British View of Atomic Energy* (San Francisco: W. H. Freeman, 1957), p. 135.
[79] Lillyman, Ernie to S. F. Johnston, interview, 18 September 2007, Thurso, Scotland, SFJ collection.

facilities, he decided that female workers—married or not—were to be actively encouraged at the production sites: 'the houses there are likely to be of small modern types and will not provide full time housewifery work for wives and certainly not for daughters. It is therefore highly desirable that work for as many women as possible should be found jobs in the Factory.' But from animal and human case studies it was known that exposure to certain chemicals, and presumably radiation, during pregnancy could cause birth defects. Hinton correspondingly decided that, 'at least as an initial step', women should not be employed 'in process or other buildings where there are minor risks of exposure to mild radiation in the ordinary course of routine duties'. Not surprisingly, this meant that certain environments became male preserves: process buildings, laboratories, and the Pile, Process and Separation groups. But altering gender roles, it also meant that women were excluded from laundry work, which involved a comparable radiation exposure risk from contaminated clothing.[80]

7.2.3 Risk and identity

Amplified concerns about nuclear safety by the mid-1970s led to a redefinition of the nuclear engineering profession itself. Indeed, Sir Monty Finniston, a former nuclear worker then responsible for a government study of the engineering professions, could state in 1978, 'nuclear engineering, if I come to think about it, is basically concerned with not making the reactor work, but making sure that it fails safe'.[81] Nevertheless, the falling public support for nuclear power was reflected in steadily growing attention to nuclear safety. Publicity peaked periodically concerning the health of nuclear workers, either because of particular accidents [82] or long-term exposure.[83]

The 1979 Three Mile Island accident (discussed in Section 8.2.4) reawakened public concerns about large-scale nuclear accidents. The new focus on reactor accidents is suggested by a shift of terminology: the term 'criticality', which at the time of the Windscale accident meant the operating point at which a reactor achieved self-sustaining and constant fission, came to mean any radiation-release incident, and was the word emblazoned on emergency exit signs at Dounreay.

During the 1990s, most operational procedures were documented and risk-assessed, with severity ranging from 'slightly harmful' ('superficial injuries; irritation from dusts; ill

---

[80] 'Factory planning for production pile 1947', NA AB 7/284. Female nuclear engineers remained as uncommon as in other engineering professions, however, for broader cultural reasons and overt sexism. When the first female graduate member joining the INucE in 1973, the announcement of her entry was accompanied by the information that 'this charming and attractive scientist in our midst is already married' ['From the Secretary', *Journal of the Institution of Nuclear Engineers* 13 (1972): 34].

[81] Finniston, Montague, 'Opening Address, Institution of Nuclear Engineers 1977–1978', NA BT 251/198. On reaction of the European Nuclear Society to public concerns, see Kenward, Michael, 'The nuclear backlash', *New Scientist*, 1 May 1975: 263–4.

[82] 'Plutonium and uranium contamination of atomic workers', NA PIN 20/788.

[83] 'UKAEA Press release: report confirms atomic workers healthy', Dounreay 16071/DJC/SAF(85) P9-1; 'Mortality of employees in the UKAEA, 1946–79 (British Medical Journal 291, 17 August 1985)', Dounreay 16071/DJC/SAF(85)P6-1. See also Walker, J. Samuel, *Permissible Dose: A History of Radiation Protection in the Twentieth Century* (Berkeley: University of California Press, 2000).

Fig. 7.6 Black humour from an isolated occupation: Cartoon from the 'Unclear Times' feature in the Christmas issue of Nuclear Engineering.[84] (Reprinted with permission of © Nuclear Engineering International 2011.)

health leading to temporary discomfort') to 'extremely harmful' (involving 'amputations; major fractures; poisonings, multiple injuries; fatal injuries; life shortening diseases; acute fatal diseases'). Even so, acceptability was determined by the likelihood of occurrence; even 'extremely harmful' activities were not prohibited unless they were likely to reoccur within a year.[85] Medical records of UKAEA employees have been of enduring importance, and are among the longest-preserved and most confidentially maintained in the organization's archives.[86]

Table 7.1 'Glossary of terms specially compiled for the uninitiated'.

Atom	Property of the A.E.A.
Consortia	Uneasy bedfellows
Degradation	Applied research
Diffusion kernel	Army information officer
Enrichment factor	Development contract
Feasibility study	Scheme for raising money
Health monitor	Factory inspector
Information, **classified**	May only be left in locked car
Information, **unclassified**	May be left in unlocked car
Manipulator	Department head
Screening	Contamination detection

Source: *Nuclear Engineering* **1** (1956), 367.

---

[84] *Nuclear Engineering* 3 (1958), 527.

[85] 'Rolls Royce Nuclear Engineering Services Ltd—Plant Operating Instruction—Pressurized Suit Working—Operator', Dounreay 19034/D8550/POI/P006-2.

[86] Personal communication, Ian Pearson to S. F. Johnston, 17 September 2007.

Safety concerns were thus an enduring factor in the design, operation, and forecasting of nuclear facilities. They affected not only the types of specialist and their activities, but also their self-perceptions. A rare practitioners' view was provided by the *Unclear Times* supplement in the first three Christmas issues of *Nuclear Engineering* (Figure 7.6). Its 'Glossary...for the uninitiated' hinted at the occupational, economic, security, and political dimensions that went unmentioned in their popular portrayals (Table 7.1).

Of course, this disjuncture between public representations and workplace perceptions is not unique. As Abbott suggests, public representations of professionals frequently bear little resemblance to their internal hierarchies and occupational problems, permitting 'large sections of a profession [to] be ghettoized'.[87] The cartoon and Christmas articles, published during the brief window between the high security of the early 1950s and the subsequent commercial secrecy of nuclear power, reveal a neglected profession stuck between dangerous working conditions and hierarchical and bureaucratic management.

---

[87] Abbott, Andrew D., *The System of Professions: An Essay on the Division of Expert Labor* (Chicago: University of Chicago Press, 1988), p.121. He cites psychiatrists in mental hospitals and house staff in medicine as examples.

PART D

Representations

'I found a lot of serious-minded young men slipping quietly around in white smock coats writing mathematical hieroglyphics down in books or peering into glass boxes or attentively eyeing electric needles and indicators on vast control boards.'
Ottawa Journal, 1953[1]

'The men and women of the [Atomic Energy] Authority...have, besides all their contributions to national strength and scientific advance, been the first people anywhere to harness atomic energy on an industrial scale to the service of us all.'
Sir Edwin Plowden, Chair, UKAEA, 1956[2]

'[T]he scientists and engineers at Argonne National Laboratory are...engaged in energy-related research and development in many areas *of which nuclear energy is only one.*'
John R. O'Fallon, Argonne National Laboratory, 1982[3]

---

[1] Smith, I. Norman, 'The magic and reality of nuclear energy: a layman takes a look', *Ottawa Journal*, 1953, in LAC MG40 B59.
[2] Plowden, E., Foreword, in Jay, Kenneth, *Calder Hall: The Story of Britain's First Atomic Power Station* (London: Methuen, 1956).
[3] O'Fallon, John R. to B. Piatkowski, letter, 18 October 1982, UI box 71 (emphasis added).

8

UNSTABLE IMPRESSIONS

8.1 Popular atomics and public recognition

Tracking the professional gestation, incubation, and growth of nuclear specialists, as we have seen, is frustrating. Their qualities were variously defined, and even their existence was sometimes disputed. Their circumscribed activities kept them apart from the public and often from each other. Given their cloistered childhood and cosseted upbringing, they were misunderstood and increasingly mistrusted by their contemporaries. Their circumstances were documented poorly, too: media portrayals, like rushed family snapshots, captured odd expressions and uncharacteristic behaviours. Reflecting wider culture, nuclear specialists were fractured into jarring images.

The construction of professional identity is dependent on factors unique to each speciality and closely linked to the cultural embeddings of a field. Previous chapters have shown how the occupational, disciplinary, and professional aspects of nuclear workers were powerfully shaped by the context of their origins. Much of this was indiscernible to those outside the field. By contrast, the wider public entertained a series of perceptions. Some were actively constructed by official information sources; others, paradoxically, were pulled from the vacuum of practical experience. Not surprisingly, these constructed identities clashed. Crucially, though, these public faces acted as a proxy that substituted for the nuclear workers themselves and represented a series of straw men to be successively vaunted, challenged, and discarded in popular culture.

One of the notable features of nuclear specialists, in fact, is that they remained peculiarly voiceless. During the Manhattan Project, nuclear workers were part of a top-secret wartime enterprise: their goals, products, achievements, and identity were hidden. After the Japanese bombings, the sketchy details of the atomic bomb were accompanied by even less information about their creators, apart from the rather unrevealing memoirs of scientists who had conceived its principles. In the post-war period, administrators portrayed nuclear specialists as scientists, and alternately as either guarantors of national status or as powerful but unreliable experts.[1] Popular understandings consequently constructed them as abstruse, enigmatic, and potentially untrustworthy.

Classification meant that nuclear engineers, technologists, and technicians were peculiarly silent during this period. When nuclear *scientists* found a voice—through calls for disarmament and declassification in the American *Bulletin of the Atomic Scientists* and the

---

[1] On portrayals of nuclear scientists rather than their engineering colleagues, see Badash, Lawrence, 'From security blanket to security risk: scientists in the decade after Hiroshima', *History and Technology* 19 (2003): 241–56.

British *Atomic Scientists' News*, for example—they proved easy targets for politicians to make hay in the era of McCarthyism. Not surprisingly, then, popular understandings of the scientists and engineers owed much to governmental views, and vacillated between praise for the hidden problem-solvers who had built the bomb, to mistrust of their shadowy successors.

A closer look is needed, though, to characterize the mixed representations of nuclear experts. As the curtain of secrecy rose during the mid-1950s, these specialists shifted in the popular mind from scientists to engineers. Later still, they were represented as efficient, if largely anonymous, modern technologists. And during the 1970s and beyond, they increasingly were understood as surreptitious and imperfect technicians, often tainted by conspiratorial association with the organizations making up the nuclear complex. Each of these depictions—cascading gradually downwards in intellectual esteem but periodically boosted by expressions of ingenuity—had distinct cultural expressions.

8.1.1 Hidden heroes: The atomic scientists

Public attitudes about nuclear workers immediately after the war were dominated by positive depictions. They owed much to government press releases and magazine stories that followed the Japanese bombings. In the week after Nagasaki, for example, Canada dubbed the members of the Montreal Laboratory 'the largest and most distinguished group of scientists ever assembled for a single investigation in any British country', and backed up the claim by releasing the names of 104 Canadian contributors and the biographies of thirteen senior administrators, scientists, and engineers.[2] The Smyth Report, released a day earlier, had provided details to inform the American general public about what it identified as the 'professional group' of scientists and engineers.[3] Soon published as a book, it remained on the *New York Times* best-seller list for over three months and spawned a flood of popular articles. Picking up the cue of the Smyth report, most stressed the scale of the wartime project and its mysterious expertise, often focusing on key scientists at the crest of this herculean enterprise.[4]

Official attention vaunted other nuclear workers, too. Those who had contributed to the atomic bomb project were feted at least briefly. 'A-bomb' pins were authorized by the War Department for all employees of the Manhattan District, and workers at the most secret of American installations, the Hanford plutonium facility, all received the highest commendation for civilians, the Army-Navy 'E' award, from General Groves. The neighbouring town

---

[2] Department of Reconstruction (Canada), 'Biographical notes', http://www.cns-snc.ca/history/history.html/1945Aug13PressReleasePart3, accessed 23 April 2009, Department of Reconstruction (Canada), 'Scientists who probed atomic secrets', http://www.cns-snc.ca/history/history.html/1945Aug13PressReleasePart2, accessed 23 April 2009.

[3] Smyth, Henry D., *Atomic Energy for Military Purposes: The Official Report on the Development of the Atomic Bomb under the Auspices of the United States Government, 1940–1945* (Princeton, NJ: Princeton University Press, 1945), p. i.

[4] A typical example is 'Manhattan Project: Its scientists have harnessed nature's basic force', *Life*, 20 August 1945: 91–5, 100, 102, 105–6. See also Hecht, David K., 'The atomic hero: Robert Oppenheimer and the making of scientific icons in the early Cold War', *Technology and Culture* 49 (2008): 943–66.

of Richland, Washington attracted the name 'Atom Town', and renamed its annual summer festivities 'Atomic Frontier Days'.[5]

The following year, the Manhattan Project was portrayed in cinema. Benefitting from approval by President Truman and scientific advisors, *The Beginning or the End* exemplified the new expertise through characters playing prominent scientists (Einstein, Fermi, Oppenheimer, Compton, Lawrence), administrators (Conant, Bush, Groves), and unnamed industrialists (representing Du Pont, General Electric, and Union Carbide). Given the sparse historical narrative sanctioned by the Smyth report, such accounts focused unavoidably on the top-down story of key participants marshalling America's industrial might.[6] This perspective was sharpened with the later publication of autobiographical accounts by several of those players.[7]

British nuclear specialists, too, were eager to promote the potential applications of atomic energy. The Atomic Scientists Association organized a travelling 'Atom Train' exhibition in 1948. Visiting twenty-five UK towns over a six-month period, the organizers were surprised by the scale of public interest in the atomic energy exhibition and its accompanying guide book, which became the most widely available publication on the topic in the country. The Ministry of Supply arranged for the loan of the exhibition to the United Nations Educational, Scientific, and Cultural Organization (UNESCO) for a tour of the Middle East.[8] The exhibit presenters became the public interpreters of the technoscientific mysteries, if not the centre of attention themselves. Canadian publicity of the period, such as the National Film Board documentaries *Inside the Atom* (Dir. Jack Olson, 1948) and *Canada's Atom Goes to Work* (Dir. Roger Blais, 1952) provided sober surveys of scientific work and biological applications such as the 'cobalt bomb' for cancer treatment.

These depictions were broadly consistent with the contemporary AEC perspective presented in the American Museum of Atomic Energy at Oak Ridge, set up in 1949 under the

---

[5] Gerber, Michele S., *On the Home Front: The Cold War Legacy of the Hanford Nuclear Site* (Lincoln: University of Nebraska Press, 1997), p. 59; Hope, Nelson W., *Atomic Town* (New York: Comet, 1954).

[6] Metro-Goldwin-Meyer, 'Facts about the making of MGM's remarkable motion picture 'The Beginning or the End', booklet, 1946, LAC MG30-E533 Vol. 1. Unsurprisingly, one reviewer characterized the film as more 'a sentimental glorification of American science and the Army Air Forces' than an objective record of the wartime development of the atomic bomb ['Review: *The Beginning or the End*', *Atomic Scientists' News* 1 (1947): 40]. See also Reingold, Nathan, 'Metro-Goldywn-Mayer meets the atom bomb', in: T. Shinn and R. Whitley (eds.), *Expository Science: Forms and Functions of Popularisation* (Dordrecht: Reidel, 1985), pp 229–45; Haynes, Roslynn, *From Faust to Strangelove: Representations of the Scientist in Western Literature* (Baltimore: Johns Hopkins University Press, 1994).

[7] Fermi, Laura, *Atoms in the Family: My Life with Enrico Fermi* (Chicago: University of Chicago Press, 1954); Compton, Arthur Holly, *Atomic Quest: A Personal Narrative* (Oxford: Oxford University Press, 1956); Groves, Leslie R., *Now it can be Told: The Story of the Manhattan Project* (New York: Harper & Row, 1962).

[8] 'The Atom Train: a successful experiment', *Atomic Scientists' News* 2 (1948): 4–8. Public reactions of the 146,000 visitors (50,000 of whom bought the guide book) were gauged by a random survey. Over 1100 completed questionnaires enabled the organizers 'to allow us to draw conclusions, with an accuracy not much smaller than that of a nuclear experiment' [p.5].

wing of the Oak Ridge Institute for Nuclear Study (ORINS). There, visitors could learn of potential applications but almost nothing about the specialists responsible for them.[9] They were equally compatible with the post-war visions of even the wary Atomic Scientists of Chicago:

> With heat and power supplied by a pile, the rich Alaskan areas might be made liveable and highly productive. Ores need no longer be shipped long distances in their crude forms but their valuable constituents may be refined at the mines...Many economies in transportation of both materials and electric power might be effected by strategic location of atomic power plants.[10]

As the war receded, the Manhattan Project scientists faded further from view, replaced by accounts of their hardware. Thus *Life* magazine based a 1950 article (which, it assured, 'reveals no secrets') on the five-year-old Smyth report and conjectural drawings by its artists. In it, scientists went unmentioned, and illustrations of workers at the Hanford piles were dwarfed by their installations.[11]

Alongside such popularizations were writings for narrower but non-specialist audiences. The American army's *Atomic Energy Indoctrination* (1950), for instance, was an instructors' manual for training soldiers and officers about atomic energy. While classified as 'restricted', much of its sober information concerning blast and biological effects of radiation seeped out eventually to magazine articles and editorials.[12] Five years after the war, then, nuclear specialists were ever more remote.

8.1.2 *A mistrusted elite*

There was consequently a growing taint to representations of nuclear knowledge. Alice Kimball Smith has written of the evanescent political potency of the 'atomic scientists', the term first used by headline writers but quickly appropriated by the specialists themselves.[13] Early popularizations by the Atomic Scientists of Chicago, for example, although only weakly political and widely supported by Met Lab personnel, recommended actions that

---

[9] ORINS, *The Peaceful Atom: American Museum of Atomic Energy* (Oak Ridge: ORINS, 1954). The facility was renamed the 'American Museum of *Science and* Energy' in 1978. This faceless representation of atomic energy is typical of the genre and is paralleled, for example, by its British equivalent at the Sellafield Visitor Centre at Windscale, UK [Jackson, Duncan, 'Bringing technology to the community: Sellafield Visitors' Centre', in: J. Durant (ed.), *Museums and the Public Understanding of Science* (London: Science Museum, 1992), pp 103–7].

[10] Atomic Scientists of Chicago, *The Atomic Bomb—Facts and Implications* (Chicago: Atomic Scientists of Chicago, 1946), p.30.

[11] 'The atomic bomb: how weapon that launched a new age was produced; here is what Americans can and must know about it', *Life*, 27 February 1950, 90–100; quotation p. 90.

[12] Department of the Army, *Atomic Energy Indoctrination* (Washington, DC: US Army, 1950).

[13] She notes, too, that the pre-emption of this label by technical personnel of the Manhattan Project, 'was the subject of caustic comment from other scientists, who pointed out that while many pioneer students of the atom were working on radar or the proximity fuse during the war scores of physicists, chemists, biologists, metallurgists, and engineers with only a peripheral understanding of the nucleus were acquiring distinction as 'atomic scientists' [Smith, Alice Kimball, *A Peril and a Hope: The Scientists' Movement in America, 1945–47* (Chicago: University of Chicago Press, 1965), p. ix].

soon deviated from government policy.[14] In wide-ranging popular articles and appeals to their governments, this elite group argued for the internationalization of atomic energy and the rejection of its military applications.

At the beginning of the new decade, the first Soviet atomic bomb, communist successes in China and Korea, and evidence of espionage within the Canadian, American, and British nuclear programmes raised fears of military use and suspicions about the allegiances of nuclear workers. As detailed in Chapter 4, security—as high as it had been during the war—was increased further. To many contemporaries, their warnings about the dangers of atomic energy and calls for international sharing appeared both naïve and unpatriotic. An example of the fictionalization of this stance was the film *Seven Days To Noon* (Dir. John Boulting and Ray Boulting, UK, 1950), in which a British atomic scientist steals a small nuclear weapon and threatens the destruction of central London if the Prime Minister does not renounce military uses. The motivations of the scientist, judged unbalanced and misguided by his colleagues, are ascribed to his having to work in isolation while carrying untenable responsibilities. Thus scientists' idealism and ideology were more easily accommodated as inexplicable deviations from the norm during a period when social and political conformity were increasingly important.

Already popularly characterized as unfathomable in their intellectual abilities, by the early 1950s the atomic scientists were also castigated as politically undependable and unpatriotic. Nuclear culture of the period was shaped by perceptions of weapons and the fears they engendered.[15] Perceptions of emerging nuclear experts suggested that they had conflicting motivations. Robert Oppenheimer himself faced the House Un-American Activities Committee (HUAC) in 1954, losing his security clearance when links were sketched to communist organizations during the 1930s.[16] This public interrogation— symbolically expressing

---

[14] Atomic Scientists of Chicago, *The Atomic Bomb—Facts and Implications* (Chicago: Atomic Scientists of Chicago, 1946). See also Strickland, Donald A., *Scientists in Politics: The Atomic Scientists Movement, 1945–46* (Lafayette: Purdue University Studies, 1968), which analyses the conflicts within the movement in the USA.

[15] Paul Boyer, *By the Bomb's Early Light: American Thought and Culture at the Dawn of the Atomic Age* (New York: Pantheon, 1985); Paul Loeb, *Nuclear Culture: Living and Working in the World's Largest Atomic Complex* (New York: New Society, 1986); B. C. Taylor, '"Our bruised arms hung up as monuments": Nuclear iconography in post-Cold War culture', *Critical Studies in Media Communication* 20 (2003): 1–34.

[16] Army security mistrust of Oppenheimer had existed since 1943, but the revocation of his clearance owed much to his differences with his former colleague Edward Teller, who had pressed for development of a hydrogen bomb during the war and helped found Lawrence Livermore Laboratory in 1951 to carry this forward. Teller's testimony to HUAC—suggesting that Oppenheimer was loyal but of questionable judgement—advanced Oppenheimer's exclusion from nuclear secrets, but also Teller's own isolation from his scientific peers. From the large industry of Oppenheimer biographies, good accounts are Thorpe, Charles, *Robert Oppenheimer: The Tragic Intellect* (Chicago: University of Chicago Press, 2006); Pais, Abraham and Robert P. Crease, *J. Robert Oppenheimer: A Life* (New York: Oxford University Press, 2005) and Herken, Gregg, *Brotherhood of the Bomb: The Tangled Lives and Loyalties of Robert Oppenheimer, Ernest Lawrence, and Edward Teller* (New York: Henry Holt and Company, 2002).

Fig. 8.1 Guilt by association for atomic workers.[17] (Reprinted with permission of L. M. Norris estate.)

the unspoken—was atypical, though. Oppenheimer's contemporaries, voiceless in public, more frequently faced scrutiny behind closed doors. And fiction mirrored reality: the film *The Thief* (Dir. Russell Rouse, USA, 1952), starring Ray Milland as an American nuclear physicist passing secrets to foreign agents, contains no dialogue at all.[18] Public mistrust, stoked during the McCarthy era, was reinforced by recent memories of the brief period of influence that the atomic scientists had held immediately after the war.

A different public slant on the unreliable scientist was the suspicion that atomic experts were *too* patriotic, in the sense of fulfilling government goals for an arms race. Ironically, this connection was a common one in Canada—which had shunned nuclear weapons development—as it was in the USA and Britain, which were then actively developing hydrogen bombs (Figure 8.1).

At the heart of both perceptions were the suspicions engendered by secrecy, and the worryingly impenetrable attributes of a deep intellect. An early novel by English physicist

---

[17] Len Norris, *Vancouver Sun*, 10 November 1954. LAC 1988–243 DAP 00030.

[18] Because nuclear engineers developed in a corporatist arrangement, a strong and independent voice for the profession never developed. Cinematic depictions retained this focus on silent (or silenced) nuclear experts even through changing professional contexts, as discussed below for the films *The China Syndrome* and *Silkwood*.

C. P. Snow straddled this public/private divide and expressed the strong optimism prevailing in the early 1950s about the new specialists. While hinting at the matter-of-fact security measures and heavy burden of knowledge for those involved in the subject, *The New Men* focused on Cambridge scientists as the creative and reflective breed typical of the new field. Indeed, the book's narrator muses on what he identifies as a scientific–engineering dichotomy and the implications for unreliable science:

> It struck me that all the top scientists... were present, but none of the engineers. As an outsider, it had taken me years to understand this rift in technical society. To begin with, I had expected scientists and engineers to share the same response to life. In fact... [t]he engineers,... who use existing knowledge to make things go, were in nine cases out of ten, conservatives in politics, acceptant of any regime in which they found themselves, interested in making their machine work, indifferent to long-term social guesses.
>
> Whereas the physicists, whose whole intellectual life was spent in seeking new truths, found it uncongenial to stop seeking when they had a look at society. They were rebellious, questioning, protestant, curious for the future and unable to resist shaping it. The engineers buckled to their jobs and gave no trouble... it was not from them, but from the scientists, that came heretics, martyrs, traitors.[19]

In these depictions, and particularly in the American and British contexts, engineers retained a relatively low status and visibility in relation to scientists. The internal categorization of engineers and technicians as supporting actors in these projects was underscored by their relative obscurity in the administration of their organizations and lack of roles as spokespersons for their facilities or the field. These social hierarchies were mirrored imperfectly in wider representations.[20]

Scientists, as the typical administrators and leaders of development groups, more frequently met reporters, explained goals and directed limited tours. Such organizational cues flavoured a wave of popular narratives focused on pen portraits of individuals, usually little-known or selectively anonymous, working in atomic energy. For the uninitiated reporter, the new nuclear sites could be slotted into cinematic stereotypes.[21] Typical of the genre was a Canadian reporter's description of Chalk River, where he found

---

[19] Snow, C. P., *The New Men* (London: MacMillan, 1954). Snow (1905–1980), a physicist and novelist, was best known for identifying a gap between scientific and literary modes of thought [Snow, C. P., *The Two Cultures and the Scientific Revolution: The Rede Lecture, 1959* (Cambridge: Cambridge University Press, 1959)], arguing that this was a long-established cultural division that reinforced mistrust of scientists and engineers by non-specialists.

[20] On the two-way influence between mass media and the public, see Strinati, Dominic, *An Introduction to Theories of Popular Culture* (London: Routledge, 1995); Fiske, John, *Understanding Popular Culture* (Boston: Unwin Hyman, 1989); Klapper, Joseph T., *The Effects of Mass Communication* (New York: Free Press, 1960). On its relationship to science, see Nelkin, Dorothy, *Selling Science: How the Press Covers Science and Technology* (New York: W. H. Freeman, 1995) and Lewenstein, B. V., 'Science and the media', in: S. Jasanoff, G. E. Markle, J. C. Peterson and T. J. Pinch (eds.), *Handbook of Science and Technology Studies* (Thousand Oaks, CA: Sage, 1995), pp 343–60.

[21] See Weart, Spencer R., *Nuclear Fear: A History of Images* (Cambridge, Mass.: Harvard University Press, 1988).

a lot of serious-minded young men slipping quietly around in white smock coats writing mathematical hieroglyphics down in books or peering into glass boxes or attentively eyeing electric needles and indicators on vast control boards. Others were poring over textbooks and still others, more power to them, were just sitting thinking.[22]

8.1.3 Ingenious engineers

The common characteristic of such portrayals is the identification of scientists as the drivers—for good or ill—of atomic energy developments. There is clearly some truth in this, as novel and unintuitive scientific concepts motivated the fission reactor, separation of radioactive isotopes, and the creation of a military explosive. Of equal importance, though, even during the war years, were engineering skills and innovation and the enduring reliance of nuclear work on the empirical characteristics of materials. Indeed, even in the distinct domain of weapons design, computer codes embodying theoretical knowledge still struggle to simulate weapons tests, and have not convincingly replaced them.[23] The boundary between scientists and engineers was, of course, artificial and permeable—supported, as we have seen, by job categories, educational specialities, and occupational prejudices—but enduringly important in public representations.

The public recognition of nuclear engineers grew as secrecy declined. Following the first Geneva Conference on the Peaceful Uses of Atomic Energy in 1955, the most senior of the shadowy nuclear specialists were for the first time identified and feted in Britain (e.g. Figures 5.2 and 8.2) and even the nameless technicians and engineers behind the first British research centre and power station at Harwell and Calder Hall, respectively, found themselves the subjects of publicity photographs and praise.[24] The new journals and popular magazines began to publish what soon became the standard iconography depicting besmocked nuclear workers tending to their specialized equipment—fuel assemblies, radiation monitors, remote actuators, and control panels. Contemporary newsreel and television images reinforced such descriptions by illustrating unidentified white-coated nuclear workers as backdrops for voice-over commentary.

After the starvation diet of nuclear news over the previous decade, a wave of magazine articles consumed the information now flowing from official sources.[25] International

---

[22] Smith, I. Norman, 'The magic and reality of nuclear energy: a layman takes a look', *Ottawa Journal*, 13 November 1953, in LAC MG40 B59.

[23] Masco, Joseph, 'Lie detectors: on secrets and hypersecurity in Los Alamos', *Public Culture* 14 (2002): 441–60.

[24] E.g. Bertin, Leonard, *Atom Harvest: A British View of Atomic Energy* (San Francisco: W. H. Freeman, 1957), an account based on 'the words of the scientists and engineers interviewed' and 'a host of lesser known people who have played vital parts in various ways' [dustjacket].

[25] Not all were impressed, however. As one British trade article reported in 1956, 'After ten years of rather awesome mystery the gates of Harwell were opened to the Press.... In general the reaction of the lay Press has been one of profound disappointment.... it must be remembered, however, that Harwell is essentially a research organization and that it is concerned entirely with matters highly technical in which the general public would have little interest and even less understanding'. [Editorial, 'The press at Harwell', *Nuclear Engineering* 1 (1956): 138].

MEN OF CALDER

Many thousands of men, both from the United Kingdom Atomic Energy Authority and from British industry, contributed in amounts large and small to the building of Calder Hall power station. It would be wrong to single out any one, for throughout they worked as a team. Nevertheless, NUCLEAR POWER is proud to record the contributions of some of the leading scientists and engineers of the Authority from the Industrial and Research Group.

Atomic Energy Research Establishment Harwell

 SIR JOHN COCKCROFT, KCB, CBE, FRS, Director, Atomic Energy Research Establishment

 DR J. V. DUNWORTH, CBE, Head of Reactor Division

 DR H. M. FINNISTON, Chief Metallurgist. Responsible for basic work on fuel element design

 DR P. FORTESCUE, formerly in charge of the Engineering Laboratory. Responsible for basic work on heat transfer problems in reactor design

 MR B. L. GOODLET, OBE, formerly Deputy Chief Engineer. Responsible for supervision of feasibility study

 MR C. A. RENNIE, formerly Deputy Head of Reactor Physics Division. Responsible for theoretical physics advice on reactor design

 DR R. SPENCE, CB, Chief Chemist. Responsible for co-ordination of basic chemical studies involved in reactor design

 DR D. TAYLOR, Head of Electronics Division. Responsible for development and production of nuclear instruments for reactor control

 MR H. TONGUE, CBE, formerly Chief Engineer

Industrial Group Risley

 SIR CHRISTOPHER HINTON, FRS, Managing Director, Industrial Group. Overall responsibility for design, construction and operation

 MR W. L. OWEN, CBE, Director of Engineering. Responsible for the overall design and construction under Sir Christopher Hinton

 MR P. T. FLETCHER, Deputy Director, Engineering. Mr Owen's Deputy. He has, for some time, acted as Chairman of the Progress Meetings held monthly to review construction

 MR J. B. W. CUNNINGHAM, Deputy Director, Reactors. Was, for a considerable period, in charge of the design office which designed the Calder Hall plant

 MR R. V. MOORE, GC, Assistant Director, Civil Reactors. Was concerned at Harwell with the early studies on power producing reactors. During a long period of illness of Mr Cunningham, he took charge of the project.

 MR T. L. VINEY, Deputy Director, Works and Buildings. Responsible for the architectural aspects of the design and recently for all construction work

 MR E. L. ASHLEY, Chief Engineer, Construction. Since 1954 he has had the responsibility for liaison between design and construction staff

 MR J. D. GLANVILLE, OBE, Chief Engineer, Inspection and Progress. Responsible for the inspection of equipment and ensuring that it was delivered to site in conformity with the programme

 MR K. B. ROSS, Director of Operations. Has overall responsibility for all factories

 MR D. W. COLE, Deputy Director of Operations, Engineering. Is responsible under Mr Ross particularly for the engineering side of the factories

 MR H. G. DAVEY, Works General Manager, Northern Area, has charge of the factories at Windscale, Calder Hall and Chapelcross

 MR K. L. STRETCH, Works Manager, Calder Hall. Is responsible to Mr Davey for the commissioning and operation of the plant.

 MR L. ROTHERHAM, Director of Research and Development. Is responsible for all research and development work carried out in the Industrial Group

 MR L. GRAINGER, Chief Metallurgist. Was until recently in charge of the laboratory in which the development work on fuel elements for Calder Hall was carried out

Fig. 8.2 The hierarchy of a profession: The two dozen 'Men of Calder', with the Harwell Research group 'over' the Risley Industrial group. The title consciously evokes the virtues of perseverance and nationalism expressed in the stirring Welsh military march *Men of Harlech*.[26] (Reprinted with permission of the publisher, David Rowse, *Nuclear Power*.)

[26] *Nuclear Power* 1 (1956), 275.

cooperation and civilian applications of atomic energy were the new focus; as a result, the public identities of nuclear specialists were rehabilitated. A 1955 *Life* magazine feature described 'young engineers' as the 'brains of the atomic future', uniting 'old hands and young geniuses'.[27] Along similar lines was the British popular book *Atomic Energy for All*, and its chapter 'Harwell: City of Giants'. Like other popular accounts, it narrated the rise of the new industrial specialists. Vaunting the research centre as the source of the 'bright ideas' behind Windscale, Calder Hall, Dounreay, and British nuclear submarines, it stressed the need for 'training in atomic engineering' if research was to do any good.[28] As noted by Paul Josephson, Soviet journalists delivered the same message, but with a different interpretation: the first generation power reactors at Beloiarsk and Novovorenezh were part of a 'great industrial experiment' founded on the cooperation of scientists and engineers, where 'science and construction, research and production, were joined in a scale and energy impossible in capitalism'.[29]

Accounts of Atoms for Peace also foregrounded not science, but technological innovation:

> Every artisan and technician in the atomic field will have to be more skilled and better informed than artisans are now. He will have been to special schools... for the demands of atomic energy are great and the penalty for failing them may be death.
>
> Engineers with a flair for "cooking up" new machines will invent better methods of mining and refining the atomic metals, new ways to make heavy water, new cooling, new heat exchanging, new pumps. Let there be a great discovery... and the field will become unlimited.[30]

In a similar vein and also with official underpinning, *Our Friend the Atom*, launched as a 1957 Walt Disney book and television film, beckoned to a world transformed by the civilian atomic age.[31] The original cartoon project had been suggested by Abbott Washburn, Director of the United States Information Agency, in late 1955, but was founded on strong corporate support. The film was produced with the cooperation of General Dynamics (responsible for the *USS Nautilus* nuclear submarine) and the US Navy, and was followed by an 'atomic submarine' ride in the new 'Tomorrowland' section of Disneyland. The collaboration proved profitable for Disney and simultaneously promoted a beneficent corporate image of applied (and commercialized) atomic energy. As Sophie Forgan has underlined, a recurring theme of the 1950s was the presentation of pure science as detached from its application.[32] By extension, such presentations provided a clear demarcation between scientist-creators and engineer-maintainers.

On the other hand, a more democratic view was provided by leisurely journalistic accounts. Typical among them were the stories of Daniel Lang, who portrayed a collage of

---

[27] 'Young engineers usher in a new era: atomic industry is already big', *Life*, 8 August 1955, 17–22.

[28] Chapman, Robert, *Atomic Energy for All: A Layman's Guide to the Atom and its Uses* (London: Oldhams Press, 1960); quotation p. 90.

[29] Josephson, Paul R., *Red Atom: Russia's Nuclear Power Program from Stalin to Today* (New York: W. H. Freeman, 1999), p. 32.

[30] Woodbury, David O., *Atoms for Peace* (New York: Dodd, Mead & Co., 1955), quotation pp. 246–7.

[31] Haber, Heinz, *The Walt Disney Story of Our Friend the Atom* (New York: Simon & Schuster, 1957).

[32] Forgan, Sophie, 'Atoms in Wonderland', *History and Technology* 19 (2003): 177–96.

earnest but pragmatic scientists, engineers, production workers, and collateral population working at AEC sites such as Oak Ridge, all adapting quickly to new occupations and responsibilities.[33] Among them were seeming incongruities such as a physicist who became an ordained Episcopal deacon, and an engineer content to ask no questions about the goals of his work. Lang's narratives sought to rehabilitate those specialists widely regarded as 'a remote and menacing breed of intellectual' into 'latter-day Merlins...burned by the strange, inextinguishable fire they have learned to create'.[34] While these sketches breathed life into nuclear specialists, they often framed them as components in a complex and hidden system, and offered identities constructed from binary opposites and paradoxes. A commonly presented perspective was that of bemused and inexplicable routine in a fairy-tale world, echoing the sentiments of Manhattan Project administrator James Conant, who had described it as a 'rather strange journey through "Alice in Wonderland"'.[35]

These representations were mirrored increasingly by official views, too. The growing prominence of engineers after the Geneva Conference shifted attention away from the contentious post-war atomic scientists; the promotion of nuclear energy vaunted the safe and routine application of nuclear knowledge. Norman Hilberry, successor to Walter Zinn as Director of ANL, stressed teamwork, rather than scientific genius alone: 'just as one achieves large power outputs by resorting to dilution within the reactor core, so in a laboratory one supplements the endeavours of the brilliant with the contributions of highly qualified yet less creative scientists and engineers'.[36]

David O. Woodbury, a popular writer specializing in American science, more positively identified ingenuity and courage as necessary attributes of the new nuclear engineers. He observed of the first experimental civilian reactor being built at Shippingport, Pennsylvania, that

> the plant will be engineered and supervised by experts who have had experience with atomic piles, but it will be full of innovations nevertheless and will be operated by technicians and artisans who are learning as they go. If an emergency arises no conventional procedure will be there to fall back on; remedies will have to be invented on the spot.[37]

---

[33] Lang, Daniel, *Early Tales of the Atomic Age* (New York: Doubleday, 1948); Lang, Daniel, *The Man in the Thick Lead Suit* (Oxford: Oxford University Press, 1954).

[34] Lang, Daniel, *From Hiroshima to the Moon: Chronicles of Life in the Atomic Age* (New York: Simon & Schuster, 1959), quotation p. 247. See also Lang, Daniel, *Early Tales of the Atomic Age* (New York: Doubleday, 1948) and Lang, Daniel, *The Man in the Thick Lead Suit* (Oxford: Oxford University Press, 1954).

[35] Conant, J. B. to C. J. Mackenzie, letter, 24 Aug 1945, Library and Archives Canada RG77 Vol. 283. For example, Hope, Nelson W., *Atomic Town* (New York: Comet, 1954) describes the adaptation of locals to the Hanford plutonium production project. In the same vein but adopting a more critical tone were Caulfield, Catherine, *Multiple Exposures: Chronicles of the Radiation Age* (London: Secker & Warburg, 1989); Hall, Jeremy, *Real Lives, Half Lives* (London: Penguin, 1996).

[36] Quotation from Whyte Jr, William H., *The Organization Man* (New York: Simon & Schuster, 1957) p.235, reproduced in Westwick, Peter J., *The National Labs: Science in an American System, 1947–1974* (Cambridge, Mass.: Harvard University Press, 2003), p. 65.

[37] Woodbury, David O., *Atoms for Peace* (New York: Dodd, Mead & Co., 1955), p. 23. See also Hewlett, Richard G. and Jack M. Holl, *Atoms for Peace and War, 1953–1961* (Berkeley: University of California Press, 1969), pp. 196–8.

Woodbury praised the bravery and resourcefulness that Canadian workers at the Chalk River NRX reactor had shown in 1952, after it had disastrously overheated and released thousands of gallons of radioactive heavy water 'loaded with...seven times the total activity of all the radium produced since the Curies discovered it'.[38] Indeed, public awareness of nuclear specialists—as engineers and technicians rather than as scientists—was made more acute by accidents. But even these incidents of engineering failure could serve to vaunt professional identity: selfless heroes emphasized the values of professionalism, and narratives of quick-thinking workers evoked public admiration.[39]

In Britain, where nuclear installations were unusually close to population centres, Windscale repeatedly altered public perceptions. An engineering problem in the spring of 1955 involving a broken fixture and stuck fuel elements was recounted a year later in a series of local newspaper articles as a tale of heroism, cool thinking, and British pragmatism.[40] A similar portrayal of 'tense and methodical' American atomic engineers disarming an atomic weapons test at the Nevada Proving Ground thrilled readers of *Life* magazine in 1957 (Figure 8.3).[41] This convention—portraying nuclear workers as cooly and even heroically competent—fitted neatly with the serious nature of safety concerns surrounding radiation. The career trait of heroism proved to be an enduring stereotype, merging well with post-war public perceptions of nuclear patriotism but sitting more awkwardly alongside notions of genius, impenetrability, and mistrust. In the following years, new contexts periodically revived each identity to cloak nuclear specialists anew.[42]

At times, these popular understandings were actively constructed. After the much more dramatic core fire at Windscale sixteen months later, American and British engineers paid close attention to the public relations aspects at their first post-accident briefing. The meeting minutes noted that 'this incident is probably the first serious blow to the public's picture of the AEA as an invincible group of scientists' but that press conferences had been 'important in controlling any hysteria in the country'.[43] Media representations, they noted, shaped the public interpretations of nuclear specialists and their field.

---

[38] Ibid., p. 139.

[39] See Abbott, Andrew D., *The System of Professions: An Essay on the Division of Expert Labor* (Chicago: University of Chicago Press, 1988), p.121; professional strains and workplace realities are commonly absent from public discourse.

[40] Chapman, Robert, 'Crisis at Windscale', *Manchester Evening News*, 26–7 June 1956: 8–9. The four-part serial on the episode was No. 221 of a series titled 'The World's Strangest Stories'. The episode, downgraded to a 'mishap', was praised because 'not one of the 251 men who went behind the biological shield suffered in any way from the effects of radiation', in Chapman, Robert, *Atomic Energy for All: A Layman's Guide to the Atom and its Uses* (London: Oldhams Press, 1960); quotation p. 104.

[41] Wainwright, Loudon S., 'The heroic disarming of Diablo—Atomic engineers make a suspense-laden climb', *Life*, 16 September 1957: 133–6, 141–4.

[42] On heroism and mistrust at Three Mile Island, Chernobyl and Fukushima, see Sections 8.2.4 and 8.2.6.

[43] Healy, J. W., 'Report of trip to Harwell to review health and safety aspects of the Windscale incident', memo, 12 November 1957, DOE DDRS D8553051.

Fig. 8.3 'Atomic engineers' in the 'heroic disarming of Diablo', *Life* magazine, 1957.[44] (Reprinted with permission of the Walter du Bois Richards estate.)

[44] Illustration by Walter du Bois Richards, in Wainwright, Loudon S., 'The heroic disarming of Diablo—Atomic engineers make a suspense-laden climb', *Life*, 16 September 1957: 133–6, 141–4.

8.1.4 Role models

Portrayals of dedicated, patriotic, and brave nuclear workers had the potential not merely to calm public concerns, but to inspire and recruit a new generation. Some of the earliest popular expressions of the new skills were more frankly fictional. From the late 1940s, the science fiction writer Robert Heinlein planned a series of novels for boys modelled on the *Tom Swift* adventure series.[45] The first, *The Young Atomic Engineers and the Conquest of the Moon*, was published as *Rocket Ship Galileo* by Charles Scribners. Although stripped of the label 'atomic engineers', Heinlein's juvenile series published between 1947 and 1958 nevertheless detailed the adventures of technology-aware boys of the future. Like the *Tom Swift* stories, which featured father and son inventors, Heinlein's protagonists portrayed an adventurous breed of practical men.[46] Reflecting the end-of-war context, the teenaged characters drew upon advances in rocketry, the electrical engineering and 'practical atomics' experience of one boy's father, and the nuclear physics of another's uncle. Atomic energy was easily allied not only to the wartime technologies of rockets and radar, but also to futuristic hopes for transportation, miracle agriculture, and clean cities.[47]

Following the mid-1950s security thaw, it became possible to sell nuclear enthusiasms as fact rather than fiction. American children's books began to communicate the adventure of the new field as a modern profession. A 1957 book from the publishers of *Popular Mechanics* magazine, for example, guided its young readers to scrape the luminous radium paint from alarm clock dials to undertake nuclear experiments suitable for science fairs. A later chapter, 'Shall I be an atomic scientist?', provided a two-page listing that gave 'a run-down on the different jobs in the atomic energy field' ranging from technologists to research engineers.[48] For young American men, the more pragmatic *Atomic Energy in Industry*

---

[45] The original Tom Swift series (1910–41), devised by American Edward Stratemeyer, focused on a young inventor and the adventures with exciting transport and electrical technologies. A second series (1954–71) featured his son, Tom Swift Jr, and updated inventions. Its 33 titles included plots based on an 'atomic earth blaster' (1954), 'outpost in space' (1955), 'caves of nuclear fire' (1956), 'space solartron' (1958), and 'electronic retroscope' (1959). See Von der Osten, Robert, 'Four generations of Tom Swift: ideology in juvenile science fiction', *The Lion and the Unicorn* 28 (2004): 268–83. See also Szasz, Ferenc M. and Issei Takechi, 'Atomic heroes and atomic monsters: American and Japanese cartoonists confront the onset of the nuclear age, 1945–80', *The Historian* 69 (2007): 728–52.

[46] Heinlein's novel plans suggest his own technological forecasting that merged atomic energy with space travel: *Young Atomic Engineers on Mars, or Secret of the Moon Corridors; Young Atomic Engineers in the Asteroids, or The Mystery of the Broken Planet; Young Atomic Engineers in Business, or The Solar System Mining Corporation*. Incidentally, the term 'atomic engineer' never became popular in the North America or Britain but has retained some currency in other countries, notably India and the Middle East. The label has been applied increasingly since the 1990s to the distinct field of nanotechnology.

[47] On the public face of the technology, as opposed to its specialists, see Del Sesto, Stephen L., 'Wasn't the future of nuclear energy wonderful?' in: J. J. Corn (ed.), *Imagining Tomorrow: History, Technology, and the American Future* (Cambridge, Mass.: MIT Press, 1986), pp. 58–76 and Forgan, Sophie, 'Atoms in Wonderland', *History and Technology* 19 (2003): 177–96.

[48] May, Julian, *There's Adventure in Atomic Energy* (New York: Popular Mechanics Press and Hawthorne Press, 1957), p. 156. This was republished for British audiences by Bailey Bros & Swinfen, London, in 1963.

offered a first glimpse of the variety of craft trades and technical careers that were becoming available. It highlighted the current shortage of labour and the adaptability of existing skills—radiography, metalwork, instrument mechanics, and drafting—to the new field. The mysteries of radioactivity and potential heroism were sidelined as career motivations: the book laid out the practical facts of radiation exposure in just the way that the dangers of electrical current had been described to earlier generations of *Tom Swift* readers.[49] Such matter-of-fact attitudes were quickly absorbed; cultural geographer Joy Parr argues that, for Canadian nuclear power workers, a sense of radiation dangers became inculcated and unremarkable as their confidence grew.[50]

By the early 1960s the UKAEA, too, sought to represent nuclear power as a career that was both modern and accessible to young Britons. The most ambitious project, an extended career guidance book, leads its adolescent readers through a visit to a UKAEA exhibition ('under the guidance of a smiling young Authority scientist'), after which its fifteen year-old principal character, Alan, decides to leave school to pursue an industrial route, combined with college courses, as a technical apprentice in the field. Guiding his decision were the confident and inspirational members of the first generation of nuclear specialists. Listening to the UKAEA demonstrator explain the 'fail safe' principle of design,

> Alan smiled to himself. There was something novel about this justification of ugly jargon which appealed to him. It was the voice of the practical man, of the engineer. And he knew what he wanted to be. A nuclear engineer. A man who helped to build atomic power stations... He saw already that it was like being a surgeon, where you had to learn to be a doctor first. He would have to learn to be an engineer and then specialize in atomic energy. But he had no doubts now. Alan Martin, Reactor Engineer.[51]

The British slant on nuclear expertise as a topping-up exercise, rather than a discipline in its own right, was reiterated through the book. Interviewed for the apprenticeship, Alan explains:

> I wanted to become a nuclear engineer after I saw an exhibition... it was a fine exhibition, sir, it made it all jolly interesting.
> Well, there you are. You want to come here. It's not just that you want a job in engineering. You want to be a nuclear engineer. I suppose you know that it's quite a long road ahead of you. You've got to be an ordinary engineer before you can specialize and become a nuclear engineer. In fact, you have to become a mechanical engineer first under the training system we run here.[52]

The story explains two other practical routes to 'working with atoms': one was via a *craft* apprenticeship, to learn traditional trades such as electrician, machinist, or technician relevant to the nuclear industry, or, like Alan's chum Bill Symes, via a *student* apprenticeship, taking a larger number of college and university courses to become an engineer who 'would

---

[49] Meyer, Leo A., *Atomic Energy in Industry: A Guide for Tradesmen and Technicians* (Chicago: American Technological Society, 1963).

[50] Parr, Joy, 'Working knowledge of the insensible: an embodied history of radiation protection in Canadian nuclear power stations, 1962–92', *Comparative Studies in Society and History* 48 (2006): 820–51.

[51] Makepeace-Lott, S., *Alan Works With Atoms* (London: Chatton & Windus, 1962), p. 19.

[52] Ibid. p. 33.

most probably get higher up the scale and would have more complicated work to do than ever Alan himself could aspire to...'.[53] Some five years later, Alan has attained a Higher National Diploma in Mechanical Engineering and has completed the short course at the Harwell Reactor School and further studies in chemical engineering, leaving the technician apprentices behind. Bill, stronger in maths, is to head the reactor control team at the company; Alan will head the fuel elements team.

Significantly, the book promoted a non-scientific and non-academic route to nuclear proficiency. None of the options in the story included a university degree followed by entry into industry: the training was to be collaborative from the start.[54] From Christopher Hinton's wartime experiences in industrial organization, then, children of the baby boom generation were being shaped for a British industry that was both new and old.

8.2 The neutron's grandchildren

My focus has been on the first generation of nuclear workers, during which new fields were born and the greatest disparity existed between countries, programmes, and goals. But this peculiar context of emergence had consequences for subsequent generations. The real-life counterparts of the fictional Alan Martin and Bill Symes found themselves entering an industry in the 1970s very different from that portrayed in recruitment literature a decade earlier.

8.2.1 *The troubled generation: Nuclear engineers and commercial power*

The public identity of nuclear specialists, as we have seen, was linked closely to their sites. But as the veil of secrecy rose and the field moved toward commercial power generation, their facilities, products, and by-products gradually came under increasing scrutiny. Indeed, sober evaluation took place within the ranks of nuclear engineers years before critical public attention. In 1959, for example, Chauncey Starr, General Manager of Atomics International and then President of the American Nuclear Society, published an engineering critique of optimistic forecasts of the atomic age, focusing on the challenges of reliability, maintenance costs, fuel reprocessing, breeder reactor feasibility, and operating economy.[55] The earlier associations of post-war security, central government direction, and national interest still shrouded the activities of nuclear specialists. But the rise of negative perceptions about the field and its workers owed much—and also contributed much—to wider cultural changes.

Historian Jon Agar's analysis of the 'long 1960s' (the period from the mid-1950s, marked by declining security concerns, to the mid-1970s, when the energy crisis drastically altered western economies) is pertinent here.[56] He suggests that the period can be understood as the overlapping of three waves, each reinforced by the Cold War. Agar's first wave is one of institutional change and accountability, during which the authority of experts was increasingly opened to public view. The second wave involved the more widely recognized social

---

[53] Ibid. p. 25.

[54] Would-be university applicants for UKAEA posts were targeted by publications such as 'Harwell: Careers in Nuclear Engineering 1959 Feb.' NA AB 17/231.

[55] Starr, T. Chauncey, 'The MYTHS of N-Energy', *Nuclear Energy Engineer* 13 (1959): 508–9.

[56] Agar, Jon, 'What happened in the sixties?' *British Journal for the History of Science* 41 (2008): 567–600.

movements of the time and the constituents and changes that they attracted, and the third wave was a shift from authority and hierarchy towards self-defined perspectives. Agar cites a history of the American debate on nuclear power as key evidence for his synthesis.[57] Indeed, his approach is fertile for assessing changes in attitude towards what had been 'atomic energy' and became 'nuclear power'; its broad sweep is discernable in the evolution of nuclear engineering described in the preceding chapters.

For the identity of nuclear specialists, the 'long 1960s' had particular relevance. Public protests against H-bombs were important in galvanizing opinion about the biological risks of atomic energy, and of the insidious nature of secrecy in the field.[58] Less obvious were the implications of the secrecy, scale, and monopolistic nature of the growing nuclear establishment. Eisenhower, in his outgoing speech as President, identified the self-perpetuating features of this post-war momentum by what he dubbed 'the military-industrial complex'.[59] The formation in 1959 of the Campaign for Nuclear Disarmament (CND) in the UK provided a template for citizens' resistance movements, mistrust of big government, and, in particular, opposition to undemocratic aspects of science and technology.[60] By the end of the 1960s, a raft of criticisms about government-funded science, free-market excesses, and apparent corporate collusion combined to challenge the public status of engineers and scientists in the Anglo-Saxon countries. Positive depictions of nuclear engineers were now more likely to be aimed at children than at increasingly cynical adult readers.[61]

There was also an enduring perception that the civilian and military aspects of nuclear energy could not be neatly separated. The Eisenhower administration's Atoms for Peace initiative had emphasized civilian applications, seeking to limit access to weapons by exporting only reactor technologies that used enriched uranium, too expensive for most countries to create themselves. At the international level, nuclear resources—either in the form of fissile material, weapons, or the technologies that made them possible—were limited by the Nuclear Nonproliferation Treaty (1968).[62]

---

[57] Balogh, Brian, *Chain Reaction: Expert Debate and Public Participation in American Commercial Nuclear Power, 1945–1975* (Cambridge: Cambridge University Press, 1991).

[58] Winkler, Allan M., *Life Under a Cloud: American Anxiety About the Atom* (Oxford: Oxford University Press, 1993). See also Kinchey, Abby J., 'African Americans in the atomic age: postwar perspectives on race and the bomb, 1945–1967', *Technology and Culture* 50 (2009): 291–315 and Scheibach, Michael, *Atomic Narratives and American Youth: Coming of age with the Atom, 1945–1955* (Jefferson, NC: McFarland, 2003).

[59] Eisenhower, Dwight D., 'The Military-Industrial Complex', in: *Public Papers of the Presidents* (Washington, DC: Government Printing Offcie, 1960), pp 1035–40.

[60] Minnion, John and Philip Bolsover (eds.), *The CND Story* (London: Allison & Busby, 1983).

[61] For example, the painted illustrations by Robert Ayton in a Ladybird children's book [Childs, E. H., *The Story of Nuclear Power* (Loughborough: Wills & Hepworth, 1972)] had first appeared in a popular introduction for adults [Gaines, Matthew J., *Atomic Energy* (London: Feltham Hamlyn, 1969)]. Gaines was a technology and science writer for the UKAEA and editor of *Atom*, its monthly magazine.

[62] The Nuclear Nonproliferation Treaty, initially signed by the USA, USSR, and UK, now includes some 189 countries and two more declared 'nuclear weapons states', France and China. Members agree to oppose proliferation of weapons and weapons expertise; to reduce international tension by negotiating the scaling-down of weapons stocks; and to support peaceful nuclear energy programmes. The division between civilian and potentially military uses of reactors remains contentious, however, as evidenced by 1998 weapons tests by Pakistan, American opposition to reactor facilities in Iran in the early 2000s, and North Korean tests in 2006 and 2009.

By the late 1960s, it was also becoming equally apparent that an academic presence for nuclear engineering was still unstable: student numbers in nuclear engineering courses could not be sustained. For example, at Imperial College in 1968, estimates of student recruitment for nuclear studies in the Mechanical and Chemical Engineering departments were downscaled from 30–40 to 20, but still deemed 'wildly optimistic' by several reviewers. Administrators planned to merge the teaching of the Nuclear Engineering Department at Queen Mary College (QMC) with that of Imperial to create a joint MSc course in Reactor Science and Engineering. However, a survey of industry and the UKAEA revealed almost no support there for a general postgraduate course. Instead, the Authority still preferred conventional engineers from existing disciplines who could be trained on the job for specific roles. Only a few industries promised to employ a handful of graduates annually, and deflected financial commitment.[63] The combined Imperial/QMC course nevertheless went ahead, with promises from the Science Research Council to rate it 'Q', making it 'the premier course in Nuclear Engineering in the UK'.[64]

American experiences were not dissimilar. By 1966—less than a decade after the launch of most American curricula—it was clear that most nuclear engineering programmes were growing slowly in competition with other disciplines. A handful prospered, owing to the convergence of government equipment grants and regional outreach programmes at the national laboratories. A few universities, such as Illinois, Kansas State, Pennsylvania State, Michigan, Arizona, and Texas A&M, had been able to expand their facilities through matching fund grants to establish large programmes. These programmes had grown to include up to ten faculty and thirty to forty graduate students, and some of these schools had, or were planning, undergraduate programmes in nuclear engineering. Several universities had strengthened their nuclear engineering curricula through cooperation with nearby government laboratories. These included the University of Tennessee with Oak Ridge, the University of New Mexico with Sandia Corporation and Los Alamos Scientific Laboratory, the University of California with the Lawrence Radiation Laboratory, and Iowa State University with Ames Laboratories.[65]

On the other hand, some of the universities that had obtained nuclear engineering facilities under early AEC educational grants decided to withdraw their programmes. By 1966, Rice University had returned the fuel from its training reactor to the AEC and was trying to sell the reactor; Oklahoma State University had ceased operating its training reactor; and the University of Wyoming had reduced support for their nuclear engineering degrees. Owing to staff losses, the University of Oklahoma was struggling to run its Nuclear Engineering programme single-handedly and soon decided to merge its Nuclear Engineering, Chemical Engineering, and Metallurgical Engineering programmes into a single School of Chemical Engineering and Materials Science.[66]

---

[63] 'Reactor Science and Engineering: Rector's Correspondence 1968–9', letters, 1968, ICA GB 0098 KNR/1/1.

[64] Board of Studies, Imperial College, 'Item 8161: Nuclear Reactor Science and Engineering', minutes, 17 February 1969.

[65] UI Box 148, 'Universities research associations'.

[66] Crosser, Orrin K. to G. M. Nordby, letter, 9 February 1966, UI Box 148.

Military support—a crucial early impetus to American nuclear programmes and university curricula—also proved diffident. A 1967 Argonne Laboratory enquiry to the Air Force Institute of Technology to encourage closer links yielded a frank reply from its head of physics:

> Over the past five years, the Air Force's interest in ground based, power producing reactors has been declining exponentially. (Rightly so, I would add. It is the function of the power industry to provide the electrical power, not the function of the armed services). In view of this decline our nuclear engineering program here at AFIT has changed dramatically to emphasize nuclear weapons effects and space nuclear applications.[67]

Representations of nuclear expertise were increasingly contested. By the early 1970s, nuclear engineering institutions and university faculties recognized that their subjects were challenged both internally and externally. The Atomic Energy Commission's Division of Nuclear Education and Training (DNET) struggled to define national standards for the profession of nuclear engineering in the face of conflicting opinions at national labs and universities. At Argonne and Brookhaven, the building of particle accelerators—the province of nuclear scientists more than engineers—continued to excite more interest than exploration of nuclear reactor designs. Argonne's Director continued to baulk at an engineering orientation, complaining about DNET categories, 'the term technologist we do not understand, for it is used with different connotation by different people'.[68] By contrast, educators sought to promote their graduate engineers. The Head of Nuclear Engineering at Kansas State opined, 'I have been sensitized because of the increasing number of attacks on the nuclear industry and in many instances, specifically upon nuclear engineering education. These are mixed emotions in that we are somewhat smug in the realization that our undergraduates and master's level students are receiving a dramatic upswing in acceptance as engineers [as] a new breed'.[69] But the components of this new breed were still disputed. The minutes of the Nuclear Engineering Education Committee of the AEC in 1971 noted the old rivalries between a physics-led profession and one built from engineering and social science disciplines:

> R. V. Laney, [Associate Director at ANL for Engineering Research and Development] expressed the thought that, whereas nuclear power was once thought of as a successor to other forms of power it is now considered as a complementary source. He questioned whether nuclear engineering as a discipline is perhaps too narrow – has nuclear engineering reached a stage of maturity where its uniqueness has become secondary, when it is now just another energy discipline? He stressed the need for an interdisciplinary approach to nuclear engineering education, bringing in such factors as economics, quality control, environmental impact and political science.

A disgruntled university administrator wrote in the margin of his copy, 'This guy ought to be told that nuclear engineering is [already] more than reactor physics in some schools'.[70]

Nuclear engineering competed increasingly for curricular and governmental attention. This change of emphasis for the national laboratories—the crucible of nuclear engineering

---

[67] Bridgman, C. J. to R. C. Taecker, letter, 3 March 1967, UI Box 100.
[68] Duffield, Robert B. to C. C. McSwain, letter, 27 July 1971, UI Box 56.
[69] Chezem, Curtis G. to A. F. Yanders, letter, 18 July 1971, UI Box 56.
[70] 'NEEC Minutes of 63rd Meeting, Argonne National laboratory', 3 May 1971, UI Box 56.

in the USA—was reinforced by the dismantling of the Atomic Energy Commission and replacement by the Energy Research and Development Administration (ERDA) during the administration of Gerald Ford in 1975. ERDA itself was replaced by the Department of Energy (DOE) in 1977 during the Carter administration. Both ERDA and DOE refocused attention from nuclear energy (in the form of the now conventional fission reactors and the more complex breeder reactor schemes) to funded research on fusion energy and a growing variety of alternative energy sources, notably solar and wind power. At the same time, public attitudes were being galvanized by exposés of what was increasingly seen as a secretive and dangerous nuclear industry.[71] The link between nuclear power reactors and plutonium production for weapons remained in the public eye, too. India tested its first nuclear weapon in 1974, having developed facilities based on the CANDU heavy-water reactor exported from Canada. The commercialization of domestic nuclear power was also proving slower, more complex, and more costly than anticipated, and public disenchantment was rising.

In response to a disaffected public, industry promotion increasingly replaced government optimism. In the USA, the Atomic Industrial Forum (AIF, 1953) became increasingly active during the 1970s. Its literature described AIF as an 'association of over 600 corporate members who share an interest in the commercial development or utilization of nuclear energy. Representing the broad spectrum of the nuclear industry, AIF's members include manufacturers, architect-engineer-constructors, electric utilities, uranium producers, financial organizations, legal and consulting firms, government agencies, labour unions, universities and others'.[72] Their aim was frankly propagandistic:

> Because nuclear power has become a highly charged and controversial subject with a growing number of critics and accompanying misinformation, our Public Affairs and Information Program is playing an increasingly important role within the AIF. We have recently expanded PAIP's activities with more services for members in connection with state and local problems; intensified efforts in dealing with special interest groups outside the industry; increased media projects to counter inaccurate news coverage and promote the reporting of positive nuclear developments; upgrading our audio/visual support services.[73]

Argonne administrators, too, identified their role in supporting the industry. Adverse publicity had prompted a study of training needs of nuclear specialists. 'The AUA Nuclear Engineering Education Committee', wrote the Vice President for Administration, 'undertook a survey of the AUA member universities to determine what actions were being taken,

---

[71] From the late 1970s, see, for example, Fuller, John G., *We Almost Lost Detroit* (New York: Ballantine Books, 1976); Knelman, Fred H., *Nuclear Energy: The Unforgiving Technology* (Edmonton: Hurtig, 1976); Bailey, Richard, *Energy: The Rude Awakening* (London: McGraw-Hill, 1977), Chapter 6; Pringle, Peter and James Spigelman, *The Nuclear Barons* (London: Joseph, 1982); McKay, Paul, *Electric Empire: The Inside Story of Ontario Hydro* (Toronto: Between the Lines, 1983); Hall, Tony, *Nuclear Politics: The History of Nuclear Power in Britain* (Harmondsworth: Penguin, 1986); John, Brian S., *Nuclear Power and Jobs: The Tranwsfynydd Experience* (Newport: Cilgwyn, 1986); Hu, Howard, Katherine Yih and Arjun Makhijani, *Nuclear Wastelands: A Global Guide to Nuclear Weapons Production and its Health and Environmental Effects* (Cambridge, Mass.: MIT Press, 1995); Bolter, Harold, *Inside Sellafield: Taking the Lid off the World's Nuclear Dustbin* (London: Quartet, 1996).

[72] Atomic Industrial Forum, 'The Atomic Industrial Forum', promotional literature, 1977, UI Box 22.

[73] Ibid.

or could be taken, by the universities to assist the nuclear power industry in addressing new education and training needs'.[74]

Even so, The Department of Energy, having recently superseded the AEC, had a declining commitment to fission reactors and to its predecessor's historical precedent of decision-making dominated by scientists. A 1978 DOE memo underlined that

> *A notable weakness is the tendency to view obstacles to adoption of new energy systems as purely technological.* Important obstacles to the adoption of new energy systems or expansion of existing ones will increasingly be recognized to be to some degree political, sociological, economic, institutional and environmental in character. Research which could assist in addressing these issues is virtually nonexistent within the DOE.[75]

Instead, it recommended establishing a research programme in the social sciences. Topics would include cost/benefit and risk-assessment methodologies; economic equity; economics and systematics of dispersed energy systems as contrasted with conventional systems; determinants of demand; energy systems siting; and social impact assessment—none of which was within the current capabilities of the Argonne National Laboratory or its technical staff. A year later, a site review by the DOE called for a reduction in nuclear reactor research and development. Against a background of a 'firm national commitment to nuclear power', it identified problems of 'waste, safety, cost and proliferation' for the fission option. Argonne administrators could merely respond impotently, 'On fission we are, of course, driven by what happens [but] we do think that the country needs to keep the nuclear option viable'.[76]

By mid-1980, Argonne and the other National Laboratories had undergone an 'Institutional Planning Process' to redefine their lead missions. Argonne was now classed as a 'multiprogram laboratory'. Its post-war specialization in nuclear engineering had been trimmed and diluted. It would now focus on community energy and urban waste programmes, thermal energy storage systems development and novel electric batteries, a shift in intellectual domain, and public engagement.[77] The nucleus of a new profession, begun at Arthur Compton's Met Lab and Walter Zinn's reactor research centre at Argonne National Lab, had lost its way.

8.2.2 *Unwelcome voices*

Piggy-backed on industry, professional identity faltered. During this transition towards rising public concern and declining government support, nuclear industry spokespersons

---

[74] O'Fallon, John R., 'Education and training needs of the nuclear power industry', memo, 15 December 1980, UI Box 71.

[75] 'Report of the Office of Science and Technology Policy Working Group on Basic Research in the Department of Energy', memo, June 1978, UI Box 47 (italics in original). This was not entirely accurate: As of 1970, an Associated Midwest Universities education committee had included a 'Sociological Effects Task Group' to evaluate the social ramifications of nuclear and related engineering applications, and to compare their effects with possible alternatives resulting from applications of other technologies ['National Engineering Education Committee, 1958–1971', memo, 1970, UI Box 100].

[76] 'Argonne National Laboratory on-site review', memo, 23 May 1979, UI Box 49.

[77] Department of Energy, 'Institutional planning process—US DOE order 25', memo, 2 June 1980, UI Box 47.

rather than working nuclear specialists came to represent the field. A handful of companies were now the major employers of nuclear engineers alongside the respective national labs. Nuclear engineers worked in the nuclear divisions of General Electric and Westinghouse, the major players in commercial reactor systems, and at a wide range of American utility companies.

The distribution of skills was similar in the UK and Canada, but the number of organizations far fewer. Institutions employing nuclear specialists included the UKAEA and British Nuclear Fuels Limited (BNFL), founded as a government-owned company in 1971 from the UKAEA Production division.[78] In Canada, the province of Ontario continued to be the primary customer for the nuclear industry, with early links between AECL, the public electricity provider Ontario Hydro, and Canadian General Electric. AECL and Ontario Hydro were crown corporations (i.e. government companies), and all three organizations employed significant numbers of nuclear engineers.[79]

At these sites, nuclear specialists remained insulated from public exposure. A collective voice for nuclear workers had been seen as problematic, particularly by the American and British governments. Moves to internationalize civilian nuclear expertise remained embroiled with the still contentious issues of proliferation of nuclear weapons.[80]

As civilian nuclear programmes came increasingly under public scrutiny and criticism for cost, safety, and security reasons, there were efforts by the US government to curtail public statements of dissent. At Argonne National Laboratory in 1972, for example, the Board of Governors fumed at pressures exerted by the AEC to remove outspoken staff. The Commission was seeking contract revisions for the operation of ANL, and the Board discussed rumours of a 'go list' of individuals identified as problematic to AEC interests. Minutes identified a range of provocations from vocal scientists and engineers having political, as well as occupational, dimensions:

> Actions known to be irritating to the AEC were (1) the Argonne Senate visit to the AEC, JCAE, other; (2) the occupation of the NIKE site by the Chicago Indian Village, with encouragement from Concerned Argonne Scientists; (3) the letter, dated Aug 11, 1971, opposing the evacuation of the Indians from the

---

[78] A private reactor industry comparable to Westinghouse and GE was slow to develop in the UK. In 1996 the country's more advanced—and commercially attractive—pressurized Water Reactor (PWR) and Advanced Gas Cooled (AGR) reactor sites were acquired by British Energy, a private company, with the older Magnox sites remaining the remit of BNFL. In 1999, BNFL acquired the nuclear business of Westinghouse and, a year later, the nuclear interests of the ABB Group of Switzerland, thereby consolidating an international presence in the industry—although the Westinghouse operation was sold again in 2006 and planned to shed its principal subsidiary, British Nuclear Group. As a result, BNFL had declining operational activities.

[79] In 1999 Ontario Hydro was divided into five companies—two of them private—to create an energy market in the province.

[80] On the history of public concern about nuclear weapons, see Boyer, Paul, 'From activism to apathy: the American people and nuclear weapons, 1963–1980', *Journal of American History* 70 (1984): 821–44; Weart, Spencer R., *Nuclear Fear: A History of Images* (Cambridge, Mass.: Harvard University Press, 1988); Smith, Jeff, *Unthinking the Unthinkable* (Bloomington: Indiana University Press, 1989); Boyer, Paul, *Fallout: A Historian Reflects on America's Half-Century Encounter With Nuclear Weapons* (Columbus: Ohio State University Press, 1998).

NIKE site; (4) the suit filed in federal court against the AEC and GSA by Goodman et al (two ANL staff members as plaintiffs) and (5) a letter to President Nixon from a group of ANL staff members.[81]

Within two months, the Board was forced to discuss the policy implications of firing Division directors at Argonne, precipitated by removal of the Director of the Chemistry Division, seemingly with pressure from the AEC. Others leaving included the former Director of Reactor Engineering, three other Division Directors, and the Business Manager. As one Board member complained, 'all these people are old-timers, some going back to World War-II days. The outright firing of three of them has caused a reaction of cynicism'.[82]

While nuclear workers and their programmes were closely scrutinized under the Nixon administration, such policies were not the product of a single American presidency. By 1977, under the Carter administration, the Brookhaven Scientific Staff Manual included an admonition that employees must not embroil the Laboratory in political activities and that 'political activities on site, whether that of individuals or groups, may in some circumstances contravene Laboratory policy by appearance, or even in substance, if carried to extremes'.[83] Individual expression gave way to Lab policy.

Yet internal wrangles over academic status and political freedom were being replaced by more potent external forces for the profession. Nuclear engineers were swept along by changing public attitudes concerning the new nuclear industries. As concerns grew about economic viability and safety, nuclear engineers found themselves recategorized. In both the USA and UK, political priorities shifted from state-promoted nuclear power during the 1960s to other forms of energy provision during the early 1970s.[84]

Prominent nuclear specialists could play a role in shaping public perceptions. Alvin Weinberg, who had headed the Oak Ridge National Laboratory since 1954, became a vocal proponent of improved reactor safety during the 1960s, arguing that attention had been focused on the economics of reactors rather than their dangers. Public concerns about the hazards of radioactivity were necessitating greater government regulation and rising costs. He identified the 'Faustian bargain' of nuclear energy as being its technical advantages—notably affordable and relatively clean energy—in exchange for long-term vigilance and stable social institutions.[85] As the American nuclear industry was being subjected to increasing regulatory scrutiny to ensure that power stations were failsafe, the AEC tended to support their contractors; Weinberg's outspokenness and dogged pursuit of novel reactor concepts became unacceptable, and in 1973 under the Nixon administration he lost his 18-year post as Director.[86]

---

[81] Minutes, Board of Governors, ANL, 'Tripartite contract problem', 28 January 1972, UI Box 110.

[82] Nevertheless, nuclear specialists remained unusually voiceless through the 1960s and beyond, and unlike their peers in other fields; see, for example, Wisnioski, Matt 'Inside "the system": engineers, scientists, and the boundaries of social protest in the long 1960s', *History and Technology* 19 (2003): 313–33.

[83] Argonne National Laboratory, 'Scientific Staff Manual', UI Box 35.

[84] In Canada, nuclear power had been adopted as a viable option only in the relatively energy-starved, densely populated, and industrially developed provinces; provincial energy policy in Ontario and Quebec thus substituted for a national policy.

[85] Weinberg, Alvin M., 'Social institutions and nuclear energy', *Science* 177 (1972): 27–34.

[86] Weinberg, Alvin M., *The First Nuclear Era: The Life and Times of a Technological Fixer* (New York: AIP Press, 1994), Chapter 10.

8.2.3 *Mistrust and responsibility*

Weinberg's focus on professional responsibility had first been foregrounded in a 1962 Stanford Research Institute study which suggested that professionalization would be promoted by the 'element of awe', 'apprehension', and 'aura of mystery' that set nuclear workers apart. Its authors argued that the enduring 'halo of secrecy' and 'professional secrets' would be the key to the special esteem anticipated for nuclear specialists, and suggested that they would attain high status akin to that of railway engineers or airline pilots.[87] Identifying nuclear specialists as a privileged profession was reiterated by Weinberg to be a matter of necessity, if not yet reality:

> The chief operator of a nuclear plant bears a heavier burden of direct responsibility than does the president of his utility, yet his salary is about 20% that of the president. This practice stems from the mistaken perception that a nuclear plant is just another generating plant. This perception is wrong: nuclear plant operators should be regarded and compensated like airline pilots – in short, a fully professional, elite cadre that carries a heavy burden of responsibility for the operation.[88]

Weinberg promoted an engineering perspective that combined the *technical* and the *social*: a new generation of reactors having 'passive' or 'inherent' safety would be designed from scratch; a relatively small number of such large reactor clusters would be sited in remote 'nuclear parks' and protected by heavy security; the nuclear power industry would be relegated to a small number of operators; public education about the new safety regime would be promoted; and nuclear specialists themselves would become a more responsible professional cadre.[89]

In Britain, the theme of professional responsibility also rose in prominence as the Institution of Nuclear Engineers continued to struggle to assert identity. Its membership had reached 2000 in 1971, but fell gradually thereafter.[90] In 1977, Jeffery Lewins of University

---

[87] Vollmer, Howard M. and Donald L. Mills, 'Nuclear technology and the professionalization of labor', *American Journal of Sociology* 67 (1962): 690–6, p. 696. The industry continued to be hamstrung by secrecy in the early 1960s. As a democratic congressman reported, the quantity of documents classified in the post-war period was three times greater than that between 1907 and 1945 [Moss, John E, 'The crisis of secrecy', *Bulletin of the Atomic Scientists* 17 (1961): 8–11].

[88] Weinberg, Alvin M., *The First Nuclear Era: The Life and Times of a Technological Fixer* (New York: AIP Press, 1994), p. 229.

[89] Weinberg, Alvin M., 'Salvaging the atomic age', *Wilson Quarterly* (Summer) (1979): 88–112. In the Soviet Union, reactor designers came to the same conclusion [Josephson, Paul R., *Red Atom: Russia's Nuclear Power Program from Stalin to Today* (New York: W. H. Freeman, 1999), p. 44].

[90] From the early 1960s, the INucE had competed with the British Nuclear Energy Society (BNES) for a smaller pool of specialists, with the BNES increasingly emphasizing learned society activities while the INucE focused on its role as a qualifying body. The two organizations collaborated increasingly via joint conferences, secretariat, branches, and journals until finally merging in 2009 to form the Nuclear Institute. Ironically, the NI continued to highlight its role in accrediting expertise in nuclear engineering and nuclear science in a country that still had muted support for these specialisms, while the American Nuclear Society and Canadian Nuclear Society downplayed this aspect of their organizations. Memberships as of 2011 were roughly 11,000 for the ANS, 2800 for the NI, and 1200 for the CNS, although it should be noted that the CNS is not a qualifying body.

College London, the incoming President, assessed the status of the British nuclear engineer. He argued that there had been a time lag between the development of nuclear engineering and the development of the nuclear engineer, but that engineering had developed to a point 'where the separate discipline "nuclear engineer" has emerged, justifying the distinction of nuclear engineer, just as mechanicals developed a separate identity from the civils'. Lewins contrasted the lack of a British professional identity with the situation in the USA, arguing that nuclear engineers there had developed as a distinct discipline because of the regulatory environment of state professional licensing and small energy utility firms, which encouraged colleges to compete to provide the most suitable first degrees for the industry. In the UK, by contrast, two major employers—the UKAEA and CEGB—had not acknowledged the value of nuclear engineering curricula or encouraged their availability at British universities. Instead, they 'preferred to take those qualified in other disciplines and provide them with in-house specialist training. They have rejected the postgraduate as much as the undergraduate nuclear engineering degree'.[91]

The President of the INucE portrayed his members' position as trapped between an unresponsive and monopolistic employer, on the one hand, and a suspicious public on the other, quoting Rudyard Kipling's poem 'The Hymn of the Breaking Strain' (1935), which begins:

> The careful text-books measure
> (Let all who build beware!)
> The load, the shock, the pressure
> Material can bear
> So, when the buckled girder
> Lets down the grinding span
> The blame of loss, or murder,
> Is laid upon the man.
> Not on the stuff, the man.

In response to such public scapegoating and consequent loss of morale, Lewins suggested emphasizing the duties of professional conduct and the supply of disinterested advice to the wider public—both newly valued attributes in an industry that was viewed with increasing suspicion by the public. Like Weinberg in the USA, he recast the identity and *esprit de corps* of the British profession in terms of its public role:

> The special considerations of nuclear radiation call for a depth of education, training and experience far beyond a superficial job application of other disciplines. The Institution must be concerned not only with initial qualifications but must also encompass a technical and ethical commitment to the development of the nuclear engineer throughout his professional life...morale is of importance in civilian affairs and touches a fundamental crisis in the current political nature of nuclear engineering. It seems to me that as a nation we now know how to make superb nuclear power stations, but we have forgotten why.[92]

A collective expression of responsibility—predicted by the Stanford sociologists and urged by Weinberg and Lewins—is a common feature of professionalization, particularly in the

---

[91] 'Is there a nuclear engineer in the house?' *Journal of the Institution of Nuclear Engineers* 18 (1977): 35–45; quotation p. 35.

[92] Ibid. pp. 43 and 45.

USA, where it is often espoused in a code of ethical practice.[93] Lewins bemoaned the lack of shared moral compass as a failing for the self-perception of nuclear engineers; Weinberg cited it as a necessary professional attribute to garner respect in wider society. As technological failures revealed, wider assessments of the ethics of nuclear workers are at the root of declining public confidence and sympathy.

8.2.4 Identity by accident: Three Mile Island and Chernobyl

The shift of public perceptions of nuclear specialists—from post-war boffins and brave innovators to collaborators in a dangerous and secretive industry—dampened professional aspirations. The morale of the British profession itself, as reflected in the ranks of the Institution of Nuclear Engineers, declined from the 1970s in the face of falling reactor orders, growing public antipathy, and government moves to rationalize the technical professions.[94] But nuclear incidents sharpened these public assessments.

Growing public concern about nuclear weapons, environmental pollution, and the accountability of government and its technical experts were punctuated by the accidents at Three Mile Island (1979), Chernobyl (1986), and Fukushima (2011). The playing out of these events, as much as any long-term management by governments and industry, altered the perceptions of nuclear energy and its specialists.

At Three Mile Island, near Harrisburg, Pennsylvania, a malfunction had led to a fairly routine 'scram' event (insertion of control rods to stop the chain reaction) in one of two pressurized water reactors. The heat still being generated by the radioactive decay of the fuel was, however, not correctly managed. A stuck valve led to the loss of cooling water, a fault that initially went unrecognized by the operators, in part because of inadequate design and awareness about the limitations of the monitoring instruments. After over an hour of overheating, increasing amounts of the water within the reactor core were converted to steam; after some two hours, cooling water no longer covered the top of the core, allowing the fuel pellets to burn, releasing radioactivity inside the containment building and causing a small hydrogen explosion. There was a shift change, but the operators had not yet recognized the existence of a serious problem. Radiation alarms finally were triggered nearly three hours after the valve failure, followed shortly after by declaration of a site emergency. It is estimated that, when the Nuclear Regulatory Commission (NRC) was informed four hours after the valve failure, nearly half the uranium fuel had melted, and the coolant water

---

[93] Abbott, Andrew D., *The System of Professions: An Essay on the Division of Expert Labor* (Chicago: University of Chicago Press, 1988), pp. 60–1.

[94] During the late 1970s, the Finniston Commission had investigated the training and categorization of engineering expertise, and during the early 1980s the Engineering Council was seeking to rejuvenate British engineering through the merger of professional bodies. In this environment, the INucE appeared likely to be subsumed into the Institute of Energy, or at least to be denied full status as a chartered institution. See Callow, J. H., 'What is a nuclear engineer? Whither INucE?' *Nuclear Engineer* 23 (1982): 129. A decade later, a council member discouraged attempts at forecasting: 'With the present uncertain economic conditions in much of the world, it is not easy to see how the future of nuclear power generation will evolve. At present, there is so little growth that trends are not clear…' [Claridge, A., 'Foreword', *Nuclear Engineer* 32 (1991): 154]. Along the same lines is Thomas, S., 'Young nuclear engineers—an endangered species', *Nuclear Engineering International* 46 (2001): 36.

was about 300 times more radioactive than normal. Some sixteen hours after the original failure, operators were able to resume the water circulation in the core and its temperature began to drop. Steam and hydrogen gas were removed gradually over the following week, some of it by release directly to the atmosphere.[95]

The incident had dramatic consequences for the American and British nuclear industries, government agencies, and public perceptions of their nuclear specialists. Initial controversy surrounded the clarity and openness of information about the accident, and the adequacy of emergency planning. The company operating the nuclear plant, Metropolitan Edison, had tardily informed the NRC of the incident, and public announcements by Met Ed and state officials were incomplete and sometimes contradictory. In the days following the accident, residents were informed initially that there was no danger, but then advised to stay indoors and, later, to evacuate pregnant women and children living within five miles (8 km) of the plant. The monitoring and management of the incident were condemned, with estimates of the radiation release remaining controversial even after the hearings of a Presidential Commission set up by then-President Jimmy Carter, who had participated as a US Navy technician in the clean-up at Chalk River's NRX reactor 27 years earlier. The Commission itself noted that training standards, control room design, and site maintenance had been inadequate. The reactor designers, Babcock & Wilcox, were cited as having communicated too little information about engineering flaws to their customers.[96] The TMI accident was a turning point for the American industry and profession: the authorization of nuclear power stations was halted in the USA, and those that had received approval faced increasing public opposition and slow construction to meet tightened safety rules.[97]

The incident proved equally troubling in Britain, where it provoked navel-gazing by nuclear specialists. On the untimely twentieth anniversary of the Institution of Nuclear Engineers, its President cautioned that TMI had 'proved the point that the first generation of power reactors cannot be considered as technically viable and socially acceptable'.[98] Lord Hinton, its Honorary Vice President, commented that 'A Harrisburg incident was bound to happen; ... [it] will make people remember as clearly as they should, that unremitting care, skill and experience alone make nuclear power safe'. He suggested that nuclear power 'had been made to run before it could walk', and that 'hurried optimism' meant the industry was heading for a fall.[99] Recognizing the 'trend against nuclear power' and linking

---

[95] Walker, J. Samuel, *Three Mile Island: A Nuclear Crisis in Historical Perspective* (Berkeley: University of California Press, 2005).

[96] Kemeny, John G., 'The President's Commission on The Accident at Three Mile Island. The need for change: the legacy of TMI', The President's Commission, Oct 1979, consulted at http://www.threemileisland.org/downloads//188.pdf.

[97] Osif, Bonnie Anne, Anthony J. Baratta and Thomas W. Conkling, *TMI 25 Years Later: The Three Mile Island Nuclear Power Plant Accident and Its Impact* (University Park, Penn.: Penn State Press, 2004).

[98] 'Anniversary Message from the President, Dr J. Lewins', *Journal of the Institution of Nuclear Engineers* 20 (1979): 130.

[99] Ibid. In particular, Hinton claimed that the Suez crisis of 1956 had encouraged an inter-government committee to set unrealistic targets for nuclear power generation even before Calder Hall had been commissioned, and that misplaced optimism for reactor sales had led to the proliferation of too many types of reactor.

the traumatic accident to the future of the profession itself, the INucE renamed its journal *Nuclear Engineer*, devoting an entire issue to a post-mortem on TMI and public relations.[100]

Uncharacteristically, the Institution published the minutes of a special meeting of their Council convened after the American accident. In their discussion, the council members sought to identify different training standards in the USA as the crux of the problem. Its president suggested that 'operators for utilities are for example very often drawn after three years' experience in the American nuclear submarine or surface navy' and 'the whole thrust of setting up undergraduate courses in nuclear engineering in America was as a source of operators to utilities who did not train their own engineers'; by contrast, British operators received substantial training at the nuclear plants. Another member proferred that American plant operators were low on the scale of technical experts: 'there are chartered engineers, technician engineers and engineering technicians. We are mainly interested in the first, with some peripheral interest in the second and not really concerned with the third, whereas in this case it may be the third level who are really the key people'.

Despite these attempts to distinguish the British context from the American one, INucE Council members agreed that both countries had inadequate emergency planning, and that the public and nuclear workers were vulnerable as a result. 'Did anyone have an emergency plan at all?', mused one member. Another responded, '... I think always that these disasters are so unexpected that nobody has ever really thought what you do in that first moment, when you've got to start telling people about it'. A third related disasters back to occupational status: 'Well it's high on our list now, that's for sure... How to protect the people dealing with the emergency?'[101]

The social potency of Three Mile Island had been primed by the release of a Hollywood drama twelve days earlier (16 March 1979) and its continuing cinema run through the crisis. *The China Syndrome* (Columbia Pictures, Dir. James Bridges, USA, 1979) portrayed the quandary of an ethically minded nuclear engineer vacillating between company loyalty, on the one hand, and concerns about engineering safety on the other. In opposition to this evocative individual were the company administrators and security forces willing to conspire to silence his public statements.

The film drew on accounts then circulating about the suspicious death of nuclear process worker Karen Silkwood (1946–1974) five years earlier, who had been scheduled to speak to a reporter about safety violations at the Kerr–McGee plutonium processing facility in

---

[100] 'TMI—one year on', *Nuclear Engineer* 21 (1980): special issue.

[101] 'TMI—one year on', *Nuclear Engineer* 21 (1980): special issue; quotations pp. 71, 74. A year later, the incoming INucE president concluded that reactor designs were already intrinsically safe, but that human fallibility and training required more attention. Aspects that had been considered as externalities in the past—a nuclear installations inspectorate, comprehensive emergency plans, a solution for disposing of nuclear waste, and proliferation of different technical systems – were all, he argued, components of an integrated engineering vision, and vital to the future of the industry. Most significantly, the field needed to become more open: 'the public deserves and needs much more nuclear education together with easy access' [Edwards, J., 'Nuclear power—learning from the lessons of the past', *Nuclear Engineer* 22 (1981): 98–106].

Oklahoma. Her story, as portrayed in the subsequent film *Silkwood* (ABC Motion Pictures, Dir. Mike Nichols, USA, 1983), emphasized corporate conspiracy, worker paranoia, and the insidious dangers of radiation. The popular representations of the period, then, characterized nuclear workers as pawns of a system centred on corporate profit rather than public benefit and worker safety, and their proficiencies as inherently deadly and poorly controlled. Both *The China Syndrome* and *Silkwood* illustrate the cloistered identity of nuclear specialists. The whistle-blowers in both cinematic portrayals, hinting at a mediaeval apostate, suggest a loss of faith and the questioning of professional ideals. By breaking their monastic vow of silence to voice misgivings about their occupations, these workers were, it seemed, revealing the conflicts within their profession.

The Chernobyl accident, in May 1986, provided a significantly more serious event to consolidate public criticism of nuclear power and its specialists. The four graphite-moderated, water-cooled reactors near Pripyat, about 110 km north of Kiev in the Ukraine, had been built during the 1970s and had what contemporary Soviet engineers felt were entirely adequate safety systems.

Ironically, the accident was precipitated by the testing of an emergency cooling system intended to prevent the overheating of the core that could occur if a 'scram' was followed by a failure of the electrical supply. The existing diesel-powered emergency pumps required about a minute to begin pumping cooling water to the core. The plant director approved tests to evaluate the reactor's steam turbine as a means of temporarily generating electrical power for the pumps as it slowed down following a scram. To perform the experiment, operators gradually reduced the reactor power with some safety systems disabled. The control rods were inserted too far, however, and to increase the power the operator consequently again withdrew the rods. The steam turbine proved to produce inadequate water flow; the water overheated and boiled, and the resulting steam created voids in the reactor which allowed the fission rate and heat generation to increase dramatically. A scram was begun, but before the control rods could be inserted the reactor power increased so quickly that steam pressure ruptured water pipes and fuel channels and blew off the lid of the reactor. A few seconds later a hydrogen gas explosion occurred, probably triggered by the dissociation of water into oxygen and hydrogen at the hot graphite surfaces. The explosion caused the graphite itself to catch fire and to open channels for airflow through the hot core. The result was a fire-borne release of radioactive materials akin to the Windscale pile fire 29 years earlier, but on a vastly more catastrophic scale.[102]

The radiation levels initially were underestimated, and reactor personnel assumed that the core was intact; as a result, most of the reactor crew and fire fighters received fatal exposures, with 31 deaths within hours or weeks. Raging for days, the fire itself was eventually snuffed out by helicopter-dropped material. In the meantime, the plume of radioactive smoke travelled across Europe. The nearby town of Pripyat was evacuated permanently, and thousands of clean-up workers and residents are estimated to have received radiation dosages likely to produce long-term health effects or death.

Nuclear identities were sacrificed, too. In the aftermath of the disaster, the Chernobyl experience generated a variety of assessments. Some analysts vaunted claims of superior

---

[102] Medvedev, Zhores A., *The Legacy of Chernobyl* (Oxford: Basil Blackwell, 1990).

reactor designs in the West (and, by extension, the greater competence of Western nuclear engineers), and even contrasted the limited effects of the Three Mile Island accident with its Ukrainian counterpart. A larger fraction, however—including the official report—criticized the training and capabilities of the Chernobyl nuclear operators.[103]

One attempt to reassure the public, and perhaps unconsciously to tap into earlier public sentiments of heroic professionals, was the initiative of Sir Frederick 'Ned' Warner (1910–2010), a senior British chemical engineer and leader of several industrial accident investigations, to form a team of experts for future emergencies. His organization Volunteers for Ionizing Radiation (VIR) was to consist of retired engineers and scientists willing to enter potentially radiatioctive environments to assess, act, and possibly sacrifice their lives during nuclear accidents. While Warner recruited several dozen senior technical volunteers, the popular press in Britain portrayed his plan largely in comic terms, and the group was never actioned.[104]

The effects of the Chernobyl accident on Western nuclear engineers were nevertheless muted. Where TMI had engendered accusations of government and industrial incompetence and collusion in underplaying and covering up safety breeches, Chernobyl attracted accusations of a bureaucratic and authoritarian state, and failed to link that country's engineering failures with Western practitioners (although the failed reactor shared design features with some of the still-operating Hanford production reactors). Together, however, TMI and Chernobyl landed a knock-out blow. The number of new reactors approved continued to fall in all countries, ending what Alvin Weinberg dubbed 'the first nuclear era'.[105] A public face for such decaying academic and institutional optimism was the satirical cartoon television series *The Simpsons* (Matt Groening, 1987–) with its principal character, Homer, cast as a poorly trained and bumbling nuclear engineer. Nuclear specialists, by turns voiceless, progressive, heroic, and mistrusted, had been reduced to a joke.

8.2.5 *Fukushima and its fallout*[106]

Twenty-five years after Chernobyl, these varying assessments of nuclear expertise were again reinforced. On the afternoon of 11 March 2011, an unusually strong earthquake off the eastern coast of Japan caused a series of unanticipated consequences for the Fukushima Dai-ichi nuclear power plant, leading to an ominously escalating accident. The outcome for the plant, Japan, and international perceptions of nuclear safety were dramatic.

---

[103] Medvedev, Grigori and Evelyn Rossiter (translator), *The Truth About Chernobyl* (New York: Basic Books, 1991).

[104] Warner, Frederick to S. F. Johnston, interview, 19 December 1997, London, England, SFJ collection; Agar, Jon, 'Dad's radioactive army: scientists, engineers (and philosophers) as volunteers after Chernobyl', presented at *British Society for the History of Science Annual Meeting*, Aberdeen, 2010. BJHS paper forthcoming.

[105] Weinberg, Alvin M., *The First Nuclear Era: The Life and Times of a Technological Fixer* (New York: AIP Press, 1994).

[106] The term 'fall-out' was coined circa 1950 to refer to the airborne radioactive particles spread by winds after the detonation of a nuclear weapon, but within four years had been used as a general synonym for 'unanticipated side effects', the meaning intended here [*Oxford English Dictionary*].

As intended following an earthquake, the three operating reactors of the six on site were shut down by an automatic fuel-rod insertion system.[107] After the scram, generators operated pumps to circulate cooling water, removing the heat that they continued to generate by radioactive decay. But the earthquake had also caused a tsunami which swept inland over an inadequate protective wall and flooded the facility. As a result, less than an hour after the quake, electricity distribution ceased and emergency diesel generators were inoperable; the pumps and their flow of cooling water stopped.

The following day, as the three reactor cores gradually overheated, fuel rods began to melt. Reactor operators began spraying seawater into the reactors through the fire extinguisher system, but cooling water vapourized, increasing pressure within the containment vessels. The hot, exposed zircaloy jackets of the fuel elements reacted with steam: the dissociated water oxidized the alloy and released hydrogen gas. As with Three Mile Island, reactor operators elected to vent steam and hydrogen to prevent an uncontrolled breach of the core, but the roof and cladding walls of one building were destroyed by a hydrogen explosion. The release of the vented radioactive gas led authorities to define a 20 km evacuation zone.

On 14 March, hydrogen in a second reactor building exploded and damaged the water supply for a third; that building's containment was breached the following day by another explosion. The same day, a fire started with spent fuel rods from a fourth inactive reactor because the water of their cooling pool, located on an upper floor of the reactor building, had evaporated, allowing them to overheat while exposed to air. The spent-fuel pools, located outside the reactor containment areas, contained the majority of radioactive material on the site. Helicopters, fire engines, and police water cannons attempted to spray water into the reactors and, a few days later, into some of the spent-fuel pools.

During the second week, electrical power was gradually restored to several installations on the site, but a few workers were contaminated by standing in radioactive water that flooded some buildings; the need for protective boots had not been foreseen in safety procedures. The operators countered the detection of highly radioactive water in some areas by reducing the water flow into one reactor, trading off a consequential rise in its temperature again. The radioactive waste water from the reactors accumulated in underground trenches carrying power cables and piping and, as they filled, one was found to be leaking directly into the sea. Repairs with concrete and other materials failed for several days, and the flooded trenches hampered the restoration of closed cooling systems for the stricken reactors. Administrators decided to release several thousand tons of slightly contaminated water into the sea so that the more intensely radioactive water in the leaking trench could be accommodated. The seawater itself—an ad hoc choice of cooling by operators—was leading to a secondary problem: the deposited salts now also had to be removed from the reactors. Within weeks of the accident, officials noted publicly that the site would have to be

---

[107] The stricken facilities were boiling-water reactors of a type designed by General Electric during the 1960s. The Fukushima events were less dramatic than those at Chernobyl, which had no containment vessel to prevent the sudden release of intense radioactivity after a core explosion and fire, but much more serious than Three Mile Island, which had suffered an explosion and venting, but no breach of its containment vessel.

decommissioned, and that entombment of the reactors in concrete, as at Chernobyl, was an option.[108]

As this narrative hints, the cascading effects, described gradually and sometimes inconsistently in a public drip-feed over the following weeks, made the impact of the event more sustained. The events at the plant were subsumed into a broader narrative of the national catastrophe. Public fears were fanned in Tokyo and neighbouring cities two weeks after the earthquake when radiation levels in water were detected that exceeded legal limits for infants. Radioactive material was later detected in milk and agricultural produce from areas near Fukushima and in seafish from the nearby coast, some exceeding regulatory limits.

The events tapped into the legacy of nuclear engineering familiar to a mistrustful public. Hardware and social systems had both failed. As at Three Mile Island, information flow had proved unreliable. In the weeks after the accident, news of radiation levels, technical developments, and outcomes jostled, and speculation even by nuclear experts failed to clarify the situation and judgements of its adequacy.

Indeed, expert voices were discordant. The early assessments were notable for their national allegiances and lack of professional cohesion. Among the enduring critiques were lack of engineering foresight, and inadequate solutions to the unexpected problems. Internet discussion lists and news feeds from the industry, often centring on particular reactor technologies, criticized the design of safety systems and consequences of ill-considered responses.[109] Not surprisingly for a beleaguered industry, criticisms of media mis-information and fearmongering were also common. Condemnation focused initially on the Tokyo Electric Power Company (TEPCO), responsible for the Fukushima plant, but also on the international nuclear industry and organizations such as the Nuclear Regulatory Commission in the USA, argued to have been too compliant to resistance by the American nuclear industry regarding implementing heightened safety measures in the USA.[110] The International Atomic Energy Agency (IAEA), established by the UN in 1957, also faced criticism for not having better policed nuclear power plants around the world or contributed more directly to the Fukushima responses. While the IAEA was able to recommend safety standards, it was unable to force member countries to comply. And the competing goals of the organization—promotion *plus* monitoring of civilian nuclear power *plus* verifying compliance of the nuclear non-proliferation treaty for nuclear weapons—reduced its effectiveness.[111]

Predictably, portrayals of nuclear workers were mixed. As with Three Mile Island and Chernobyl, news coverage flitted briefly from the technical and administrative disaster to accounts of heroic workers. Five days after the earthquake, Japan publicized a group of

---

[108] Government of Japan Nuclear Emergency Response Headquarters, 'Report of Japanese Government to the IAEA Ministerial Conference on Nuclear Safety—The Accident at TEPCO's Fukushima Nuclear Power Stations', http://www.kantei.go.jp/foreign/kan/topics/201106/iaea_houkokusho_e.html, accessed 20 June 2011.

[109] E.g. the London-based *World Nuclear News* (http://www.world-nuclear-news.org/).

[110] E.g. Collins, Robin, 'Reactor design in Japan has long been questioned' and Franta, Jaro, 'Media meltdown', both at cdn-nucl-l@mailman1.CIS.McMaster.CA, 15 March 2011.

[111] Kurczy, Stephen, 'Japan nuclear crisis sparks calls for IAEA reform', *Christian Science Monitor*, 17 March 2011.

control-room engineers, maintenance technicians, fire fighters, and other emergency workers dubbed the 'Fukushima Fifty'. Information was sparse. It remained unclear whether the Fifty had volunteered or submitted to TEPCO orders; the company itself, now working under close government supervision, had been directed to be resolute in ending the crisis, and directed its workers accordingly. The Health Ministry raised the legal limit for short-term, whole-body radiation exposure of Japanese nuclear workers from 100 millisieverts (10 rem) to 250—five times that allowed for American workers and twelve times the British annual limit.[112] Perhaps in recognition of this sacrifice, its rotating group of workers, actually numbering in the hundreds, was described as 'courageous' and 'stoic', but, characteristically for nuclear workers, also as 'silent' and 'faceless'.[113]

Like Chernobyl a quarter-century earlier, the repercussions from Fukushima were profound. Support by governments for nuclear energy, already strained by public reaction to Chernobyl and to rising concerns about nuclear terrorism, was further strained in defending the safety of existing power plants. Within days of the earthquake, international announcements sought to quell concerns. In the UK and Canada, regulatory organizations began to review their emergency procedures in parallel with public assurances. The German government initially declared a three-month moratorium on operation of eight reactors and soon decided to phase out its nuclear programme, as did Switzerland and Italy. Two months after the accident, the Japanese prime minister announced that the country's energy policy would deemphasize nuclear power in favour of renewable sources and creation of an energy-saving society.[114] Plans for new initiatives, such as another generation of reactors and the expert personnel needed to design and operate them, were placed on the defensive. And nuclear specialists, having dwindled over the past generation, saw declining optimism even within their own ranks.

8.2.6 *Nuclear generations*

In the seven decades between 1940 and the first decade of the twenty-first century, scarcely three generations of nuclear engineer had practised their art. The 'neutron's children'—the first cohort of professionals—have been the primary focus of previous chapters. Their successors, the 'neutron's grandchildren', university-trained and working at nuclear power stations from the mid-1960s, faced the challenges of sustainable nuclear energy through their careers. But the 'neutron's great-grandchildren' from the 1980s represented a lost generation.

Professional identities had been increasingly contested, as new nuclear projects evaporated after Three Mile Island and then Chernobyl. Sustaining the existing power stations

---

[112] The acceptable exposure limit in the early Manhattan Project was some 36 rem/year [Weinberg, Alvin M., *The First Nuclear Era: The Life and Times of a Technological Fixer* (New York: AIP Press, 1994), p.182].

[113] Branigan, Tania and Justin McCurry, 'Fukushima 50 battle radiation risks as Japan nuclear crisis deepens', *Guardian*, 15 March 2011; Bradsher, Keith and Hiroko Tabuchi, 'Last Defense at Troubled Reactors: 50 Japanese Workers', *New York Times*, 15 Mar 2011; Gilligan, Andrew, 'Japan earthquake: the Fukushima 50 fight to stave off disaster', *Daily Telegraph*, 16 March 2011, retitled 'Fukushima fifty on whose bravery all hope rests' by the *Montreal Gazette* and other newspapers that picked up the story.

[114] http://www.guardian.co.uk/world/2011/may/11/japan-nuclear-power-expansion-plans-abandoned.

became increasingly problematic as a hole grew in the nuclear workforce. The engineers and scientists trained between the 1950s and 70s were aging and not being replaced. In 2010, a quarter-century after those traumatic events, over 40% of the nuclear experts at a new Finnish installation were over the age of 50; the French power utility Électricité de France (EDF) reported that half the employees of its nuclear branch would retire by 2015. At American sites, the peak age of nuclear specialists was about 50; three-quarters of the nuclear personnel at Department of Energy labs were elegible to retire by 2010, and the enrolment of undergraduate students of engineering had fallen by two-thirds over the previous thirty years.[115]

Indeed, the small intake was more sobering than the inevitable attrition by retirement. As early as 2000, the Organization for Economic Co-operation and Development (OECD) and the Nuclear Energy Agency (NEA) had warned that the numbers of higher education programmes, faculty members, and students in nuclear subjects were declining precipitously in Europe, North America, and Japan. The remaining programmes were being diluted to attract a broader range of students, and a significant fraction of the graduates were opting not to enter the nuclear industry. Given this 'downward spiral of low enrolment and budgetary cuts', the organizations recommended a renewed strategic role for governments in supporting education, manpower, and infrastructure.[116]

University and college programmes are sensitive indicators of the health and viability of a profession because of their dependence on a continuing supply of students and jobs in industry. As student numbers fell at Imperial College, for example, nuclear engineering staff moved to the Earth Science group within the Geology department: the nuclear modellers, adept at devising elaborate computer codes for reactors, could apply themselves efficiently to new problems relevant to environmental remediation.[117]

There is an irony in this shifting focus. Waste disposal, the issue that was to rise to greatest prominence during the 1980s and 1990s, had always been a secondary focus for nuclear specialists. As Alvin Weinberg of ORNL observed, the original Clinton laboratory had been the first site at which megacuries of radioactivity had been produced:

> We realized even then that proper disposal of radioactive wastes was necessary if nuclear energy was to succeed, but instead of waste disposal becoming a central focus for the laboratory, the matter was always a side issue – one that never commanded the attention of the most sophisticated people.

Consequently, he described the Clinton Laboratories' mission of developing chemical recycling methods and basic research in health physics as a 'dismal decision', commenting 'after all, who wanted to work at a place whose primary purpose was garbage— nuclear garbage, to be sure, but garbage nevertheless'.[118] There was a sense in which

---

[115] 'Nuclear industry in midst of skills crisis', *Reuters*, 19 December 2010; Wogman, N. A., L. A. Bond, A. E. Waltar and R. E. Leber, 'The nuclear education and staffing challenge: rebuilding critical skills in nuclear science and technology', *Journal of Radioanalytical and Nuclear Chemistry* 263 (2005): 137–43.

[116] Nuclear Energy Agency and Organization for Economic Co-Operation and Development, 'Nuclear Education and Training—Cause for Concern? A Summary Report', NEA #02428, 2000; quotation p. 21.

[117] Goddard, Tony to S. F. Johnston, interview, 2 July 2008, London, England, SFJ collection.

[118] Weinberg, Alvin M., *The First Nuclear Era: The Life and Times of a Technological Fixer* (New York: AIP Press, 1994), pp. 63 and 68.

disposal problems were nothing new: for a century the expanding chemical industry had dealt with its waste products by employing tall smokestacks and pipes feeding effluent into large waterways or reservoirs, thereby diluting noxious products and their dangers. The occasional radioactive effluent evacuated inadvertently by the Windscale piles was meant to be captured by the smokestack filters or vented and diffused in the surrounding air; liquid waste from Dounreay cooling systems was, with equal confidence in traditional methods, pumped into the fast-flowing waters of the adjacent Pentland Firth. Post-war designers concerned with radioactive waste focused on the short-term problems of dispersal rather than the long-term problems of accumulating waste, which was relegated to ad hoc holding tanks or pits having unknown contents and unpredictable lifetimes.[119] Rather than concentrate on such unappealing and intractable concerns, Weinberg and his colleagues focused on reactor design rather than on reactor products, describing the ORNL mission as a national lab 'established to try things too difficult or too risky for private industry to undertake'. Noting that he had 'never acquired the instinctive, detailed understanding of the waste problem that I had of the reactor problem', Weinberg later regretted not having elevated waste disposal to the top of the agenda at ORNL.[120]

Through accumulating experience such as Weinberg's, the nascent experts who built and ran the production reactors from the 1940s and the experimental reactors of the 1950s had been transformed into power specialists from the 1960s, assuming increasing responsibility for the environmental dangers and side-effects. An illustration is the creation of BNFL Ltd in 1990 with the remit of decontaminating and decommissioning nuclear sites. The creation of this government-owned company changed the trajectory of what had once been Christopher Hinton's Production Group in the UKAEA.[121] But the term 'waste disposal' retained its deprecating connotations for subsequent generations. Some participants adopted more popular and neutral labels such as the phrases 'nuclear legacies' and 'radiological engineering'.[122]

A more diffident example of future specialization for nuclear experts was that of nuclear security. Following the rise of terrorism against Western countries exemplified by the 11 September 2001 al-Qaeda attacks, attention focused on protecting nuclear facilities from intentional sabotage and theft of fissile materials. These measures often concentrated on vehicle barriers, personnel scanners, and security checks, representing nuclear sites as

---

[119] Harwell and Brookhaven researchers of the period investigated atmospheric pollution and the absorption of effluent by ground and vegetation [Sutton, O. G. to W. B. Lewis, letter, 26 April 1949, QU Lewis fonds Box 1 File 13].

[120] Weinberg, Alvin M., *The First Nuclear Era: The Life and Times of a Technological Fixer* (New York: AIP Press, 1994), quotations pp. 117 and 183.

[121] Similarly, the UKAEA, the organization from which BNFL had fissioned in 1971, formed UKAEA Ltd as a subsidiary dedicated to nuclear clean-up of its remaining sites responsible for the production of nuclear fuels and weapons materials. In 2009, the UKAEA announced the creation of Research Sites Restoration Limited (RSRL) to decommission the Harwell and Winfrith facilities.

[122] Wogman, N. A., L. A. Bond, A. E. Waltar and R. E. Leber, 'The nuclear education and staffing challenge: rebuilding critical skills in nuclear science and technology', *Journal of Radioanalytical and Nuclear Chemistry* 263 (2005): 137–43.

potent sources of danger and instability. Training courses, too, satisfied this sudden demand for engineering skills.[123]

But the turn of the twenty-first century offered another change of identity for nuclear workers, and more palatable work than mere clean-up operations and anti-terrorism planning. Evidence for a link between fossil fuel consumption and climate change became strong by the end of the twentieth century, with the United Nations and national governments paying increasing attention to the findings of a series of reports by an international scientific panel as early as 2001.[124] The tardily acknowledged link proved an opportunity to nominate nuclear power as a desirable alternative. Proponents of the 'nuclear renaissance' mooted the nuclear fission fuel cycle of energy production to generate considerably lower quantities of carbon than alternatives such as coal- or gas-fired power plants, although the full-cycle costs—like those of the original fission reactor and breeder reactor programmes—remained unclear and contentious.[125] Thus, for the first time since the 1950s, nuclear engineers were cautiously portrayed not merely as competent technicians but as providers of technical solutions of tangible social and environmental benefit. The popular 1950s dream of clean cities could now be recast in the even more seductive vision of a sustainable planet. For the Russian Federation, in fact, this approached technological utopianism akin to that of mid-century.[126] For nuclear specialists, it represented a new opportunity—if a relatively brief and little acknowledged one before Fukushima.

The representations of these now-hidden, now-revealed specialists has altered repeatedly. Depicted for a decade as diffident and enigmatic Cold Warriors, and for a further decade as shadowy creators of progressive technologies by their governments, the public image of nuclear workers was constructed largely from a vacuum of fact. From the 1950s nuclear engineers had been presented as a voiceless collection of experts, their achievements trumpeted second-hand by their governments. Wider cultural perceptions remained enduringly flavoured by those secretive early years; divorced from direct evidence, stereotypes and clichés were sustained and episodically revived.

Over seven decades, then—the span of two or three working careers—nuclear specialists were transmuted from heroic geniuses to inventive technical experts, to ever more distant engineering technicians characterized as the pawns of corporate interests or even as potential environmental saviours. Their own voices remained unusually muted and, on occasion, actively stifled by their governments and employers. As a 1990s popular account—midway between Chernobyl and Fukushima—summarized it, 'they would love to be loved again'.[127]

---

[123] Hall, H. L., Jr Dodds, H. L., J. P. Hayward, L. H. Heilbronn, J. W. Hines, H. Liao, G. I. Maldonado, L. F. Miller, R. E. Pevey, A. E. Ruggles, L. W. Townsend and B. R. Uphadhyaya, 'Nuclear engineering and nuclear security: a growing emphasis at the University of Tennessee', presented at *Proceedings of the Pacific Northwest International Conference on Global Nuclear Security–the Decade Ahead*, 2010.

[124] The International Panel on Climate Change (IPCC), reporting in 1990, 1992, 1995, 2001, and 2007.

[125] For an early discussion see, for example, Weinberg, Alvin M., *The First Nuclear Era: The Life and Times of a Technological Fixer* (New York: AIP Press, 1994), pp. 233–7.

[126] Josephson, Paul R., 'Technological utopianism in the twenty-first century: Russia's nuclear future', *History and Technology* 19 (2003): 277–92.

[127] Bolter, Harold, *Inside Sellafield: Taking the Lid off the World's Nuclear Dustbin* (London: Quartet, 1996), p. 30.

This story of shifting identities suggests little trace of stability. Nuclear engineers have seen their public identities successively revealed, vaunted, and criticized, usually shaped by others in a changing context that melded technical know-how with politics, economics, social goals, and capricious events. Their shifting cultural representations reveal the fickle trajectory of a state managed discipline.

9

CONCLUSIONS: CAREERS FROM THE MANHATTAN PROJECT TO FUKUSHIMA

The Anglo-Saxon atom experts have been an elusive target in this book, tracked over challenging historical terrain. Born during wartime secrecy, the viability of these ad hoc specialists was disputed; in secure post-war environments, their growing expertise was hidden between the lines of promotional press releases and the archived reports of government and industry. And—less secure and vocal than their cousins, the atomic scientists—they were represented second-hand by their employers and via their contentious products. Constructing a stable identity proved to be a perennial battle.

Andrew Abbott's identification of *jurisdiction* as a key to understanding professional identity has provided a fruitful perspective for the preceding chapters. Jurisdictional disputes about boundaries—how jobs, expertise, and responsibilities are apportioned—defined the technical workplace and gave momentum to institutional developments. The history of these clashes is the real history of technical specialisms such as nuclear engineering. To explain the nature of this knowledge, then, requires an assessment of the ecology of professions in which it is embedded. And by intercomparing dissimilar environments and their ecosystems, generic evolutionary influences may be discerned.[1]

This final chapter delineates the forces that shaped the mutable identities, and examines what a comparative approach reveals. As detailed in the preceding chapters, those factors include (a) the wartime butting-up of a burgeoning scientific field and pre-existing industrial cultures; (b) post-war concentration at secure institutional sites, where categorizations of expertise set enduring precedents; (c) a pervasive role for the state and its institutions in mediating a staged emergence and in controlling occupational conditions; and (d) public representations constructed largely by employers and by perceptions of engineering successes and failures.[2]

9.1 Fertile environments

To a remarkable degree, the evolution of nuclear specialists was played out in environments shaped and directed by government funding, political edicts, and international agreements. But, as I have argued, a bottom-up perspective is needed to appreciate the consequences for their occupations, disciplines, and professions. The nascent cadres were simultaneously

---

[1] Abbott, Andrew D., *The System of Professions: An Essay on the Division of Expert Labor* (Chicago: University of Chicago Press, 1988), pp. 2, 124.

[2] These influences correspond to chapter coverage: (a) Chapters 2 and 3; (b) Chapters 4 and 5; (c) Chapters 6 and 7; (d) Chapter 8.

cosseted and isolated; their work was both free-running and rigidly constrained; their goals and products were intimately wedded to government policies and yet divorced from public awareness. The evolution of nuclear engineering proved to be highly contingent on circumstances, shaped in isolated environments in which local context and key individuals played decisive roles. As a result, social accommodations were, to adopt Abbott's words, 'fluctuating and geographic'. Jurisdictional problems were identified and solved to suit the players in each locale.[3]

Time, as well as place, was critical. Technical identity was defined in discrete phases and experienced sudden ruptures at well-defined phase transitions. Security demarcated the early field, with three periods identifying distinct threats from the circulation of knowledge and personnel: (i) the wartime Manhattan Project; (ii) the post-war decade between the 1946 Atomic Energy Act and the 1955 Geneva Conference; and (iii) the subsequent beginnings of the commercialization of nuclear power. National governments played an exceptional role, defining the nature of the subject by filtering the workforce, subsidizing and segregating development, categorizing occupational identity, influencing disciplinary categories in higher education, and supporting particular forms of professional identity. As 'a form of government regulation', quoting Daniel Moynihan, secrecy shaped these specialists and fostered nationally distinct versions.[4]

But in these settings, the new national labs fostered innovation and a divergence of expertise. The pent-up desire to investigate new applications led to a proliferation after the war of new reactor designs nurtured in distinct environments. Staff at the Argonne and Oak Ridge National Laboratories vied for novelty in reactor development, and both sites attracted powerful sponsors; the urgent demands of the UK Ministry of Supply's Atomic Energy Division similarly focused efforts at Harwell and Risley on air-cooled reactors. Chalk River's contrasting focus on a single technology—the heavy-water reactor—for some time identified no clear objective for Canada's expensive Atomic Energy Project.

The importance of historical contingency, even in what is often assumed to have been a richly supported endeavour, is supported strongly by the reflections of Alvin Weinberg, the prototypical American nuclear engineer. Weinberg argued that the eventual world-wide prevalence of light-water reactors could be traced to the adoption of this technology for the *Nautilus* submarine power-plant of 1954 by Admiral Hyman Rickover. This snowballing of a particular technological solution into a system of mutually reinforcing components was dubbed *technological momentum* by American historian Thomas Hughes.[5] The rapid

---

[3] Abbott, Andrew D., *The System of Professions: An Essay on the Division of Expert Labor* (Chicago: University of Chicago Press, 1988), p. xiv. On the relevance of local perspectives over national ones, see Pyenson, Lewis, 'An end to national science: the meaning and extension of local knowledge', *History of Science* 40 (2002): 251–90.

[4] Commission on Protecting and Reducing Government Secrecy, 'Secrecy: Report of the Commission on Protecting and Reducing Government Secrecy', Government Printing Office, 1997, p. xxi.

[5] Hughes, Thomas Parke, 'Technological momentum', in: M. R. Smith and L. Marx (eds.), *Does Technology Drive History? The Dilemma of Technological Determinism* (Cambridge, Mass.,: MIT Press, 1994), pp. 101–14. See also Cowan, Robin, 'Nuclear power reactors: a study in technological lock-in', *The Journal of Economic History* 50 (1990): 541–67.

development of know-how surrounding this design at Oak Ridge, and the organization's seminal role in training American nuclear engineers, provided commercial firms with the background needed to successfully export their designs and to support the training of further specialists with comparable experience. At the same time, nuclear workers in the government organizations learned from their industrial counterparts: forays from Hanford, Oak Ridge, and Argonne gleaned knowledge about commercial operations. These interconnected technological, educational, corporate, and institutional networks generated momentum that subsequently proved difficult to redirect.[6]

As their products were consolidated, so too were professional identities. Nuclear reactor design, unlike some other aspects of the Manhattan Project, offered relatively clear career potential. Reactors provided novel and open-ended opportunities for exploring design concepts and as generators of phenomena. Dedicated to power generation, their design richness gradually diverted their specialists towards more frank engineering and production concerns. The early secrecy surrounding nuclear expertise facilitated this, keeping the distinct national flavours of nuclear engineering bottled up and free of wider social or commercial taint, unlike other nascent professions. Reactor engineering arguably held a Goldilocks position in terms of security, being more secretive and hence more regionally delineated than other new forms of nuclear knowledge having wider application such as radiochemistry and nuclear chemical engineering, and yet less classified than weapons development.

This technical divergence developed alongside other professional qualities. Abbott's focus reveals the relatively weak professional status of nuclear specialists, especially in terms of autonomy and the ethical criteria that often follow from it. Their failure to identify and adopt recognized responsibilities set them apart from many twentieth-century professionals, for whom ethical standards conferred public status and occupational distinctiveness. They also lacked a well-developed and stable expert–client relationship, unlike lawyers, psychiatrists or, indeed, chemical engineers. The context of employment was a key factor: early nuclear engineers were institutionally based rather than entrepreneurial, and served their governments more often than customers on clearly conceived projects.[7]

While professional aspirations were often unachieved, occupational security was also uneven. Jurisdictional disputes were played out differently in the three countries: in Britain, by the unsuccessful battles to form distinctive labour unions, a high-status profession represented by the Institution of Nuclear Engineers, and a stable discipline represented by academic programmes of nuclear engineering; in the USA, by the long-running challenges of Du Pont engineers to usurp the dominance of Met Lab and National Lab scientists, and the eventual establishment of nuclear engineering curricula and licensing by industry–university–government collaborations; and in Canada, with the relatively placid shepherding by the National Research Council, the avoidance of jurisdictional conflicts, and the flowering of occupational labels.

---

[6] Weinberg, Alvin M., *The First Nuclear Era: The Life and Times of a Technological Fixer* (New York: AIP Press, 1994), p. 133.

[7] Abbott, Andrew D., *The System of Professions: An Essay on the Division of Expert Labor* (Chicago: University of Chicago Press, 1988), pp. 4–8, 15–16; quotation p.8.

More widely successful was the recognition of the abstract knowledge that underpinned nuclear engineering. Abbott identifies the 'path to abstraction' as an essential feature in the ecology of professions, allowing a group to find a sustainable niche by its mastery of generic skills.[8] The American focus on radioactivity as the basis of the field—unlike the British attention to constituent disciplines—provided a basis for consolidating and extending the intellectual territory of nuclear engineers.

By the early 1960s, civilian nuclear power provided new working environments. In 1962, the first pilot Canadian nuclear power station became operational at Rolphton, Ontario, and nuclear workers began the gradual expansion into industrial firms. In Britain, the Berkeley Power Station became operational and represented a break with military origins. This second-generation design, optimized for power generation rather than plutonium production, offered the prospect of a British discipline cleaved more cleanly from its wartime origins. And in the USA that year, nuclear engineering as a discipline and nuclear engineers as a profession were consolidated by the ANS/ASEE report agreeing university curricula and accreditation. With these developments, a viable occupational status became established in each country. The aspirational label 'atomic engineer', circulating intermittently during the 1950s, was trumped by the steady rise of 'nuclear engineer' in contemporary texts and a growing number of job advertisements.[9] In each of the three founding countries, these technical specialists had achieved varying degrees of institutional stability—if not public perception—that were to endure for another two decades.[10]

This later phase for the profession was thus characterized by the proliferation of commercial nuclear power plants during a period of profound social shifts. The authority of central governments—and, with them, the scientific authority that had been state-directed since the war—fell precipitously. Public challenges to the assurances of nuclear facility administrators were increasingly directed at their designers and operators, too. The link between governments and science that had developed with the atomic bomb proved to taint the identity of civilian nuclear workers by the early 1970s. And with the decline of national nuclear programmes after Three Mile Island and Chernobyl, nuclear engineers fell in status and numbers—a phase of their existence that has been underlined by Fukushima in the early twenty-first century.

---

[8] Ibid., p. 93.

[9] Similarly, the term 'atomic engineering' was used increasingly in English-language technical sources such as the *Bulletin of the Atomic Scientists* until about 1960, but attained much less currency than either 'nuclear engineering' or 'nuclear science' [Google Ngram viewer, http://ngrams.googlelabs.com/graph?content=atomic+engineering%2C+nuclear+engineering%2C+nuclear+science&year_start=1930&year_end=2011].

[10] Paul Josephson asserts that nuclear engineers played an enduring role in Soviet culture, being identified as the key to economic and social progress, and embodying the ideals of mass production [Josephson, Paul R., 'Atomic-powered communism: nuclear culture in the postwar USSR', *Slavic Review* 55 (1996): 297–324]. Even more strongly, Dolores Augustine argues that engineers became closely identified with the East German dictatorship and its socialist ideals [Augustine, Dolores, *Red Prometheus: Engineering and Dictatorship in East Germany, 1945–1990* (Cambridge, Mass.: MIT Press, 2007)].

9.2 Fragile ecosystems

These discrete geographical and temporal environments shaped the identities of nuclear engineers. Bringing together different technical cultures, workplaces imposed an accommodation on the juxtaposed disciplines. In some contexts, a sustainable equilibrium of coexisting professionals was established; in others, nuclear engineers dominated the ecosystem or, alternatively, occupied a precarious occupational niche.

The earliest traces of a distinct specialism of nuclear engineering emerged at the working interface between scientists and industrial engineers. Three groupings were significant during the war: the British MAUD Committee, which brought together academic scientists with ICI engineers; the University of Chicago Metallurgical Laboratory (and its programme offshoots, the Clinton laboratory and Hanford Engineer Works), where physicists and Du Pont engineers were forced to interact; and the Montreal Laboratory (and its successor, Chalk River), where ICI and the National Research Council melded an assortment of scientists under engineer administrators. At each site, professional identities were shaped by renegotiating power relations.

Not all wartime working environments proved sustainable. Senior nuclear physicists—men like Enrico Fermi, Rudolf Peierls, and Otto Frisch—had relatively little impact on the post-war directions or post-war identity of nuclear workers, apart from the brief attention accorded to the social and political views of the atomic scientists in the first years after the war. All of them, as pre-war physicists, had matured in an academic environment largely free of the concerns of external funding and, with it, external influence. Most were eager to return to 'pure' research after the war, but their recent administrative expertise yielded more optimal post-war environments. Physicist I. I. Rabi and others, for example, lobbied for a brand-new national laboratory to service universities of the north-east. At Brookhaven, scientists and engineers would study not just chain reactors, but particle accelerators and their products. Thus sustainable scientific environments, too, were made more congenial by engineering experience.

By contrast, the wartime workers lower in the hierarchy proved to be active constructors of post-war nuclear engineering. Walter Zinn, Fermi's sometime assistant, assumed direction of Argonne; Alvin Weinberg assumed a similar role at Oak Ridge; the industrial chemist Christopher Hinton, dubbed a 'nuclear baron' by contemporaries, was appointed by the UK government to build the atomic factories necessary for a British atomic bomb and, along with John Cockcroft and William Penney, he received a knighthood for the job. They and others effectively shaped the nature of nuclear specialists, their goals, and their criteria of success.

The sustainability of these managed occupational ecosystems is evident in the formation and growth of the American National Laboratories, particularly Argonne and Oak Ridge. Zinn and his colleagues made strong efforts to define nucleonics, to seed it with ex-Met Lab scientists, and to implant and nurture it at ANL. The design of atomic piles was to be their focus for this new discipline, and scientists were to be on the ascendant. At Oak Ridge, a similar post-war cohort grew, promoting a working environment richer in engineers than that of ANL.

Abbott has argued that the rapidly growing field of electrical engineering attained a similarly heterogeneous balance. He notes that radio engineering had become established via two strategies: growth/division and abstraction. The rapid expansion of low-voltage technology during the 1920s grew the population of its specialists, who eventually surpassed their counterparts in power engineering. Establishing a distinct identity and professional organization, they buttressed their position by hybridizing expertise from physics. Beginning with Vannevar Bush's population of the wartime Radiation Laboratory with scientists as well as engineers, PhD physicists continued to dominate research in the field, particularly at major research firms such as Bell Laboratories. Electrical engineers were thus transformed from practical specialists to members of an academically embedded discipline.[11] In such a rapidly evolving field, career progression of the first generation involved riding the wave of technical progress or else boosting their skills through midlife training at short courses.

His analysis is pertinent to the field of nuclear engineering and to key historical actors such as Wigner, Cockcroft, Zinn, and Weinberg. Usually labelled as physicists owing to their formal training and the subject's higher social ranking in the post-war period, their work was substantially engineering in orientation. Where Wigner and Cockcroft retained their established scientific identities, their younger counterparts such as Zinn and Weinberg built their careers through the technological innovations they managed. In their separate ways, Walter Zinn's ANL, Alvin Weinberg's ORNL, C. J. MacKenzie's Chalk River, and Christopher Hinton's Industrial Group embodied authority, inertia, and national goals. Each director fostered a distinct working culture and abstraction of the subject.

Like the technological momentum they facilitated, institutional inertia was a significant factor for each of these centres. In the USA, the early wartime siting decisions made by Crawford Greenewalt had enduring consequences, fissioning the original Met Lab scientific team to Argonne, Clinton, and Hanford. As post-war Director of Du Pont, Greenewalt also oversaw the company's adoption of the Savannah River Project, which had its own profound influence on the practice and institutionalization of American nuclear engineering. Management of the Canadian project by the NRC, and project continuity by MacKenzie, fixed its location, remit, and outlook. And the overwhelming impact of ICI's dominance in Britain meant that its working methods, personnel hierarchies, and the bulk of Britain's reactor know-how became embedded in the UKAEA. These organizations thus had a lasting impact in directing the field in the three countries.

9.3 Critical conditions

This book has explored three national experiences of atomic energy and the nurturing of their specialists. Its aim has not been to argue for unique national identities or for their seminal status, but to better correlate outcomes with historical conditions. The co-evolution of technical specialists in three countries—in the contrasting contexts of wartime collaborative development, post-war classified exploration, and subsequent commercialization—offers an advantageous analytical view. We can intercompare and, to a degree, unpick the distinct historical contingencies that shaped their professional destinies. Factors common to all three countries give credence to explanations that explain shared experiences; unique contextual factors, on the other hand, provide plausible evidence for understanding national differences in engineering distinctiveness.

---

[11] Abbott, Andrew D., *The System of Professions: An Essay on the Division of Expert Labor* (Chicago: University of Chicago Press, 1988), pp. 179–82.

Among the obvious influences is political context. Prior studies of the Soviet, East German, Spanish, and Portuguese experiences of nuclear energy have suggested common institutional features attributable to the authoritarian regimes of Josef Stalin/Nikita Khruschev/Leonid Brezhnev, Walter Ulbricht/Erich Honecker, Francisco Franco, and António de Oliveira Salazar, respectively.[12] On the other hand, as I have argued, the experiences of the Anglo-Saxon democracies—which pursued dissimilar civilian and military objectives after the war—had remarkably similar engagements with security apparatus and state management surrounding nuclear power.[13] Security framed the early maturation of the field and is integral to understanding its later development. Owing to shared concerns about domestic and foreign espionage, the severity of security procedures in Britain and Canada paralleled American changes—an occurrence somewhat at odds with Edward Shils's depiction of distinct patterns of public 'luxuriating publicity' jostling with 'hyperpatriotism' in the USA, as opposed to the dominance of deference and privacy in the UK.[14] In each country, political allegiances were scrutinized, selecting a cohort of nuclear specialists that met official guidelines but remained potentially suspect. The state monopoly on atomic energy instilled an unusual degree of common interest with its experts and industrial contractors. This commonality of corporatist elements—extending, in some respects, to the Iberian countries and arguably others—suggests that the pursuit of nuclear energy has been associated with a degree of technological determinism. As suggested by Langdon Winner, nuclear technology may require or even produce particular political environments.[15]

From a present-day perspective, of course, similarities dominate. Globalization and increasingly free international exchange of information and personnel have tended to erase national differences or at least to allow one version of nuclear engineering expertise to dominate more easily, as in the case of the light-water reactor.[16] Indeed, the export of

---

[12] Josephson, Paul R., *Red Atom: Russia's Nuclear Power Program from Stalin to Today* (New York: W. H. Freeman, 1999), especially Chapter 2; Augustine, Dolores, *Red Prometheus: Engineering and Dictatorship in East Germany, 1945–1990* (Cambridge, Mass.: MIT Press, 2007), pp. 115–18; Presas i Puig, Albert, 'Science on the periphery. The Spanish reception of nuclear energy: an attempt at modernity?' *Minerva* 43 (2005): 197–218; Gaspar, Júlia 'The two Iberian nuclear programmes: post-war scientific endeavours in a comparative approach (1948–1973)', presented at *7th Meeting, Science and Technology in the European Periphery*, Galway, Ireland, 2010.

[13] On the corresponding Soviet experience of centrally organized atomic energy research and development, and the creation of 'Big Science', see Kojevnikov, Alexei, *Stalin's Great Science: The Times and Adventures of Soviet Physicists* (London: Imperial College Press, 2004).

[14] Shils, Edward, *The Torment of Secrecy: The Background and Consequences of American Security Policies* (Glencoe, Ill.: Free Press, 1956), Sections 2-II and 2-IV.

[15] Winner, Langdon, 'Do artifacts have politics?' in: L. Winner (ed.), *The Whale and the Reactor: A Search for Limits in an Age of High Technology* (Chicago: University of Chicago Press, 1986), pp. 19–39. For a related argument in relation to alternative power sources, see Hayes, Denis, *Rays of Hope: The Transition to a Post-Petroleum World* (New York: W. W. Norton, 1977), pp. 71 and 159.

[16] A similar homogenization was seen in chemical engineering, which evinced distinct national traits in America, Britain, and Germany before the Second World War. These diminished from the post-war period, when American petroleum plant designers came to dominate European developments, as discussed in Divall, Colin and Sean F. Johnston, *Scaling Up: The Institution of Chemical Engineers and the Rise of a New Profession* (Dordrecht: Kluwer Academic, 2000).

hardware and engineering proficiencies became significant aims after the Geneva Conference of 1955. All three countries instituted internal training programmes within their new organizations. The existence and broad curricula of some of these courses was publicized internationally, even when restricted to a favoured few, thus serving as templates for other countries. The Oak Ridge School of Reactor Technology (1950), mirrored by the Harwell Reactor School (1954) and Chalk River Reactor School (1959), admitted foreign students, exporting national perspectives to the graduates who implemented them in their own countries. The same was true of the Soviet Union: from the late 1950s, Igor Kurchatov sought to establish nuclear programmes in the Soviet republics. His successors trained foreign nuclear engineers from friendly countries to establish a Soviet reactor export market.[17]

In conjunction with these shared features which suggest deterministic features in the distinguishing of nuclear expertise, there were also differences in national experience that can be attributed to their unique social contexts. National contingencies were most marked by differences in military influence, and this strongly shaped the workers in each country. The sudden termination of the Manhattan Engineer District, and its replacement by the civilian AEC, masked continuing military interests. While enthusiastic to pursue technical ideas identified during the war, scientists and engineers at the new American post-war laboratories had unclear goals. This vacuum of purpose was rapidly filled by sponsors from the armed forces, the Navy seeking propulsion power-plants for submarines, the Air Force funding development of nuclear-powered aircraft, and the Army seeking more compact, powerful, and efficient nuclear weapons, continuing the programme it had managed during the Manhattan Project. While there was a degree of separation between these sites and the experience they embodied, the domination of military funding guided their specialists. Oak Ridge chemists, dedicated to developing separation processes, influenced the reactor schemes investigated there and at Argonne. Los Alamos scientists conceiving new weapons concepts that involved novel radioactive species—uranium-233, plutonium-239, tritium, thorium, and others—encouraged reactor designers, radiochemists, and chemical engineers to explore a wide range of systems. Engineers at Hanford and Savannah River, called upon to manufacture these materials, tied together the specialist knowledge refined at these sites. Nationally defined goals for advanced weapons systems, military propulsion, and portable power systems therefore linked American workers in a socio-technical system and fostered a distinct complement of expertise.

British nuclear workers were moulded according to a different national template. They, too, worked toward the goal of a British nuclear weapon in the post-war decade, but with distinctive management. Harwell, Risley, Springfields, and Aldermaston sprang up on former military sites. Their first planners and workers were reabsorbed from the Anglo-Canadian project and from the Ministry of Supply, transplanting a working culture that had grown in the British chemical industry. The civil service tone was sustained by the relatively distant links to military needs (defined more by government policy than by strong proponents within the armed forces). Surprisingly, then—given the late start for the British

---

[17] Josephson, Paul R., *Red Atom: Russia's Nuclear Power Program from Stalin to Today* (New York: W. H. Freeman, 1999), pp. 204–5.

atomic energy programme—its notable characteristic was the post-war continuity of wartime practices.

By contrast, Canadian nuclear workers had no practical involvement with bomb development during the war and publicly shunned such aims afterwards. Those Canadians who had contributed directly to the Manhattan Project were usually tempted into the various post-war US national laboratories. As a result, Canadian nuclear workers at Chalk River soaked up a research council working culture. The budding AECL focused efforts towards newly identified customers in the mid to late 1950s: medical and biological applications of radioisotopes, and electrical utility companies. The technical focus on heavy-water reactors further honed the unique profile of Canadian nuclear specialists.

The labelling of knowledge was also nationally distinct. While 'nuclear engineer' became a viable label at American research and production sites directed by the US Atomic Energy Commission and administered by firms such as Du Pont and General Electric and consolidated by their support of the fledgling American Nuclear Society, the term remained contested in the UK.[18] And Canada's programme was equally influenced by institutional cultures: growing from the National Research Council environment of close cooperation between engineers and scientists, specialists had relative freedom to explore reactor technologies and, for the lower-ranked engineers and skilled workers, at least, to vaunt their skills through the creation of new labour unions. These distinct national experiences, much like divergent siblings, owed much to their early segregation, allowing the original complement of staff and organizational mind-set to define an independent course, and sustained by the technological momentum of their reactor technologies of choice. In each country, the trajectories for nuclear specialists served national interests rather than aims of professional identity. This had enduring implications not just for the jobs they performed, but for their professional identity and their capacity to engage effectively with a wider public.

9.4 Between anonymity and rhetoric

The origins and evolution of nuclear engineers were exceptional. They appeared suddenly: not evolving gradually in a familiar social and commercial context, but created by timely scientific discoveries and harnessed to urgent wartime goals. Sprouting from a single seductive idea—the fission chain reaction mediated by the neutron—distinct national contexts created three flavours of Anglo-Saxon nuclear expert in the decades that followed. The emergence and maturation of these specialists was contingent on their shared collective experiences of institutional cultures, national goals, and state intervention.

In the American, British, and Canadian atomic energy programmes, ample and uncritical early funding had yielded rapid growth, but in a sterile environment lacking essential nutrients, during which participants were starved periodically of information, resources, and opportunities to interact. If the field of nuclear engineering were considered in the framework of developmental psychology, the neutron's children might be perceived as suffering

---

[18] A survey of job advertisements in *New Scientist* through the 1960s, for example, reveals that *lecturers in nuclear engineering* were sought, but that the specific term *nuclear engineer* was infrequently used. The magazine increasingly used the term as a popular label in articles from the late 1970s, however.

from arrested development, peculiar idiosyncrasies and worldview, insecure self-image, weak communication skills, and poor socialization with their peers. The gradual estrangement of their governments and the traumatic experiences of Three Mile Island, Chernobyl, and Fukushima further shaped their identity. One might hope that, with redefined goals, the profession's transition to a tardy maturity during the twenty-first century may offer better social integration.

APPENDIX I: ACRONYMS, ORGANIZATIONS, AND YEAR OF ORIGIN

The nuclear field is burdened with a bewildering number of acronyms, mainly for technical terms. The following are mainly *organizational* acronyms employed in this book.

AEC	Atomic Energy Commission (USA, 1946)
AECB	Atomic Energy Control Board (Canada, 1946)
AECL	Atomic Energy of Canada Ltd (Canada, 1952)
AED	Atomic Energy Division, Du Pont (USA, 1950)
AERE	Atomic Energy Research Establishment, Harwell (UK, 1946)
AEOR	Atomic Engineers of Oak Ridge (USA, 1945)
AFL	American Federation of Labor (USA, 1886)
AIChE	American Institute of Chemical Engineers (USA, 1908)
AIF	Atomic Industrial Forum (USA, 1953)
AMU	Association of Midwest Universities (USA, 1958)
ANL	Argonne National Laboratories (USA, 1946)
ANS	American Nuclear Society (USA, 1954)
AORS	Association of Oak Ridge Scientists (USA, 1945)
AORES	Association of Oak Ridge Engineers and Scientists (USA, 1946)
APSOR	Atomic Production Scientists of Oak Ridge (USA, 1945)
ARWU	Atomic Research Workers' Union (Canada, 1952)
ASA	Atomic Scientists Association (UK, 1946)
ASC	Atomic Scientists of Chicago (USA, 1945)
ASEE	American Society for Engineering Education (USA, 1893)
AUA	Argonne Universities Association (USA, 1965)
AUI	Associated Universities Inc. (USA, 1947)
AWRE	Atomic Weapons Research Establishment (UK, 1947)
BCPMA	British Chemical Plant Manufacturers' Association (UK, 1920)
BNEC	British Nuclear Energy Conference (UK, 1955)
BNES	British Nuclear Energy Society (UK, 1960)
BNFL	British Nuclear Fuels Ltd (UK, 1971)
BWR	Boiling Water Reactor
CANDU	Canadian Deuterium reactor (Canada, 1958)
CEGB	Central Electricity Generating Board (UK, 1957)
CONE	Career Oriented Nuclear Engineering Program (USA, 1972)
CNA	Canadian Nuclear Association (Canada, 1960)
CGE	Canadian General Electric (Canada, 1892)
CNS	Canadian Nuclear Society (Canada, 1979)
CTS	Clinton Training School, Oak Ridge (USA, 1946)
DIL	Defence Industries Limited (Canada, 1939)
DNET	Division of Nuclear Education and Training, AEC (USA, 1955)

DOE	Department of Energy (USA, 1977)
DSIR	Department of Scientific and Industrial Research (UK, 1916)
ENS	European Nuclear Society (1975)
ERDA	Energy Research and Development Administration (USA, 1975)
FAS	Federation of Atomic Scientists (USA, 1946)
HAMTC	Hanford Atomic Metal Trades Council (USA, 195?)
HEW	Hanford Engineer Works (USA, 1943)
IAEA	International Atomic Energy Agency (UN, 1957)
IChemE	Institution of Chemical Engineers (UK, 1922)
ICI	Imperial Chemical Industries (UK, 1926)
IMechE	Institution of Mechanical Engineers (UK, 1847)
INucE	Institution of Nuclear Engineers (UK, 1959)
MED	Manhattan Engineer District (USA, 1942)
MIT	Massachusetts Institute of Technology (USA, 1861)
MURA	Midwest Universities Research Association (USA, 1952)
NDRC	National Defense Research Committee (USA, 1940)
NEA	Nuclear Energy Agency (International, 1958 as European Nuclear Energy Agency, and 1972 as NEA)
NEEC	Nuclear Engineering Education Committee, Association of Midwest Universities (USA, 1958)
NEPA	Nuclear Energy for the Propulsion of Aircraft (USA, 1946)
NI	Nuclear Institute (UK, 2009)
NRC	National Research Council of Canada (Canada, 1916)
NRC	Nuclear Regulatory Commission (USA, 1974)
NUAW	National Union of Atomic Workers (UK, 1958)
OECD	Organization for Economic Co-operation and Development (International, 1961)
ORAU	Oak Ridge Associated Universities (USA, 1966, successor to ORINS)
ORINS	Oak Ridge Institute of Nuclear Studies (USA, 1947)
ORNL	Oak Ridge National Laboratory (USA, 1942)
ORSORT	Oak Ridge School of Reactor Technology (USA, 1950)
OSRD	Office of Scientific Research and Development (USA, 1941)
PWR	Pressurized Water Reactor
RSRL	Research Sites Restoration Limited (UK, 2009)
SSFL	Santa Susana Field Laboratory (USA, 1948)
UKAEA	United Kingdom Atomic Energy Authority (UK, 1954)
USAEC	See AEC
WNRE	Whiteshell Nuclear Research Establishment (Canada, 1963)

APPENDIX II: NUCLEAR ENGINEERING PERIODICALS SINCE 1945

As discussed in the text, the publishers, titles, and content of British and North American trade magazines and institutional periodicals—alternately stressing technical advances, news of the industry, occupational challenges, and skills sets—reveal the shifting identities of their nuclear specialists.

Titles are listed by date of origin, but clustered to indicate succeeding titles.
Atomic Engineering (1945–6)
Atomic Power (New York: McGraw-Hill, 1946)
   Nucleonics (New York: McGraw-Hill, 1947–67)
      Nucleonics Week (New York: McGraw-Hill, 1967–)
Atomic Engineer and Scientist (AORES, 1946)
Bulletin of the Atomic Scientists of Chicago (Chicago: Atomic Scientists of Chicago, 1946)
   Bulletin of the Atomic Scientists (Educational Foundation for Nuclear Science, 1946–)
Atomic Scientists' News (ASA, 1947–53)
   Atomic Scientists' Journal (ASA, 1953–6)
      New Scientist (1956–)
Atomics (London: Leonard Hill Technical Group, 1949–53)
   Atomics and Atomic Technology (London: Leonard Hill, 1953–5)
      Atomics Engineering and Technology (London: Leonard Hill, 1956)
         Atomics and Nuclear Energy (London: Leonard Hill, 1957–8)
            Atomic World – The Journal of International Nuclear Industry (London: Leonard Hill, 1958–9)
Journal of Reactor Science and Technology (ORNL, 1951–?; classified)
Journal of Nuclear Energy (London, New York: Pergamon, 1954–73)
   Annals of Nuclear Science and Engineering (Oxford: Pergamon, 1974)
      Annals of Nuclear Energy (Oxford: Pergamon, 1975–)
IRE Transactions on Nuclear Science (IRE, 1955–62)
   IEEE Transactions on Nuclear Science (IEEE, 1963–)
Nuclear Science and Engineering (ANS, 1956–)
Reactor Technology (AEC/ANS, 1957–67; 1970–73)
   Reactor and Fuel Processing Technology (AEC/ANL, 1967–9)
Journal of the British Nuclear Energy Conference (BNEC, 1956–61)
   Journal of the British Nuclear Energy Society (BNES, 1962–77)
      Nuclear Energy (BNES 1978–2004; BNES and INucE, 2004–)
         Nuclear Future (INucE and BNES, 2005–)

Nuclear Engineering (London: Temple Press, 1956–68)
　Nuclear Engineering International (London: Temple Industrial Publications, 1968–)
Nuclear Power—the Journal of British Nuclear Engineering (London: Rowse Muir, 1956–63)
Progress in Nuclear Energy (Oxford: Pergamon, 1956–)
Combustion, Boiler House and Nuclear Review (London: Princes Press, 1957–8)
　Nuclear Energy Engineer (London: Princes Press, 1958–9)
　　Nuclear Energy (INucE, 1959–71)
　　　Journal of the Institution of Nuclear Engineers (INucE, 1971–80) *Atomic Energy News: Bulletin of the Institution of Nuclear Engineers* (1971–82)
　　　　Nuclear Engineer (INucE, 1980–2004)
Transactions of the American Nuclear Society (ANS, 1958–)
Nuclear News (ANS, 1959–)
Nuclear Safety (ORNL, 1959–79)
Advances in Nuclear Science and Technology (New York: Plenum, 1962–)
Nuclear Applications (ANS, 1965–9)
　Nuclear Applications and Technology (ANS, 1969)
　　Nuclear Technology (ANS, 1969–)
Nuclear Engineering Bulletin (ANS, 1967–8)
Nuclear Technology International (London: Sterling, 1987?–)
Nuclear Journal of Canada (CNS, 1987)
Canadian Nuclear Society Bulletin (CNS, 1988–)
BNFL Engineer (Manchester: BNFL Engineering Group, 1990–)
Fuel Cycle Review (Sutton, UK: Reed Publishing, 1990–)
International Journal of Nuclear Energy Science and Technology (Olney, UK: Inderscience, 2004–)
Science and Technology of Nuclear Installations (New York: Hindawi Publishing, 2006–)

APPENDIX III: ARCHIVAL SOURCES

The documentation of nuclear history has become accessible episodically and patchily since the Second World War. The UK government announced its intention to open records as early as 1966, well before the originally planned 50-year closure period. The transfer of 'the cream of wartime records' to the National Archives began in 1972,[1] a process that, at the time of writing, is still incomplete for the bulk of records at the UKAEA facilities at Harwell, Dounreay, and Winfrith. Such records can be sought under the UK Freedom of Information Act, but this is often a blind and inefficient process unless made easier by a direct visit, as proved possible at Dounreay, thanks to Records Manager Ian Pearson. The creation of British Nuclear Fuels Ltd from the UKAEA Production Group in 1971 exempted the new organization from subsequent public access, although it maintained its own archive. The personal files of prominent figures in the field, such as Christopher Hinton and W. Bennett Lewis, have been archived at academic and institutional repositories.

In Canada, the nuclear programme has been documented at Libraries and Archives Canada through the files of the National Research Council (responsible for the wartime Montreal Laboratory and Chalk River establishment) and thereafter Atomic Energy of Canada Ltd. As in the UK, individual universities have archived the papers of a handful of government departments, career workers, and university faculty active in the field.

In the USA, there has been a more dispersed preservation and archiving of documents. Records of the Manhattan Engineer District, superseded by the Atomic Energy Commission in 1946 and the Department of Energy in 1977, remained substantially at the relevant laboratories.

For all three countries, records of professional bodies in the nuclear industry generally have been difficult to obtain, reflecting perhaps the reservations of institutions that have been embattled from a variety of sources over the preceding half-century.

---

[1] 'Raw Materials of History – UK Atomic Energy Records Handbook', 1 July 1976, Dounreay AEA/DRO/1.

Abbreviations used for archive citations

CC	Churchill College, Cambridge, UK: Sir Harold Hartley papers.
DOE DDRS	US Department of Energy Declassified Document Retrieval System (online database), http://www5.hanford.gov/ddrs: Hanford Engineering Works archive.
DOE ON	US Department of Energy OpenNet (online database), https://www.osti.gov/opennet/: archive of AEC and DOE files declassified since 1994.
Dounreay	UKAEA, Dounreay, UK: UKAEA records of the Dounreay site.
Hagley	Hagley Museum and Archives, Wilmington, Delaware, USA: E. I. Du Pont de Nemours Atomic Energy Division, including Clinton, Hanford and Savannah River administrative records.
ICA	Imperial College archives, London, UK: nuclear engineering programme.
ICE	Institution of Chemical Engineers, Rugby, UK: membership application forms; council minutes.
IME	Institution of Mechanical Engineers, London, UK: C. Hinton papers.
LAC	Library and Archives Canada, Ottawa, Canada: Atomic Energy of Canada Ltd, Chalk River, National Research Council, A. N. Budden, D. A. Keys, C. J. Mackenzie, L. R. Thomson papers.
NA	UK National Archives (formerly Public Record Office), Kew, UK: UKAEA, Ministry of Supply and Foreign Office papers.
NLS	National Library of Scotland, Edinburgh, UK: runs of *Atomic Scientists' News*, *Atomic Scientists' Journal*, *Nuclear Energy Engineer*, *Nuclear Engineer*, *Journal of the Institution of Nuclear Engineers and Atomic Energy News*.
QU	Queen's University archives, Kingston, Canada: W. B. Lewis, B. W. Sargent and W. H. Watson papers.
UI	University of Illinois archives, Urbana-Champaign, Illinois, USA: records of the Argonne Universities Association.

BIBLIOGRAPHY

'100 Area facilities and operations', September 1945, Hagley 1957 Series III Box 58 folder 5
'1942–1945 Staff, general', NA AB 1/246
'1943 Canadian organization: personnel', NA AB 1/380
'1944 Removal of Montreal Laboratory to UK', NA AB 1/149
'1946 Organization of Engineering Division Chalk River and Montreal Laboratories—National Research Council: "Atom smasher aids research in many fields"', *Popular Mechanics*, February 1939: 238–9
'1946 Suggested re-organization of the Engineering Branch at Chalk River in the light of present and future responsibilities and the formation of a crown company to administer the Atomic Energy Project', NA AB 2/128
'1953–1955 Industrial representatives attached to atomic energy establishments', NA AB 16/1324
'1954 Industrial applications: arrangements for training British Electricity Authority and consulting engineers', NA AB 16/1303
'1955 Shortage of staff: group proposals', NA AB 16/1770
'1955–1956 Report on the recruitment of scientists and engineers by the engineering industry', NA AB 16/1772
'1956–1960 Personnel and administration: manpower statistics; Ministry of Labour surveys of qualified scientists and engineers', NA AB 16/2063
'1956–1961 Surveys of apprentice schemes in industry', NA AB 16/343
'1956–1962 Programmes: professional manpower in the atomic energy industry including Advisory Council on Scientific Policy and Ministry of Labour survey.' NA AB 16/1971
'1957 Personnel and administration: transfer of scientists and engineers from defence to civil work', NA AB 16/2083
'1960 Minister's case: security clearance for process workers', NA AB 16/3675
'1961 Minister's case: National Union of Atomic Workers, recognition by United Kingdom Atomic Energy Authority.' NA AB 16/3678
AECL-PD-323, QU W. B. Lewis fonds Box 12 file 11
'A message from Sir Leonard Owen, CBE', *Nuclear Energy Engineer* 12 (1958): 33
'Admin files—political correspondence', prospectus, Hagley 1957 Series III Box 12 folder 7
'American scientists involved in security investigations', *Atomic Scientists' News* 2 (1948): 49–54
'Analysis of lost production time, Windscale Piles, by J. L. Phillips 1 October 1957', NA AB 7/3234
'Anniversary Message from the President, Dr J. Lewins', *Journal of the Institution of Nuclear Engineers* 20 (1979): 130
'Apology for impression possibly given by Secretary of State in a BBC broadcast that the Atomic Scientists' Association are "fellow travellers"', NA FO 371/129241
'Argonne National Laboratory on-site review', memo, 23 May 1979, UI Box 49
'Atomic energy—Canada—General—US Atomic Energy Commission liaison officer at Chalk River', LAC RG25 Vol. 6675
'Atomic piles', *Discovery* 7 (11) (1946): 321–4
'Atomic Research Workers Union, No. 24291, Applicant—and Atomic Energy of Canada Limited, Respondent', LAC RG145 Vol. 114 File 766:336:52
'Atomic Scientists Association's general correspondence 1946–1951', NA AB 16/52
'B reactor front face', 1944, DOE DDRS N1D0029053
'Biography, Milton Shaw', file, 31 January 1969, UI Box 13
'Bionucleonics at Purdue', 2 December 1959, UI Box 13
'Board of Education and successors: Technical Branch and Further Education Branch: Registered Files (T Series)', NA ED 46/1062
'Board of Governors' Minutes, ANL', 2 May 1950, UI Box 44
'Britain's atomic deficiencies', *Discovery* 12 (8) (1951): 235–6

'Britain's premier nuclear training centre: Harwell Reactor School', *Nuclear Energy Engineer* 13 (1959): 246–8
'British Nuclear Energy Conference', *The Chemical Engineer* (October 1958): A47
'Canadian Association of Nuclear Energy Technicians and Technologists, Local 1568, CLC, Applicant—and Atomic Energy of Canada Limited, Chalk River, Ont., Respondent (Technicians)', LAC RG145 Vol. 168 File 766:886:58
'Careers in Nuclear Engineering at Harwell and Winfrith 1960 December.' NA AB 17/234
'Central Electricity Authority requirements of engineers with training in nuclear science', interview memorandum, Mr Bellamy, 9 May 1957, NA ED 46/1062
'Choice of site for production pile 1946', NA AB 7/157
'Commercial atomic power: how soon?' *Discovery* 8 (11) (1947): 331
'Dana engineering and design history—Girdler Corp.' September 1952, Hagley 1957 Series V Box 53 folder 7
'Dana engineering and design history—Lummus Co.' September 1952, Hagley 1957 Series V Box 53 folder 8
'Dana History, Startup through December 1952', memo, 1953, Hagley 1957 Series V Box 53 folder 10
'Draft operating policy of the Argonne National Laboratory', memo, 28 February 1950, UI Box 19
'Eve Curie writes a biography of her mother', *Life*, 22 November 1937: 75–8
'Einstein explains relativity: his new book explains science to the layman', *Life*, 11 April 1938: 48–9
'Factory planning for production pile 1947', NA AB 7/284
'Freedom in Science', *Atomic Scientists' News* 1 (new series) (1951): 2
'From the Secretary', *Journal of the Institution of Nuclear Engineers* 13 (1972): 34
'From the Secretary', *Nuclear Energy* (February 1966): 31
'From the Secretary', *Nuclear Energy* (November/December 1967): 153
'Hanford organization', Hagley 1957 Series I Box 2 folder 8
'Hanford Story, chap 7, 8', Hagley 1957 Series V Box 50
'Hanford Story, chap 12–14', Hagley 1957 Series V Box 50
'Hanford Story, chap 17', Hagley 1957 Series V Box 50
'Harwell: Careers in Nuclear Engineering 1959 February.' NA AB 17/231
'History of operations—Administrative', Hagley 1957 Box 1 folder 1
'History of the Savannah River Laboratory, Volume III—Power Reactor and Fuel Technology', June 1984, Hagley 1957 Series V Box 54 folder 12
'History of TNX', Hagley 1957 Box 1 folder 1
'Imperial College of Science and Technology (University of London) Postgraduate Course in Nuclear Power', brochure, February 1957, ICA GB 0098 KNP/3/1
'Institution requirements for membership', *Journal of the Institution of Nuclear Engineers* 19 (1978): 46–7
'International Atomic Energy Agency, International Symposium on Design, construction and operating experience of demonstration liquid metal fast breeder reactors (Bologna, Italy 10–14 April 1978)', Dounreay Superarchive Box/SR shelf/FRDC/FEWP/P(78)20, IAEA-SM-225/4–1 FRFPDC(84)P5–1
'Is there a nuclear engineer in the house?' *Journal of the Institution of Nuclear Engineers* 18 (1977): 35–45
'Keeping up with the times', *Combustion, Boiler House and Nuclear Review* 12 (1958) (119): 61
'Letters Re: Personnel—Organization at Chalk River', LAC RG77 Vol. 283
'Life story of Robert A. Millikan told in pictures', *Startling Stories*, May 1939: 29–31
'Manhattan Project: Its scientists have harnessed nature's basic force', *Life*, 20 August 1945 91–5, 100, 2, 5–6
'Minutes of Board of Governors' Meeting, ANL', 2 May 1948, UI Box 19
'Minutes of Board Meeting, ANL', 7 March 1949, UI Box 19
'Minutes of Board Meeting, ANL', 6 May 1946, UI Box 19
'Minutes of conference with District Engineer and Metallurgical Laboratory', 5 April 1946, UI Box 19
'Minutes of special meeting of ANL Board of Governors', 4 January 1948, UI Box 19
'Mortality of employees in the UKAEA, 1946–79 (*British Medical Journal* 291, 17 August 1985)', Dounreay 16071/DJC/SAF(85)P6–1
'National Engineering Education Committee, 1958–1971', memo, 1970, UI Box 100
'National Union of Atomic Workers 1958–63', NA FS 27/406
'NEEC Minutes of 63rd Meeting, Argonne National laboratory', 3 May 1971, UI Box 56
'Nuclear engineers and the trades unions', *Journal of the Institution of Nuclear Engineers* 20 (1979): 77
'Nuclear engineers in the making', *Nuclear Energy Engineer* 12 (1958): 334
'Nuclear industry in midst of skills crisis', *Reuters*, 19 December 2010,
'Nuclear power leaflets', brochure, 1958, ICA GB 0098 KNP/2
'Nuclear Science Instruction—UGC enquiry 1957 (File No. 571)', letters, 1957, ICA GB 0098 KNP/3/1

'Obituary: Sir Wallace Alan Akers', *Atomic Scientists' Journal* 4 (1955): 257–8
'Ottawa Atomic Energy Workers, Local No. 1541 (CLC), Applicant—and Atomic Energy of Canada Limited, Ottawa, Ont., Respondent (Commercial Products Division)', LAC RG145 Vol. 160 File 766:811:57
'Personnel policy—proposed new notes for guidance for promotion in the professional grades', Dounreay 07647/47/PP/1–1
'Plans for staff organization, 1946–1950', NA AB 19/56
'Plutonium and uranium contamination of atomic workers', NA PIN 20/788
'Press release', 22 December 1948, Hagley 1957 Series I Box 2 folder 9
'Production Piles—Air Commodore Rawley's History', NA AB 8/557
'Progress in declassification', *Bulletin of the Atomic Scientists* 10 (1954): 143
'Raw Materials of History—UK Atomic Energy Records Handbook', 1 July 1976, Dounreay AEA/DRO/1
'Reactor activities—100 Areas', 1944, DOE DDRS N1D0002151
'Reactor Science and Engineering: Rector's Correspondence 1968–9', letters, 1968, ICA GB 0098 KNR/1/1
'Regular output of engineers for nuclear work—Battersea College of Technology', *Nuclear Energy Engineer* 13 (1959): 136–7
'R. E. Newell', NA AB 2/123
'Report of the Office of Science and Technology Policy Working Group on Basic Research in the Department of Energy', memo, June 1978, UI Box 47
'Report of the Technical Advisory Board to the Technical Committee of the Aircraft Nuclear Propulsion Program', 4 August 1950, ANP-52
'Report of visits to the Radiation Laboratory, Berkeley, California, Clinton Laboratories, Oak Ridge, Tennessee, and the Argonne National Laboratories, Chicago', memo, 13 July 1956, DOE DDRS D197214145
'Review: *The Beginning or the End*', *Atomic Scientists' News* 1 (1947): 40
'Rolls Royce Nuclear Engineering Services Ltd—Plant Operating Instruction—Pressurized Suit Working—Operator', Dounreay 19034/D8550/POI/P006–2
'Secrecy in nuclear engineering: comments on a paper by Dr. J.G. Beckerley, USAEC 1952 11/1/7/2(17)', NA AB 6/1063
'Security procedures in the USA', *Atomic Scientists' News* 1 (1948): 162–4
'Statement on US censorship policy', memo, 13 August 1945, LAC RG77 Vol. 283
'Teaching nuclear engineering in the mile end road', *Nuclear Energy Engineer* 12 (1958): 335–6
'The atomic bomb: how weapon that launched a new age was produced; here is what Americans can and must know about it', *Life*, 27 February 1950, 90–100
'The Association of Scientific Workers and Atomic Energy', *Atomic Scientists' News* 1 (1947): 37–8
'The Atom Train: a successful experiment', *Atomic Scientists' News* 2 (1948): 4–8
'The Atomic Scientists Association Ltd: policy and associated correspondence 1950–1951: 1954–1959', NA AB 27/6
'The birth of the Institution', *Nuclear Energy Engineer* 13 (1959): 579–802
'The Federation of Atomic Scientists', *Bulletin of the Atomic Scientists* 1 (1945): 2
'The Hanford Atomic Project and Columbia River pollution', memo, 20 December 1957, DOE DDRS D8413586
'The journal that spawned an institution', *Journal of the Institution of Nuclear Engineers* 20 (1979): 140–2
'The National Reactor Testing Station', *Nuclear Energy Engineer* 13 (1959): 200
'The New Government Employee Security Program', *Bulletin of the Atomic Scientists* 9 (1953): 175
'The Oak Ridge School of Reactor Technology 1953–1954', prospectus, 1 September 1952, Hagley 1957 Series III Box 12 folder 5: ORSORT 1953–57
'The Oak Ridge School of Reactor Technology 1954–1955', prospectus, 3 September 1953, Hagley 1957 Series III Box 12 folder 5: ORSORT 1953–57
'The Oak Ridge School of Reactor Technology 1957–1958', prospectus, April 1957, Hagley 1957 Series III Box 12 folder 5: ORSORT 1953–57
'The purge in Britain', *Discovery* 11 (6) (1950): 202
'The Scottish Research Reactor Centre, East Kilbride, Glasgow, Scotland', *Journal of Radioanalytical Chemistry* 6 (1970): 273–83
'The supply and distribution of chemical engineers in Great Britain', *The Chemical Engineer* (October 1958): A41–A5
'TMI—one year on', *Nuclear Engineer* 21 (1980): special issue
'Training nuclear engineers at Southampton University', *Nuclear Energy Engineer* 13 (1959): 305–6

'Training of SRP Personnel at Hanford', 1952, Hagley 1957 Series I Box 2 folder 9
'UKAEA Press release: report confirms atomic workers healthy', Dounreay 16071/DJC/SAF(85)P9–1
'Universities of Manchester and Liverpool Joint Research Reactor', *Nature* 203 (1964): 348
'University: nuclear engineering 1954–1957 Train/1', NA AB 19/84
'Visits to scientific establishments—Oak Ridge Institute', 1948, LAC RG25-B-2 Vol. 2143
'Young chemists needed for atomic energy factory', *Discovery* 8 (6) (1947): 192
'Young engineers usher in a new era: Atomic industry is already big', *Life*, 8 August 1955, 17–22
Abbott, Andrew D., *The System of Professions: An Essay on the Division of Expert Labor* (Chicago: University of Chicago Press, 1988)
AECL, 'The university graduate and Atomic Energy of Canada Limited', 1959, LAC MG30 B59 Vol. 8
Agar, Jon, 'What happened in the sixties?' *British Journal for the History of Science* 41 (2008): 567–600
——, 'Dad's radioactive army: scientists, engineers (and philosophers) as volunteers after Chernobyl', presented at *British Society for the History of Science Annual Meeting*, Aberdeen, 2010
Agar, Jon and B. Blamer, 'British scientists and the Cold War: The Defence Research Policy Committee and information networks, 1947–1963', *Historical Studies in the Physical and Biological Sciences* 28 (1998): 209–52
Akers, Wallace to M. W. Perrin, letters, 29 and 31 August 1946, NA AB 16/52
Alberty, Robert A., 'Farrington Daniels, March 9, 1889–June 23, 1972', *National Academy of Sciences Biographical Memoirs* 65 (1994): 109–21
Allison, Samuel B. to R. Williams, 13 March 1944, Chicago, Hagley 1889—Greenewalt Manhattan Project Diary, Vol. III
American Nuclear Society, 'Summer institutes on nuclear energy for 1964', QU Sargent fonds, Series III Box 4
American Nuclear Society, American Society for Engineering Education, 'Report on objective criteria in nuclear engineering education', QU Sargent fonds, Series III Box 4
Anderson, E. E., F. C. Vonderlage and R. T. Overman, 'Education and training in nuclear science and engineering in the United States of America', presented at *International Conference on the Peaceful Uses of Atomic Energy*, Geneva, 1955
Anonymous to C. J. Mackenzie, memo, 24 December 1945, LAC RG77 Vol. 283
Argonne Laboratory, 'Broad Policy on National Laboratories Recommended By General Groves' Advisory Committee on Research and Development', December 1945, UI Box 44
Argonne National Laboratory, 'Subcontractor contract template', July 1946, UI Box 44
——, 'Inter-institution committee for considering a cooperative effort for advanced nuclear engineering education', minutes, 20 December 1955, UI Box 100
——, *School of Nuclear Science and Engineering* (Lemont Ill.: ANL, 1955)
——, 'Scientific Staff Manual', UI Box 35
Arnold, Lorna, *Britain and the H-Bomb* (London: Palgrave MacMillan, 1979)
——, *Windscale 1957: Anatomy of a Nuclear Accident* (London: Macmillan, 1992)
Association of Scientific Workers, *Peaceful Uses of Atomic Energy* (London: Labour Research Department, 1955)
Atomic Energy Commission, telex, 31 October 1953, Department of Energy, CD 59–5–20/FORM 189. Accession No. NV0702100 [OpenNet http://www.osti.gov/opennet/detail.jsp?osti_id=16289444]
——, 'Oak Ridge Operations Information Manual, Budget and Reports Division', Department of Energy Accession No. NV0714712 [on OpenNet as http://www.osti.gov/opennet/detail.jsp?osti_id=16111668], 31 October 1953
——, 'SRP Fact Book', 25 November 1960, Hagley 1957 Series V Box 54 folder 15
Atomic Industrial Forum, *A Growth Survey of the Atomic Industry, 1955–1965* (New York: Atomic Industrial Forum, 1955)
——, 'The Atomic Industrial Forum', promotional literature, 1977, UI Box 22
Atomic Scientists of Chicago, *The Atomic Bomb—Facts and Implications* (Chicago: Atomic Scientists of Chicago, 1946)
Auger, P. V., H. H. Halban, R. E. Newell, F. A. Paneth and G Placszek, 'Research programmes for development of heavy water boiler', memo, 30 December 1943, LAC RG77 Vol. 283
Augustine, Dolores, *Red Prometheus: Engineering and Dictatorship in East Germany, 1945–1990* (Cambridge, Mass.: MIT Press, 2007)
Badash, Lawrence, 'From security blanket to security risk: scientists in the decade after Hiroshima', *History and Technology* 19 (2003): 241–56

Badash, Lawrence, Joseph O. Hirschfelder and Herbert P. Broida (eds.), *Reminiscences of Los Alamos, 1943–1945* (Dordrecht: Reidel, 1980)

Bagley, K.Q. and J. A. Gatley, 'The application of core and fuel performance experience in British reactors to commercial fast reactor design', presented at *International Symposium on Design, construction and operating experience of demonstration liquid metal fast breeder reactors*, Bologna, Italy, 1978

Bailey, Richard, *Energy: The Rude Awakening* (London: McGraw-Hill, 1977)

Balogh, Brian, *Chain Reaction: Expert Debate and Public Participation in American Commercial Nuclear Power, 1945–1975* (Cambridge: Cambridge University Press, 1991)

Barca Salom, Francesc X., *Els Inicis de L'Enginyeria Nuclear a Barcelona: La Càtedra Ferran Tallada (1955–1962)*, PhD thesis, Departament de Matemàtica Aplicada 1, Universitat Politècnica de Catalunya (2002)

——, 'Nuclear power for Catalonia: The role of the official Chamber of Industry of Barcelona, 1953–1962', *Minerva* 43 (2005): 163–81

Bauer, S. G. and J. Diamond, 'Note on piles for the production of useful power', memo, 4 June 1945, LAC RG77 Vol. 283

Beck, Clifford K., 'Undergraduate nuclear engineering curriculum at North Caroline State College', *Nucleonics* 8 (1951): 54–9

Beckerley, J. G., 'Declassification of low-power reactors', *Nucleonics* 8 (1951): 13–6

Behrens, D. J., 'Life at Harwell', *Atomic Scientists' News* 2 (1953): 173–6

Bernstein, Jeremy, *Oppenheimer: Portrait of an Enigma* (London: Duckworth, 2004)

Bertin, Leonard, *Atom Harvest: A British View of Atomic Energy* (San Francisco: W. H. Freeman, 1957)

Board of Directors, ORINS, 'Annual Report of the Board of Directors to the Council of the Oak Ridge Institute of Nuclear Studies Inc', 30 June 1947, DOE ON NV0707737

——, 'ORINS Annual Reports Excerpts: Special Training Division 1947–1958', 30 June 1958, DOE ON NV0712404

Board of Studies, Imperial College, 'Item 8161: Nuclear Reactor Science and Engineering', minutes, 17 February 1969

Bolter, Harold, *Inside Sellafield: Taking the Lid off the World's Nuclear Dustbin* (London: Quartet, 1996)

Bolton, B. K. to C. J. Mackenzie, letter, 18 August 1944, Ottawa, Ontario, LAC RG77 Vol. 283

Bormaier, R. J., 'Reports on atomic energy in relation to power', memo, 26 October 1953, DOE DDRS D8471637

Bothwell, Robert, *Nucleus: The History of Atomic Energy of Canada Limited* (Toronto: University of Toronto Press, 1988)

Boudia, Soraya, 'The Curie laboratory: radioactivity and metrology', *History and Technology* 13 (1997): 249–65

Bowman, H. J., 'Instrument Department—Procurement and Training of Non-Exempt Personnel', memo to file, 4 August 1945, Hagley 1957 Series V Box 50 folder 16

Boyce, J. C., 'To Council of Participating Institutions', memo, 26 January 1951, UI Box 134

Boyer, Paul, 'From activism to apathy: the American people and nuclear weapons, 1963–1980', *Journal of American History* 70 (1984): 821–44

——, *By the Bomb's Early Light: American Thought and Culture at the Dawn of the Atomic Age* (New York: Pantheon, 1985)

——, *Fallout: A Historian Reflects on America's Half-Century Encounter With Nuclear Weapons* (Columbus: Ohio State University Press, 1998)

Bradsher, Keith and Hiroko Tabuchi, 'Last defense at troubled reactors: 50 Japanese workers', *New York Times*, 15 March 2011

Branigan, Tania and Justin McCurry, 'Fukushima 50 battle radiation risks as Japan nuclear crisis deepens', *Guardian*, 15 March 2011

Brennan, J. B., 'Chemical engineering manpower in the chemical industry', April 1961

Bridgman, C. J. to R. C. Taecker, letter, 3 March 1967, UI Box 100

British Medical Council, *The Hazards to Men of Nuclear and Allied Radiations* (London: HMSO, 1956)

Brookhaven National Laboratory, 'Brookhaven 1960–82 and its Associated Universities', UI Box 35

Brown, C. L., 'Hanford Technology Course', personal notebook, 15 October 1948–8 October 1953, DOE DDRS D198027813

Brown, H. F., 'Atomic Energy Division', memo, 2 August 1950, Hagley 1957 Series II Box 6 File—Administrative policy correspondence, general, 1950–1963

——, 'Atomic Energy Survey Committee', memo, 11 July 1950, Hagley 1957 Series II Box 6 File—Administrative policy correspondence, general, 1950–1963

Brown, John K., 'Design plans, working drawings, national styles: engineering practice in Great Britain and the United States, 1775–1945', *Technology and Culture* 41 (2000): 195–238

Brown, L., B. Pippard and A. Pais (eds.), *Twentieth Century Physics*, Vol. I (Bristol: Institute of Physics Publishing, 1995)

Brown, Ralph S., *Loyalty and Security: Employment Tests in the United States* (New Haven: Yale University Press, 1958)

Bryce, J. C., 'Admission of uncleared scientific visitors', memo, 30 September 1953, UI Box 44

Buchanan, Nicholas, 'The atomic meal: the cold war and irradiated foods, 1945–1963', *History and Technology* 21 (2005): 221–49

Bugbee, S. J., 'Technical Department Functions and Organization to 1 July 1945', memo to file, Hagley 1957 Series III Box 58 folder 4

——, 'Pile Operations Dept, Part II, 1944–46', memo to file, Hagley 1957 Series III Box 58 folder 2a

Bush, S. H. and E. A. Evans, 'Trip to Harwell, Nov 27–28, and to Risley and Culcheth, November 29 1957', memo, 31 December 1957, DOE DDRS D8298841

Bush, Vannevar, 'Science The Endless Frontier: A Report to the President by Vannevar Bush, Director of the Office of Scientific Research and Development' (Washington, DC: United States Printing Office, July 1945)

Bushey, A. H., 'The less familiar elements of the atomic energy program', report, 12 October 1953, DOE DDRS D198149819

Buyers, William J. L., 'Neutron and other stories from Chalk River', *Physics in Canada* 56 (2000): 145–51

Cahan, David, *An Institute for an Empire: The Physikalisch-Technische Reichesanstalt 1871–1918* (Cambridge: Cambridge University Press, 1989)

Callow, J. H., 'A personal tribute to the memory of J. B. Pinkerton', *Journal of the Institution of Nuclear Engineers* 19 (1978): 48

——, 'What is a nuclear engineer? Whither INucE?' *Nuclear Engineer* 23 (1982): 129

Canadian Nuclear Association, 'Survey of Nuclear Education and Research in Canadian Universities 1961', QU B. W. Sargent fonds, Series III Box 4

Cantello, Gerald Wynne, *The Roles Played by the Canadian General Electric Company's Atomic Power Department in Canada's Nuclear Power Program: Work, Organization and Success in APD, 1955–1995*, MA thesis, Frost Centre for Canadian Studies and Native Studies, Trent (2003)

Carlisle, Rodney P. and Joan M. Zenzen, *Supplying the Nuclear Arsenal: American Production Reactors 1942–1992* (Baltimore: Johns Hopkins Press, 1996)

Caulfield, Catherine, *Multiple Exposures: Chronicles of the Radiation Age* (London: Secker & Warburg, 1989)

Central Office for Information, *Nuclear Energy in Britain* (London: HMSO, 1969)

Chadwick, James to C. J. Mackenzie, letter, 8 January 1946, LAC RG77 Vol. 283

Chalk River scientists to J. Cockcroft, memo, 27 December 1945, CCFT25/20

Chambers, F. S., 'Manpower requirements for separation work at ORNL', memo to file, 19 September 1950, Hagley 1957 Series IV Box 44 folder 1

Chandler, Alfred D. and Stephen Salsbury, *Pierre S. Du Pont and the Making of the Modern Corporation* (New York: Harper & Row, 1971)

Chapman, Robert, 'Crisis at Windscale', *Manchester Evening News*, 26–27 June 1956: 8–9

——, *Atomic Energy for All: A Layman's Guide to the Atom and its Uses* (London: Oldhams Press, 1960)

Chatzis, Konstantinos, 'Introduction: The national identities of engineers ', *History and Technology* 23 (2007): 193–6

Chezem, Curtis G. to A. F. Yanders, letter, 18 July 1971, UI Box 56

Church, G. P. to E. M. Cameron, letter, 13 December 1950, Hagley 1957 Series III Box 12 folder 3

—— to E. T. Macki, letter, 5 September 1951, Hagley 1957 Series II Box 12 folder 3

Claridge, A., 'Foreword', *Nuclear Engineer* 32 (1991): 154

Clark, R. W., *The Birth of the Bomb* (London: Phoenix, 1961)

Clayton, R. J. and J. Algar, *The GEC Research Laboratories 1919–1984* (London: Institution of Engineering and Technology, 1989)

Cochrane, Rexmond C., *Measures for Progress: A History of the National Bureau of Standards* (Washington, DC: US Department of Commerce, 1966)

Cockcroft, John, 'Montreal staff', NA AB 1/278

——, 'The development of chain reacting systems', memo, 30 July 1945, LAC RG77 Vol. 283

——, 'Power Development Programme', 7 May 1947, LAC MG30 B59 Vol. 4

——, 'Interpretation of scientific knowledge', *Chemical Age* (1949): 613–4
—— to R. E. France, letter, 6 September 1951, NA AB 27/6
—— to unlisted recipients, memo, 22 November 1954, NA AB 19/84
——, 'Foreword—The Journal of Nuclear Energy', *The Journal of Nuclear Energy* 1 (1954): 1
——, 'The Harwell Reactor School', *Nuclear Engineering* 1 (1956): 10–1
——, 'Scientific problems in the development of nuclear power', *Nuclear Power—The Journal of British Nuclear Engineering* 1 (1956): 200
Codd, J. and R. F. Jackson, 'Notes on the second American Nuclear Society conference, Chicago, 6th–8th June, 1956', NA AB 15/5081
Cole, J. E., 'Trip report, Health physics meeting, Chicago, Jan 16–18 1951', memo to file, 22 January 1951, Hagley Series IV Box 44 folder 2
Collins, H. M. and Robert Evans, 'The third wave of science studies: studies of expertise and experience', *Social Studies of Science* 32 (2002): 235–96
Commission on Protecting and Reducing Government Secrecy, 'Secrecy: Report of the Commission on Protecting and Reducing Government Secrecy' (Washington, DC: Government Printing Office, 1997)
Compton, Arthur Holly, *Atomic Quest: A Personal Narrative* (Oxford: Oxford University Press, 1956)
Condon, E. U., 'Physics gives us—nuclear engineering', *Westinghouse Engineer*, 5 November 1945: 167–73
Conn, H. G. to P. Mackintosh, letter, 2 January 1958, QU Sargent fonds Series III Box 4
Cosgrove, Denis, 'Introduction: Project Plowshare', *Cultural Geographies* 5 (1998): 263–6
Council of Representatives, 'Minutes of Meeting of Council of Representatives, Participating Institutions, ANL', 2 May 1950, UI Box 44
Cowan, Robin, 'Nuclear power reactors: a study in technological lock-in', *The Journal of Economic History* 50 (1990): 541–67
Cowell, S. F. and S. B. Dowd, 'We are what we think we are: Professionalization in nuclear medicine technology', *Journal of Nuclear Medicine Technology* 24 (1996): 336–41
Crane, P. W. to J. M. Tilley, memo to file, 20 February 1945, Hanford, DOE DDRS D8255334
Creager, Angela N. H., 'The industrialisation of radioisotopes by the US Atomic Energy Commission', in: K. Grandin and N. Wormbs (eds.), *The Science–Industry Nexus: History, Policy, Implications* (New York: Watson Publishing, 2004), pp. 141–67
Crosser, Orrin K. to G. M. Nordby, letter, 9 February 1966, UI Box 148
Curtiss, D. E. and R. E. Nightingale, 'Trip report US-UK Graphite Conference, London, England, Dec 16–20, 1957', memo to file, 9 January 1958, DOE DDRS D4763348
Daerr, R. L. to Officer in Charge: Intelligence Office, memo, 31 July 1946, Hanford, DOE DDRS D4763417
—— to Officer in Charge: Intelligence Office, memo, 17 October 1946, Hanford, DOE DDRS D5763387
Dandurand, Louise, *The Politicization of Basic Science in Canada: NRC's Role, 1945–1976*, PhD thesis, History, University of Toronto (1982)
Davis, W. K. and W. A. Roger, 'The chemical engineer and nuclear energy', in: W. T. Dixon and A. W. Fisher Jr (eds.), *Chemical Engineering in Industry* (New York: American Institute of Chemical Engineers, 1958), pp. 88–97
Del Sesto, Stephen L., 'Wasn't the future of nuclear energy wonderful?' in: J. J. Corn (ed.), *Imagining Tomorrow: History, Technology, and the American Future* (Cambridge, Mass.: MIT Press, 1986), pp. 58–76
Department of Energy, 'School of Nuclear Engineering class photo', 1954, DOE DDRS N1D0002164
——, 'Institutional planning process—US DOE order 25', memo, 2 June 1980, UI Box 47
Department of Reconstruction (Canada), 'Biographical notes', http://www.cns-snc.ca/history/history.html/1945Aug13PressReleasePart3, accessed 23 April 2009
——, 'Canada's role in atomic bomb drama', http://www.cns-snc.ca/history/history.html/1945Aug13PressReleasePart1, accessed 23 April 2009
——, 'Scientists who probed atomic secrets', http://www.cns-snc.ca/history/history.html/1945Aug13PressReleasePart2, accessed 23 April 2009
Department of the Army, *Atomic Energy Indoctrination* (Washington, DC: US Army, 1950)
DeVoss, H. G., D. C. Keck, G. V. R. Smith and E. W. Wilson, 'Report of trip to Dresden Power Station, Shippingport Atomic Power Station, Yankee Atomic Power Station, Savannah River Plant, and Combustion Engineering, Chattanooga Tennessee', letter, 9 January 1961, DOE DDRS D4763348
Dick, William E., 'The hangars hide uranium piles', *Discovery* 9 (9) (1948): 281–5
Divall, Colin and Sean F. Johnston, 'Scaling up: The evolution of intellectual apparatus associated with the manufacture of heavy chemicals in Britain, 1900–1939', in: A. S. Travis, H. G. Schröter and E. Homburg (eds.),

Determinants in the Evolution of the European Chemical Industry, 1900–1939: New Technologies, Political Frameworks, Markets and Companies (Dordrecht: Kluwer Academic, 1998), pp. 199–214
——, *Scaling Up: The Institution of Chemical Engineers and the Rise of a New Profession* (Dordrecht: Kluwer Academic, 2000)
DOE, OpenNet, 'History of the activities of the Manhattan District Research Division, October 15, 1945–December 31, 1946', 31 December 1946, DOE ON NV0714682
——, 'Remarks by Seaborg at the 25th anniversary of Argonne National Laboratory, Argonne Illinois (AEC-S-12–71)', 19 June 1971, DOE ON NV0712412
Doern, G. Bruce, *Science and Politics in Canada* (Montreal: McGill University Press, 1972)
Du Pont, 'Savannah River Plant History—All Areas—Aug 1950 through June 1954', bound report, 1954, Hagley 1957 Series V Box 51
Duffield, Robert B. to C. C. McSwain, letter, 27 July 1971, UI Box 56
Duncan, Otis Dudley, 'Sociologists should reconsider nuclear energy', *Social Forces* 57 (1978): 1–22
Dunworth, John V., 'Pay rise for 15,000', *Nuclear Power—The Journal of British Nuclear Engineering* 1 (1956): 48
——, 'It's not just ordinary engineering', *Nuclear Power—The Journal of British Nuclear Engineering* 3 (1958): 36
Editorial, *Nucleonics* 9 (1954): 3
——, *The Engineer* 200 (1955): 569
——, 'Another industrial revolution?' *Nuclear Engineering* 1 (1956): 1
——, 'The press at Harwell', *Nuclear Engineering* 1 (1956): 138
——, 'Training the nuclear engineer', *Nuclear Engineering* 1 (1956): 5
——, 'Jobs for the boys', *Nuclear Engineering* 2 (1957): 85
——, 'Secrecy wraps lifted', *Nuclear Power—The Journal of British Nuclear Engineering* 2 (1957): 5
——, 'The nuclear industry', *Nuclear Power—The Journal of British Nuclear Engineering* 3 (1958): 150
Edwards, J., 'Nuclear training facilities at the Royal Naval College, Greenwich', *Nature* 200 (1960): 545–7
——, 'Nuclear power—learning from the lessons of the past', *Nuclear Engineer* 22 (1981): 98–106
Eggleston, Wilfred, *Canada's Nuclear Story* (Toronto: Clarke, Irwin, 1965)
——, *National Research in Canada: the NRC, 1916–1966* (Toronto: Clarke, Irwin, 1978)
Egorov, Nikolai N., *The Radiation Legacy of the Soviet Nuclear Complex: An Analytical Overview* (London: Earthscan, 2000)
Eisenhower administration, 'General outline for Agronsky program', 16 December 1953, Eisenhower Presidential Library, http://www.eisenhower.utexas.edu/dl/Atoms_For_Peace/Binder14.pdf
——, 'Project 'Candor'', 22 July 1953, Eisenhower Presidential Library, http://www.eisenhower.utexas.edu/dl/Atoms_For_Peace/Binder17.pdf
Eisenhower, Dwight D., 'Atoms for Peace', *United Nations General Assembly*, 470th Plenary Meeting 8 December 1953
Evans, R. M. to C. A. Nelson, letter, 9 August 1950, Hagley 1957 Series II Box 6 File—Administrative policy correspondence, general, 1950–1963
Fanner, A. A. and J. M. Hill, 'Incidents in the life of the Windscale piles', 26 May 1953, NA AB 7/18254
Fawcett, Ruth, *Nuclear Pursuits: The Scientific Biography of Wilfrid Bennett Lewis* (Montreal: McGill-Queen's University Press, 1994)
Fermi, Laura, *Atoms in the Family: My Life with Enrico Fermi* (Chicago: University of Chicago Press, 1954)
Finniston, Montague, 'Opening Address, Institution of Nuclear Engineers 1977–1978', NA BT 251/198
Fishenden, R. M. to Editor of *Atomic Scientists' News*, letter, 19 August 1949, NA AB 16/52
Fishlock, David, 'The world's nuclear engineers compare notes', *New Scientist*, 3 September 1964: 551–2
Fiske, John, *Understanding Popular Culture* (Boston: Unwin Hyman, 1989)
Folger, J. K. and M. L. Meeks, 'Educated manpower: key to nuclear development', in: R. Sugg Jr (ed.), *Nuclear Energy in the South* (Baton Rouge, 1957), pp. 105–38
Foreword, 'Soviet Journal of Atomic Energy', *Soviet Journal of Atomic Energy* 1 (1956): 75
Forgan, Sophie, 'Atoms in Wonderland', *History and Technology* 19 (2003): 177–96
Forman, Paul, 'Behind quantum electronics: National security as basis for physical research in the United States, 1940–1960', *Historical Studies in the Physical and Biological Sciences* 18 (1987): 149–229
Forman, Paul and José M. Sánchez Ron, *National Military Establishments and the Advancement of Science and Technology: Studies in 20th Century History* (Dordrecht: Kluwer Academic, 1996)
Fox, Marvin, 'Fundamentals of reactor design', presented at *Atomic Energy in Industry, 2nd Annual Conference*, Montauk, NY, 1953

Franklin, N. L., 'The contribution of chemical engineering to the U. K. nuclear industry', in: W. F. Furter (ed.), *History of Chemical Engineering* (Washington, DC: American Chemical Society, 1980), pp. 335–66

Frenkel, Stephen, 'A hot idea? Planning a nuclear canal in Panama', *Cultural Geographies* 5 (1998): 303–9

Frisch, Otto, *What Little I Remember* (Oxford: Oxford University Press, 1979)

Fuller, John G., *We Almost Lost Detroit* (New York: Ballantine Books, 1976)

Fulling, R. W. to F. C. VonderLage, letter, 14 May 1954, Hagley 1957 Series III Box 12 folder 6

Fullmer, G. C., 'Pile Technology Course', personal notebook, 26 January 1950–30 Septembar 1960, DOE DDRS D198027813

Furman, Necah Stewart, *Sandia National Laboratories: The postwar decade* (Albuquerque: University of New Mexico press, 1990)

Furter, William F. (ed.), *History of Chemical Engineering* (Pittsburgh, Penn.: American Chemical Society, 1980)

Gaines, Matthew J., *Atomic Energy* (London: Feltham Hamlyn, 1969)

Galbreath, Ross, 'The Rutherford connection: New Zealand scientists and the Manhattan and Montreal projects', *War in History* 2 (1995): 306–21

Galison, Peter, *Image and Logic: A Material Culture of Microphysics* (Chicago: University of Chicago Press, 1997)

——, 'Removing knowledge', *Critical Enquiry* 31 (2004): 229–43

Galison, Peter and Barton J. Bernstein, 'Physics between war and peace', in: M. R. S. Mendelsohn and P. Weingart (eds.), *Science, Technology, and the Military* (Dordrecht: Kluwer Academic, 1988), pp. 47–86

Galison, Peter and Bruce Hevly, *Big Science: The Growth of Large-Scale Research* (Stanford, Calif.: Stanford University Press, 1992)

Gaspar, Júlia 'The two Iberian nuclear programmes: Post-war scientific endeavours in a comparative approach (1948–1973)', presented at *7th Meeting, Science and Technology in the European Periphery*, Galway, Ireland, 2010

Gast, P. F., 'Comment on the probability of a pile disaster', memo, 20 October 1954, DOE DDRS D8372268

Gast, P. F. and C. W. J. Wende, 'Reactivity experience and control to July 1, 1945', memo to file, 20 September 1945, Hagley 1957 Series III Box 58 folder 4

Geison, Gerald L., 'Scientific change, emerging specialties, and research schools', *History of Science* 29 (1981): 20–40

General Electric, 'Four years at Hanford', report, 1951, DOE DDRS D19803392

Genereaux, R. P., 'Object of survey', memo, 29 December 1948, Hagley 1957 Series I Box 2 folder 9

Gerber, Michele S., *On the Home Front: The Cold War Legacy of the Hanford Nuclear Site* (Lincoln: University of Nebraska Press, 1997)

Gibb, Claude, 'Industry and atomic power', *Atomic Scientists' News* 2 (1952): 98–103

Gilligan, Andrew, 'Japan earthquake: the Fukushima 50 fight to stave off disaster', *Daily Telegraph*, 16 March 2011

Gingras, Yves, 'The institutionalization of scientific research in Canadian universities: the case of physics', *Canadian Historical Review* 67 (1986): 181–94

Gispen, Kees, *New Profession, Old Order: Engineers and German Society, 1815–1914* (Cambridge: Cambridge University Press, 1989)

Glasstone, Samuel, *Principles of Nuclear Reactor Engineering* (Princeton, NJ: Van Nostrand, 1955)

Glasstone, Samuel and Milton C. Edlund, *The Elements of Nuclear Reactor Theory* (Princeton, NJ: Van Nostrand, 1952)

Glennan, T. Keith, 'The engineer in the AEC', *Bulletin of the Atomic Scientists* 8 (1952): 55

Goddard, Tony to S. F. Johnston, interview, 2 July 2008, London, England, SFJ collection

Goodman, Clark (ed.), *The Science and Engineering of Nuclear Power*, Vol. I (Reading, Mass.: Addison-Wesley, 1947)

Goudsmit, Samuel A., *Alsos: The Failure of German Science* (New York: Sigma Books, 1947)

Government of Japan Nuclear Emergency Response Headquarters, 'Report of Japanese Government to the IAEA Ministerial Conference on Nuclear Safety—The Accident at TEPCO's Fukushima Nuclear Power Stations', http://www.kantei.go.jp/foreign/kan/topics/201106/iaea_houkokusho_e.html, accessed 20 June 2011

Gowing, Margaret, *Britain and Atomic Energy, 1939–1945* (New York: St Martin's Press, 1964)

Gowing, Margaret and Lorna Arnold, *Independence and Deterrence: Britain and Atomic Energy*, Vol. I: *Policy Making, 1945–52* (London: MacMillan, 1974)

——, *Independence and Deterrence: Britain and Atomic Energy*, Vol. II: *Policy Execution, 1945–52* (London: MacMillan, 1974)

Graham, Loren, *The Ghost of the Executed Engineer: Technology and the Fall of the Soviet Union* (Cambridge, Mass.: Harvard University Press, 1993)
——, *What Have We Learned About Science and Technology from the Russian Experience?* (Stanford, Calif.: Stanford University Press, 1998)
Grayson, Lawrence P., *The Making of an Engineer: An Illustrated History of Engineering Education in the United States and Canada* (New York: John Wiley, 1993)
Greenbaum, Leonard, *A Special Interest: The Atomic Energy Commission, Argonne National Laboratory, and the Midwestern Universities* (Ann Arbor: University of Michigan Press, 1971)
Greenewalt, Crawford H., 'Manhattan Project Diary, Vol. I', 1942, Hagley 1889
——, 'Manhattan Project Diary, Vol. II', 1942–43, Hagley 1889
——, 'Manhattan Project Diary, Vol. III', 1944–45, Hagley 1889
——, 'Stine's memorandum', 27 November 1942, Hagley 1957 Series I Box 1
Gregory, Colin to S. F. Johnston, interview, 19 September 2007, Thurso, Scotland, SFJ collection
Groves, Leslie R., *Now it can be Told: The Story of the Manhattan Project* (New York, N.Y.: Harper & Row, 1962)
Guéron, Jules to C. J. Mackenzie, memo, 13 December 1945, LAC RG77 Vol. 283
Gusterson, Hugh, *Nuclear Rites: A Weapons Laboratory at the End of the Cold War* (Berkeley: University of California Press, 1996)
——, 'The death of the authors of death: prestige and creativity among nuclear weapons scientists', in: Biagioli, Mario and Peter Galison (eds.), *Scientific Authorship: Credit and Intellectual Property in Science* (New York: Routledge, 2003), pp. 281–307
Haber, Heinz, *The Walt Disney Story of Our Friend the Atom* (New York: Simon & Schuster, 1957)
Hales, Peter B., *Atomic Spaces: Living on the Manhattan Project* (Urbana: University of Illinois Press, 1997)
Hall, G. R., 'Nuclear technology in a university environment', presented at *Imperial College of Science and Technology inaugural lecture*, 1964
Hall, H. L., Jr. Dodds, H. L., J. P. Hayward, L. H. Heilbronn, J. W. Hines, H. Liao, G. I. Maldonado, L. F. Miller, R. E. Pevey, A. E. Ruggles, L. W. Townsend and B. R. Uphadhyaya, 'Nuclear engineering and nuclear security: a growing emphasis at the University of Tennessee', presented at *Proceedings of the Pacific Northwest International Conference on Global Nuclear Security–the Decade Ahead*, 2010
Hall, Jeremy, *Real Lives, Half Lives* (London: Penguin, 1996)
Hall, Tony, *Nuclear Politics: The History of Nuclear Power in Britain* (Harmondsworth: Penguin, 1986)
Harrell, W. B. to A. H. Frye Jr, letter, 9 March 1946, UI Box 19
Harrer, Joseph R., 'Briefing on atomic science for Missouri newsmen', press release, Argonne National Laboratory, 14 February 1958, UI Box 10
Hartcup, Guy and T. E. Allibone, *Cockcroft and the Atom* (Bristol: Adam Hilger, 1984)
Hartley, Harold, *Chemistry and Industry* (25 April 1953): 404–5
——, 'The place of chemical engineering in modern industry', *School Science Review* (March 1954): 199–202
Harwood, Jonathan, 'Engineering education between science and practice: rethinking the historiography', *History and Technology* 22 (2006): 53–79
Hawes, Lewis, 'Far too secret a secret', *Atomic Scientists' News* 3 (1949): 7–19
Hayes, Denis, *Rays of Hope: The Transition to a Post-Petroleum World* (New York: W. W. Norton, 1977)
Haynes, Roslynn, *From Faust to Strangelove: Representations of the Scientist in Western Literature* (Baltimore: Johns Hopkins University Press, 1994)
Healy, J. W., 'Computations of the environmental effects of a reactor disaster', memo, 14 December 1953, DOE DDRS D198160849
——, 'Report of trip to Harwell to review health and safety aspects of the Windscale incident', memo, 12 November 1957, DOE DDRS D8553051
Hecht, David K., 'The atomic hero: Robert Oppenheimer and the making of scientific icons in the early Cold War', *Technology and Culture* 49 (2008): 943–66
Hecht, Gabrielle, 'Political designs: nuclear reactors and national policy in postwar France', *Technology and Culture* 35 (1994): 657–85
——, 'Rebels and pioneers: technocratic ideologies and social identities in the French nuclear workplace 1955–1969', *Social Studies of Science* 26 (1996): 483–529
——, 'Enacting cultural identity: risk and ritual in the French nuclear workplace', *Journal of Contemporary History* 32 (1997): 483–507

——, *The Radiance of France: Nuclear Power and National Identity after World War II* (Cambridge, Mass.: MIT Press, 1998)
——, 'The power of nuclear things', *Technology and Culture* 51 (2010): 1–30
Hentschel, Klaus (ed.) and Ann M. Hentschel (translator), *Physics and National Socialism: An Anthology of Primary Sources* (Berlin: Birkhäuser, 1996)
Herken, Gregg, '"A most deadly illusion": the atomic secret and American nuclear weapons policy, 1945–1950', *Pacific Historical Review* 49 (1980): 51–76
——, *Brotherhood of the Bomb: The Tangled Lives and Loyalties of Robert Oppenheimer, Ernest Lawrence, and Edward Teller* (New York: Henry Holt and Company, 2002)
Herran, Néstor, 'Spreading nucleonics: the Isotope School at the Atomic Energy Research Establishment, 1951–67', *British Journal for the History of Science* 39 (2006): 569–86
Hewlett, Richard G., 'Beginnings of development in nuclear technology', *Technology and Culture* 17 (1976): 465–78
Hewlett, Richard G. and Francis Duncan, *Nuclear Navy, 1946–1962* (Chicago: University of Chicago Press, 1974)
Hewlett, Richard G. and Jack M. Holl, *Atomic Shield, 1947–1952* (Berkeley: University of California Press, 1969)
——, *Atoms for Peace and War, 1953–1961* (Berkeley: University of California Press, 1969)
——, *The New World, 1939–1946* (Berkeley: University of California Press, 1969)
Hilberry, Norman, 'Plan for continued operation of Argonne Laboratory', 5 December 1945, UI Box 44
——, 'Statement to the Advisory Committee on Future Argonne Laboratory Operation (draft)', memo, 24 November 1945, UI Box 134
—— to F. Daniels, letter, 5 February 1948, UI Box 19
—— to M. B. Powers, letter, 24 February 1960, UI Box 85
Hinton, Christopher to H. Hartley, letter, 4 August 1954, NA AB 19/84
——, 'Inaugural address', *Journal of the British Nuclear Energy Society* 1 (1955): 1–2
—— to T. P. Creed, letter, 23 November 1955, NA AB 19/84
—— to J. M. Kay, letter, 4 January 1955 and 22 December 1954, NA AB 19/84
—— to O. A. Saunders, letters, 5 and 9 January 1956, NA AB 7/40
——, 'Inaugural address', *Journal of the British Nuclear Energy Conference* 1 (1956): 1–2
—— to J. M. Kay, letter, 9 January 1957, NA AB 19/84
——, *Engineers and Engineering* (Oxford: Oxford University Press, 1970)
——, 'The birth of the breeder', in: J. S. Forrest (ed.), *The Breeder Reactor: Proceedings of a Meeting at the University of Strathclyde 25 March 1977* (Edinburgh, 1977), pp. 8–13
——, Hinton A.3, Institution of Mechanical Engineers Archives
——, 'unpublished memoirs Chap XI: Risley Organization', IME Hinton A.3
——, 'unpublished memoirs Chap XIV: The Diffusion Plant', IME Hinton A.4
——, 'unpublished memoirs Chap XVI: The Reasons Why', IME Hinton A.4
HMSO, *Scientific and Engineering Manpower in Great Britain* (London: HMSO, 1956)
Hoddeson, Lillian, Paul W. Henriksen, Roger A. Meade and Catherine Westfall, *Critical Assembly: A Technical History of Los Alamos During The Oppenheimer Years, 1943–1945* (Cambridge: Cambridge University Press, 1993)
Hodgson, P. E., 'The British Atomic Scientists' Association, 1946–59', *Bulletin of the Atomic Scientists* 15 (1959): 393–4
Holliday, Joe, *Dale of the Mounted: Atomic Plot* (Toronto: Thomas Allen, 1959)
Holloway, David, *Stalin and the Bomb: The Soviet Union and Atomic Energy, 1939–1956* (New Haven, Conn.: Yale University Press, 1994)
Hope, Nelson W., *Atomic Town* (New York: Comet, 1954)
Hounshell, David A. and John Kenly Smith, *Science and Corporate Strategy: Du Pont R & D, 1902–1980* (Cambridge: Cambridge University Press, 1988)
Howe, C. D. to D. C. Abbott, letter, Ottawa, Ontario, LAC RG29-F-2 Vol. 5761
Hu, Howard, Katherine Yih and Arjun Makhijani, *Nuclear Wastelands: A Global Guide To Nuclear Weapons Production And Its Health And Environmental Effects* (Cambridge, Mass.: MIT Press, 1995)
Hughes, Jeff, 'The French connection: the Joliot-Curies and nuclear research in Paris, 1925–1933', *History and Technology* 13 (1997): 325–43
——, 'Modernists with a vengeance': changing cultures of theory in nuclear science, 1920–1930', *Studies In History and Philosophy of Science Part B: Studies In History and Philosophy of Modern Physics* 29 (1998): 339–67

——, 'Plasticine and valves: industry, instrumentation and the emergence of nuclear physics', in: J.-P. Gaudillière and I. Löwy (eds.), *The Invisible Industrialist. Manufactures and the Production of Scientific Knowledge* (London: Macmillan, 1998), pp. 58–101
——, *The Manhattan Project: Big Science and the Atom Bomb* (New York: Columbia University Press, 2003)
——, 'Radioactivity and nuclear physics', in: M. J. Nye (ed.), *The Cambridge History of Science: The Modern Physical and Mathematical Sciences* (Cambridge: Cambridge University Press, 2003), pp. 350–74
——, 'Essay review—Deconstructing the bomb: recent perspectives on nuclear history', *British Journal for the History of Science* 37 (2004): 455–64
Hughes, Thomas Parke, *American Genesis: A Century of Invention and Technological Enthusiasm, 1870–1970* (New York: Viking, 1989)
——, *Networks of Power: Electrification in Western Society, 1880–1930* (Baltimore: Johns Hopkins University Press, 1993)
——, 'Technological momentum', in: M. R. Smith and L. Marx (eds.), *Does Technology Drive History? The Dilemma of Technological Determinism* (Cambridge, Mass.: MIT Press, 1994), pp. 101–14
Hull, J. W., H. T. Kelley and R. L. Daerr to Officer in Charge: Intelligence Office, memo, 29 July 1946, Hanford, DOE DDRS D4763587
Hurst, D. G. and E. Critoph (eds.), *Canada Enters the Nuclear Age: A Technical History of Atomic Energy of Canada Limited as Seen from its Research Laboratories* (Montreal: Atomic Energy of Canada Ltd, 1997)
Imperial Chemical Industries Ltd, 'Report by M.A.U.D. Committee on the Use of Uranium for a Bomb, Appendix VII: Nuclear energy as a source of power', in: Gowing, M., *Britain and Atomic Energy 1939–1945* (London: Macmillan, 1964)
Isbin, Herb, 'Licensing and regulation of university research reactors', memo, February 1963, UI Box 100
Jackson, Duncan, 'Bringing technology to the community: Sellafield Visitors' Centre', in: J. Durant (ed.), *Museums and the Public Understanding of Science* (London: Science Museum, 1992), pp. 103–7
Jammer, Max, *The Philosophy of Quantum Mechanics: The Interpretations of Quantum Mechanics in Historical Perspective* (New York: John Wiley, 1974)
Jarausch, Konrad, *The Unfree Professions* (Oxford: Oxford University Press, 1990)
Jay, Kenneth, *Calder Hall: The Story of Britain's First Atomic Power Station* (London: Methuen, 1956)
Jeffries, Zay, 'Prospectus on Nucleonics', report, University of Chicago Metallurgical Laboratory, 13 November 1944, reproduced in: Smith, A. K., *A Peril and a Hope: The Scientists' Movement in America: 1945–47*, Appendix I
Jesiek, Brent K., *Between Discipline and Profession: A History of Persistent Instability in the Field of Computer Engineering, circa 1951–2006*, PhD thesis, Princeton (2006)
Joerges, Bernward and Terry Shinn (eds.), *Instrumentation: Between Science, State, and Industry* (Dordrecht: Kluwer, 2001)
John, Brian S., *Nuclear Power and Jobs: The Tranwsfynydd Experience* (Newport: Cilgwyn, 1986)
Johnson, Charles W., *City Behind A Fence: Oak Ridge, Tennessee 1942–1946* (Knoxville: University of Tennessee Press, 1981)
Johnson, Leland and Daniel Schaffer, *Oak Ridge National Laboratory: The First Fifty Years* (Knoxville: University of Tennessee Press, 1994)
Joseph, J. Walter and Cy J. Banick, 'The genesis of the Savannah River site key decisions, 1950', presented at *50 Years of Excellence in Science and Engineering at the Savannah River Site: Proceedings of the Symposium*, May 17, 2000, Aiken, SC
Josephson, Paul R., 'Atomic-powered communism: nuclear culture in the postwar USSR', *Slavic Review* 55 (1996): 297–324
——, *Red Atom: Russia's Nuclear Power Program from Stalin to Today* (New York: W. H. Freeman, 1999)
——, 'Technological utopianism in the twenty-first century: Russia's nuclear future', *History and Technology* 19 (2003): 277–92
Jungk, Robert, *Brighter than a Thousand Suns: A Personal History of the Atomic Scientists* (San Diego: Harcourt Brace, 1956)
Kamack, Harry J., 'ORSORT curriculum', memo, 10 June 1953, Hagley 1957 Series III Box 12 folder 4 ORSORT 1953–54
——, 'Report on first term of ORSORT year', memo, 20 January 1954, Hagley 1957 Series III Box 12 folder 4 ORSORT 1953–54
——, 'Report on year of training at Oak Ridge School of Reactor Technology—1953–1954', memo, 8 September 1954, Hagley 1957 Series III Box 12 folder 4 ORSORT 1953–54

Kammash, T. to C. E. Dryden, letter, 22 February 1966, UI Box 100
Katz, Ralph and Thomas J. Allan, 'Investigating the Not Invented Here (NIH) syndrome: A look at the performance, tenure, and communication patterns of 50 R & D Project Groups', *R&D Management* 12 (2007): 7–20
Kay, John Menzies, 'Imperial College course proposal', 4 January 1957, NA AB 19/84
Keller, Alex, *The Infancy of Atomic Physics: Hercules in his Cradle* (Oxford: Clarendon Press, 1983)
Kemeny, John G., 'The President's Commission on The Accident at Three Mile Island. The need for change: the legacy of TMI', The President's Commission, October 1979
Kenward, Michael, 'The nuclear backlash', *New Scientist*, 1 May 1975: 263–4
Kevles, Daniel, *The Physicists: The History of a Scientific Community in Modern America* (New York: Knopf, 1977)
——, 'Cold War and hot physics: science, security and the American State, 1945–1956', *Historical Studies in the Physical and Biological Sciences* 20 (1990): 239–64
Keys, D. A., LAC MG30 B59 Vol. 4
——, 'Monthly report', August–September 1947, LAC MG30 B59 Vol. 5
—— to C. C. Cook, letter, 21 January 1948, LAC MG30 B59 Vol. 4.
—— to Editor of *Nucleonics*, letter, 23 April 1951, LAC MG30 B59 Vol. 4
—— to J. Cockcroft, letters, 17 July and 8 August 1951, LAC MG30 B59 Vol. 4
——, 'Monthly report', March 1954, LAC MG30 B59 Vol. 5
Kinchey, Abby J., 'African Americans in the atomic age: postwar perspectives on race and the bomb, 1945–1967', *Technology and Culture* 50 (2009): 291–315
Kinsey, Freda, 'Life at Chalk River', *Atomic Scientists' Journal* 3 (1953): 18–23
Klapper, Joseph T., *The Effects of Mass Communication* (New York: Free Press, 1960)
Knelman, Fred H., *Nuclear Energy: The Unforgiving Technology* (Edmonton: Hurtig, 1976)
Kojevnikov, Alexei, *Stalin's Great Science: The Times and Adventures of Soviet Physicists* (London: Imperial College Press, 2004)
Kopp, Carolyn 'The origins of the American scientific debate over fallout hazards', *Social Studies of Science* 9 (1979): 403–22
Kowarski, L., 'Atomic energy developments in France', *Atomic Scientists' News* 2 (1948): 16–21
Kraft, Alison, 'Between medicine and industry: medical physics and the rise of the radioisotope, 1945–1965', *Contemporary British History* 20 (2006): 1–35
Kragh, Helge, *Quantum Generations: A History of Physics in the Twentieth Century* (Princeton, NJ: Princeton University Press, 1999)
Kramer, Andrew W., 'Atomic energy is here', *Power Plant Engineering*, 49 September 1945: 74–7
Kranakis, Eda, 'Social determinants of engineering practice', *Social Studies of Science* 19 (1989): 5–70
Krenz, Kim, *Deep Waters: The Ottawa River and Canada's Nuclear Adventure* (Montreal: McGill-Queen's University Press, 2004)
Krige, John, 'Atoms for Peace, scientific internationalism and scientific intelligence', *Osiris* 21 (2006): 161–81
Krige, John and Dominique Pestre (eds.), *Science in the Twentieth Century* (London, 1997)
Kuhn, James W., *Scientific and Managerial Manpower in Nuclear Industry* (New York: Columbia University Press, 1966)
Kurczy, Stephen, 'Japan nuclear crisis sparks calls for IAEA reform', *Christian Science Monitor*, 17 March 2011
Laboratory, Oak Ridge National, 'ORSORT: Oak Ridge School of Reactor Technology', *Oak Ridge Nuclear Laboratory Review*, 25 (3 & 4)
Landa, Edward, *Buried Treasure to Buried Waste: The Rise and Fall of the Radium Industry* (Golden, Colo.: Colorado School of Mines, 1987)
Lang, Daniel, *Early Tales of the Atomic Age* (New York: Doubleday, 1948)
——, *The Man in the Thick Lead Suit* (Oxford: Oxford University Press, 1954)
——, *From Hiroshima to the Moon: Chronicles of Life in the Atomic Age* (New York: Simon and Schuster, 1959)
Lapp, Ralph Eugene and Howard Lucius Andrews, *Nuclear Radiation Physics* (New York: Prentice-Hall, 1948)
Lassman, Thomas C., 'Industrial research transformed: Edward Condon at the Westinghouse Electric and Manufacturing Company, 1935–1942', *Technology and Culture* 44 (2003): 306–39
Latour, Bruno, *Science in Action* (Cambridge, Mass.: Harvard University Press, 1987)
——, *We Have Never Been Modern* (Hemel Hempstead: Harvester Wheatsheaf, 1993)

Layton, Edwin T., *The Revolt of the Engineers: Social Responsibility and the American Engineering Profession* (Cleveland, Ohio, 1971)
Leslie, Stuart W., *The Cold War and American Science: the Military-Industrial-Academic Complex at MIT and Stanford* (New York: Columbia University Press, 1993)
Lewenstein, B. V., 'Science and the media', in: S. Jasanoff, G. E. Markle, J. C. Peterson and T. J. Pinch (eds.), *Handbook of Science and Technology Studies* (Thousand Oaks, Calif., 1995), pp. 343–60
Lewis, W. B. to D. A. Keys, 24 September 1947, LAC RG77 Vol. 283
Lewis, W. R., 'Trip report: UK—US—Canada Reactor Safety Conference', memo, 23 October 1953, DOE DDRS D8490526
Libby, Leona Marshall, *The Uranium People* (New York: Crane, 1979)
Lillyman, Ernie to S. F. Johnston, interview, 18 September 2007, Thurso, Scotland, SFJ collection
Longacre, Andrew, 'Proposal for a cooperative facility for education in nuclear engineering', memo, November 1955, UI Box 100
Lord Waverley and Alexander Fleck, 'Wallace Alan Akers. 1888–1954', *Biographical Memoirs of Fellows of the Royal Society* 2 (1956): 1–4
Mackenzie, C. J., LAC MG30-B122 Vol. 1
——, LAC RG24 Box 5002 File 3310–50/7
—— to H. G. Thode, letter, 29 December 1947, LAC RG77 Vol. 283
—— to J. Cockcroft, letter, 1 October 1947, LAC RG77 Vol. 283
——, 'War Technical and Scientific Development Committee 19th meeting', 2 July 1943, MG30 B122 Vol. 3
—— to G. L. Groves, letter, 26 December 1945, LAC RG77 Vol. 283
—— to J. Chadwick, 28 January 1946, LAC RG77 Vol. 283
—— to J. Chadwick, letter, 12 February 1946, LAC RG77 Vol. 283
—— to W. Zinn, letter, 17 April 1946, LAC RG77 Vol. 283
—— to C. D. Howe, letter, 15 August 1952, LAC MG30-B122 Vol. 3
—— to K. F. Tupper, letter, 24 April 1952, LAC MG30-B122 Vol. 3
—— to F. E. Simon, letter, 4 December 1952, LAC MG30-B122 Vol. 3
—— to J. Cockcroft, letter, 23 May 1952, LAC MG30-B122 Vol. 3
——, *The Mackenzie–McNaughton Wartime Letters* (Toronto: University of Toronto Press, 1978)
Maier, C. S., 'Between Taylorism and Technocracy: European ideologies and the vision of industrial productivity in the 1920s', *Journal of Contemporary History* 5 (1970): 27–61
Makepeace-Lott, S., *Alan Works With Atoms* (London: Chatton & Windus, 1962)
Manley, John H., 'Secret science', *Physics Today* 3 (1950): 8–11
Marshall, John, 'Plant assistance (physics) to Jul 1 1945', 17 September 1945, Hagley 1957 Series III Box 58 folder 4
Martin, W. R., 'Undergraduate education in nuclear engineering in the USA', *Journal of Radioanalytical and Nuclear Chemistry* 171 (1993): 183–92
Masco, Joseph, 'Lie detectors: on secrets and hypersecurity in Los Alamos', *Public Culture* 14 (2002): 441–60
——, *The Nuclear Borderlands: The Manhattan Project in Post-Cold War New Mexico* (Princeton, NJ: Princeton University Press, 2006)
Mather, R. E., 'Report of visit to Chalk River Ontario', memo, 22 January 1951, DOE DDRS D4193933
May, Julian, *There's Adventure in Atomic Energy* (New York: Popular Mechanics Press and Hawthorne Press, 1957)
McKay, Paul, *Electric Empire: The Inside Story of Ontario Hydro* (Toronto: Between the Lines, 1983)
McNeight, S. A., 'Hanford cohort of trainees', memo, 6 February 1953, Hagley 1957 Series I Box 2 folder 9
Medvedev, Grigori and Evelyn Rossiter (translator), *The Truth About Chernobyl* (New York: Basic Books, 1991)
Medvedev, Zhores A., *The Legacy of Chernobyl* (Oxford: Basil Blackwell, 1990)
Meier, Richard L., 'The origins of the scientific species', *Bulletin of the Atomic Scientists* 7 (1951): 169–73
Meitner, Lise and Otto R. Frisch, 'Disintegration of uranium by neutrons: a new type of nuclear reaction', *Nature* 143 (1939): 239–40
Messrs. ICI, 'Report by M.A.U.D. Committee on the Use of Uranium for a Bomb, Appendix VI: Nuclear energy as a source of power', in: Gowing, M., *Britain and Atomic Energy 1939–1945* (London: Macmillan, 1964)
Metro-Goldwin Meyer, 'Facts about the making of MGM's remarkable motion picture 'The Beginning or the End'', booklet, 1946, LAC MG30-E533 Vol. 1
Meyer, Leo A., *Atomic Energy in Industry: A Guide for Tradesmen and Technicians* (Chicago: American Technological Society, 1963)

Middleton, W. E. Knowles, *Physics at the National Research Council of Canada, 1929–1952* (Waterloo, Ont.: Wilfrid Laurier University Press, 1979)
Millar, R. N., 'Future power programme in Britain', *Nuclear Engineering* 1 (1956): 304
Minnion, John and Philip Bolsover (eds.), *The CND Story* (London: Allison & Busby, 1983)
Minutes, Board of Governors, ANL, 'Tripartite contract problem', 28 January 1972, UI Box 110
MIT, 'History of the MIT Department of Nuclear Science and Engineering', http://web.mit.edu/nse/overview/history.html, accessed 20 March 2006
Mitchell, D. S. to L. Owen, memorandum, 1 January 1958, NA AB 9/1869
Mladjenovic, Milorad, *History of Early Nuclear Physics (1896–1931)* (Singapore: World Scientific, 1992)
Moss, John E, 'The crisis of secrecy', *Bulletin of the Atomic Scientists* 17 (1961): 8–11
Mott, N. F., 'The scientist and dangerous thoughts', *Atomic Scientists' News* 2 (1949): 171–2
Mountjoy, P. B., 'Security report on Hanford Engineer Works', memo to District Engineer, Manhattan Engineer District, 4 November 1943, DOE DDRS D4851457
——, 'History of Intelligence and Security Division, HEW', memo to file, 1947, DOE DDRS D4740793
—— memo to file, 5 June 1947, Hanford, DOE DDRS D4763339
Moynihan, Daniel P., *Secrecy: The American Experience* (New Haven: Yale University Press, 1998)
Murray, Raymond L., *Introduction to Nuclear Engineering* (New York: Prentice-Hall, 1955)
National Academy of Sciences, 'The Biological Effects of Atomic Radiations' (Washington, DC: National Research Council, 1956)
Ndiaye, Pap A., *Nylon and Bombs: DuPont and the March of Modern America* (Baltimore: Johns Hopkins University Press, 2007)
Needell, Allan A., 'Nuclear reactors and the founding of Brookhaven National Laboratory', *Historical Studies in the Physical Sciences* 14 (1983): 93–122
Nelkin, Dorothy, *Selling Science: How the Press Covers Science and Technology* (New York: W. H. Freeman, 1995)
Nelson, George, 'British nuclear engineering: the next ten years', *New Scientist*, 27 October 1960: 1115–17
NES, 'Nuclear Engineering Society: Annual General Meetings; notices and nominations, 1958–1978', NA AB 65/459
Newell, R. E. to L. R. Thomson, letter, 11 November 1943, LAC MG30 E533 Vol. 1
Newman, Bernard, *Soviet Atomic Spies* (London: Robert Hale, 1952)
Newman, M. K. to J. H. Roberson, letter, 17 February 1960, UI Box 85
News, 'Locomotion', *Nuclear Engineering* 1 (1956): 3
——, 'Unclassified UKAEA documents', *Nuclear Engineering* 2 (1957): 30–2
——, 'Acton Technical College', *Nuclear Engineering* 1 (1956): 95
——, 'Atomic energy course at Birkenhead Technical College', *Nuclear Engineering* 1 (1956): 109
——, 'Course of nuclear engineering at the University of Glasgow', *Nature* 177 (1956): 508–9
——, 'London evening classes in nuclear physics at Borough Polytechnic', *Nuclear Power—The Journal of British Nuclear Engineering* 1 (1956): 142
——, 'School for Dounreay', *Nuclear Power* 1 (1956): 47
——, 'Training nuclear engineers at Oak Ridge', *Nuclear Engineering* 1 (1956): 115
Newton, G. W. A., 'Nuclear and radiochemistry training in the UK ', *Journal of Radioanalytical and Nuclear Chemistry* 171 (1993): 45–56
Nichols, C. M. and A. S. White, 'Chemical engineering and atomic energy', *Chemistry and Industry* (17 January 1963): 51–5
Nichols, Kenneth D., *Nucleonics*, July 1949, reproduced in: *Atomic Energy Indoctrination* (Washington, DC: Dept. of Army, 1950)
Noble, David F., *America By Design: Science, Technology, and the Rise of Corporate Capitalism* (New York: Knopf, 1977)
Norwood, W. D., 'Disaster plans', memo, 2 June 1948, DOE DDRS D8417395
Nuclear Energy Agency and Organization for Economic Co-Operation and Development, 'Nuclear Education and Training—Cause for Concern? A Summary Report', NEA #02428, 2000
O'Connor, D. F., 'Hanford Personnel and Training Program', memo, 31 August 1950, Hagley 1957 Series II Box 12 folder 3
O'Fallon, John R., 'Education and training needs of the nuclear power industry', memo, 15 December 1980, UI Box 71
—— to B. Piatkowski, letter, 18 October 1982, UI Box 71
Ogburn, William Fielding, 'Sociology and the atom', *American Journal of Sociology* 41 (1946): 267–75

Oliphant, M. I. E., 'How the atom can be harnessed for peace: power production by nuclear methods; with diagrammatic drawings by G. H. Davis', *Illustrated London News*, 207 13 October 1945: 399–401

Olwell, Russell B., *At Work in the Atomic City: A Labor and Social History of Oak Ridge, Tennessee* (Knoxville: University of Tennessee Press, 2004)

Operations Coordinating Board, 'A program to exploit the A-bank proposals in the President's UN speech of December 8, 1953', 4 February 1954, Eisenhower Presidential Library, http://www.eisenhower.utexas.edu/dl/Atoms_For_Peace/Binder11.pdf

ORINS, *The Peaceful Atom: American Museum of Atomic Energy* (Oak Ridge: ORINS, 1954)

ORSORT, 'Oak Ridge Operations Information Manual', DOE ON NV0714712

Osif, Bonnie Anne, Anthony J. Baratta and Thomas W. Conkling, *TMI 25 Years Later: The Three Mile Island Nuclear Power Plant Accident and Its Impact* (University Park, Penn.: Penn State Press, 2004)

Overbeck, W. P., 'Instrument Department functions and organization to July 1, 1945', memo to file, 14 August 1945, Hagley 1957 Series V Box 50 folder 16

Owen, Leonard, *Journal of the British Nuclear Energy Society* 1 (1962): 7

——, 'Nuclear engineering in the United Kingdom—the first ten years', *Journal of the British Nuclear Energy Society* 2 (1963): 23–32 and 296–8

Pais, Abraham and Robert P. Crease, *J. Robert Oppenheimer: A Life* (New York: Oxford University Press, 2005)

Parfit, Michael, *The Boys Behind the Bombs* (New York: Little and Brown, 1983)

Parker, H. M., 'What is a health physicist?' presented at *Health Physics Meeting*, Chicago, 1951

——, 'Radiation protection in the atomic energy industry: a ten-year review', draft conference presentation for meeting of the Radiological Society of North America, Los Angeles, 7 November 1954, DOE DDRS D198195963

Parker, M. M., 'Radiation protection in the atomic energy industry: a ten-year review', draft conference presentation for meeting of the Radiological Society of North America, Los Angeles, 7 November 1954, DOE DDRS D198195963

Parr, Joy, 'Working knowledge of the insensible: an embodied history of radiation protection in Canadian nuclear power stations, 1962–92', *Comparative Studies in Society and History* 48 (2006): 820–51

Patterson, H. Wade and Ralph H. Thomas, *A History of Accelerator Radiation Protection: Personal and Professional Memoirs* (Ashford, Kent: Nuclear Technology Publishing, 1994)

Peierls, Rudolf, *Bird of Passage: Recollections of a Physicist* (Princeton, NJ: Princeton University Press, 1985)

Perrot, Donald, 'Problems of the supply of scientists and engineers', UKAEA memorandum, 16 February 1956, NA AB 16/1971

Perry, C. C., 'Nuclear engineering at Wayne State University', memo, 6 April 1959, UI Box 22

Pickering, Andrew, *Constructing Quarks: A Sociological History of Particle Physics* (Edinburgh: Edinburgh University Press, 1984)

Pilkey, O. H., 'The civil engineer and atomic energy', draft for American Society of Civil Engineers Conference, 22 February 1962, DOE DDRS DA615027

Pinch, Trevor J. and Wiebe E. Bijker, 'The social construction of facts and artefacts: or how the sociology of science and the sociology of technology might benefit each other', *Social Studies of Science* 14 (1984): 399–441

Pinkerton, J. B., 'The Institution of Nuclear Energy Engineers', *Nuclear Energy Engineer* 12 (1958): 49

——, 'The Institution of Nuclear Engineers', *Nuclear Energy Engineer* 12 (1958): 197

——, 'The Institution of Nuclear Engineers: Officers and Council', *Nuclear Energy Engineer* 13 (1959):576–7

Pocock, Rowland Francis, *Nuclear Power: Its Development in the United Kingdom* (London: Unwin Brothers, 1977)

Poor, R. S., 'The Atomic Energy Commission and nuclear education', presented at *Interrelated Role of Federal Agencies and Universities in Nuclear Education Conference*, Gatlinburg, TN, 1963

Prentice, B. R., 'Development of atomic energy for subsonic plane', memo, 10 December 1950, DOE DDRS DA02114307

Presas i Puig, Albert, 'Science on the periphery. The Spanish reception of nuclear energy: an attempt at modernity?' *Minerva* 43 (2005): 197–218

Preston, Mel A. and Helen E. Howard-Lock, 'Emergence of physics graduate work in Canadian universities 1945–1960', *Physics in Canada* 56 (2000): 153–62

Pringle, Peter and James Spigelman, *The Nuclear Barons* (London: Joseph, 1982)

Proctor, Robert N., '"-Logos," "-ismos," and "-ikos": the political iconicity of denominative suffixes in science (or, phonesthemic tints and taints in the coining of science domain names)', *Isis* 98 (2007): 290–309

Pyatt, Edward C., *The National Physical Laboratory: A History* (Bristol: Institute of Physics Publishing, 1982)

Pyenson, Lewis, 'An end to national science: the meaning and extension of local knowledge', *History of Science* 40 (2002): 251–90

Rae, H. K., 'Three decades of Canadian nuclear chemical engineering', in: W. F. Furter (ed.), *History of Chemical Engineering* (Washington, DC: American Chemical Society, 1980), pp. 313–34

Ramsey, Norman F., 'Early history of associated universities and Brookhaven National Laboratory', *Brookhaven Lecture Series* 55 (1966)

Reader, W. J., *Imperial Chemical Industries: A History* (Oxford: Oxford University Press, 1975)

Redhead, Paul A., 'The National Research Council's impact on Canadian physics', *Physics in Canada* 56 (2000): 109–21

Reich, L. K., *The Making of Industrial Research, Science and Business at GE and Bell* (New York: Cambridge University Press, 1985)

Reingold, Nathan, 'Metro-Goldywn-Mayer meets the atom bomb', in: T. Shinn and R. Whitley (eds.), *Expository Science: Forms and Functions of Popularisation* (Dordrecht: Reidel, 1985), pp. 229–45

Rentetzi, Maria, 'The U.S. radium industry: industrial in-house research and the commercialization of science ', *Minerva* 46 (2008): 437–62

Reynolds, Terry S., *Seventy-Five Years of Progress: A History of the American Institute of Chemical Engineers 1908–1983* (New York: AIChE, 1983)

——, 'Defining professional boundaries: chemical engineering in the early 20th century', *Technology and Culture* 27 (1986): 694–716

Rhodes, Richard, *The Making of the Atomic Bomb* (London: Simon & Schuster, 1986)

——, *Dark Sun: The Making of the Hydrogen Bomb* (New York: Simon & Schuster, 1995)

Roberson, John H. to W. F. Stubbins, letter, 28 September 1959, UI Box 83

—— to E. L. Multhaup, letter, 11 April 1960, UI Box 83

—— to M. Carbon, letter, 3 November 1959, UI Box 86

Roberts, Arthur, 'Take away your billion dollars (the Brookhaven song)', lyrics, UI Box 35

Roqué, Xavier, 'Marie Curie and the radium industry: A preliminary sketch', *History and Technology* 13 (1997): 267–91

Rose, Hilary and Steven Rose, *Science and Society: The Chemists' War* (Harmondsworth: Penguin, 1970)

Rose, Paul L., *Heisenberg and the Nazi Atomic Bomb Project: A Study in German Culture* (Berkeley: University of California Press, 1998)

Rosenthal, Debra, *At the Heart of the Bomb: The Dangerous Allure of Weapons Work* (Reading, Mass.: Addison-Wesley, 1990)

Ross, K. B., 'The chemical engineer in industry', *The Chemical Engineer* (October 1960): A38–A9

Rothbaum, Melvin, *The Government of the Oil, Chemical and Atomic Workers Union* (New York: John Wiley, 1962)

Rowe, P., 'Obituary: Arthur Southan White', *IChemE Diary & News*, April 1987: 3

Rowse, David to S. F. Johnston, interview, 5 July 2011, telephone, SFJ collection

Salmon, A., 'Technologists for the nuclear power industry', *Atomic Scientists' Journal* 5 (1956): 280–5

Sanger, S. L. and Robert W. Mull, *Hanford and the Bomb: An Oral History of World War II* (Seattle: Living History Press, 1989)

Sargent, B. W. to G. M. Everhart, letter, 4 November 1953, QU Sargent fonds Series III Box 4 file 4.16

——, 'file 4.1 Nuclear engineering', QU B. W. Sargent fonds, Series III Box 4

Schatz, Ronald W., *The Electrical Workers: A History of Labor at General Electric and Westinghouse, 1923–1960* (Urbana: University of Illinois Press, 1983)

Scheibach, Michael, *Atomic Narratives and American Youth: Coming of Age with the Atom, 1945–1955* (Jefferson, NC: McFarland, 2003)

Schonland, Basil to J. Cockcroft, letter, 8 January 1958, NA AB 6/1971

Schwartz, Stephen I. (ed.), *Atomic Audit: The Costs and Consequences of U.S. Nuclear Weapons Since 1940* (Washington, DC: Brookings Institute, 1998)

Sclove, Richard E., 'From alchemy to atomic war: Frederick Soddy's "technology assessment" of atomic energy, 1900–1915', *Science, Technology, & Human Values* 14 (1989): 163–94

Scott, E. E., 'Atomic energy power generation', memo, 14 February 1948, DOE DDRS DA02779406

Scranton, Philip, 'None-too-porous boundaries: labor history and the history of technology', *Technology and Culture* 29 (1988): 722–43

Seidel, Robert W., 'A home for Big Science: the Atomic Energy Commission's laboratory system', *Historical Studies in the Physical and Biological Sciences* 16 (1986): 135–75

——, 'Secret scientific communities: classification and scientific communication in the DOE and DoD', in: M. E. Bowden, T. B. Hahn and R. V. Williams (eds.), *Proceedings of the 1998 Conference on the History and Heritage of Scientific Information Systems* (1999), pp. 46–60

Seitz, F., 'Report on Wigner disease', memo to file, 3 February 1945, DOE DDRS D4763348

Semenovsky, P., *Conquering the Atom: A Story about Atomic Engineering and the Uses of Atomic Energy for Peaceful Purposes* (Moscow: Foreign Languages Publishing House, 1956)

Shaw, D. F., 'Forthcoming espionage trial', memo, 7 March 1951, DOE DDRS D0983254

Shils, Edward, *The Torment of Secrecy: The Background and Consequences of American Security Policies* (Glencoe, Ill.: Free Press, 1956)

Sime, R. L., 'From radioactivity to nuclear physics: Marie Curie and Lise Meitner', *Journal of Radioanalytical and Nuclear Chemistry* 203 (1996): 247–57

Skelton, B., 'The education & training of nuclear engineers in the UK', *Nuclear Engineer* 42 (2001): 175–8

Skinner, H. W. B., 'Atomic energy and the public interest', *Discovery* 12 (9) (1951): 269–72

——, to W. B. Lewis, letter, 5 December 1950, Kingston, Ontario, QU Box 1 File 13

Smith, Alice Kimball, *A Peril and a Hope: The Scientists' Movement in America, 1945–47* (Chicago: University of Chicago Press, 1965)

Smith, I. Norman, 'The magic and reality of nuclear energy: a layman takes a look', *Ottawa Journal*, 1953, in LAC MG40 B59

Smith, Jeff, *Unthinking the Unthinkable* (Bloomington: Indiana University Press, 1989)

Smith, Merritt Roe and Leo Marx, *Does Technology Drive History?: The Dilemma of Technological Determinism* (Cambridge, Mass.: MIT Press, 1994)

Smith, Pat, 'On the trail of Drum T-7', *AECL Inter-Comm*, 2 June 1989

Smith Stahl, Margaret, *Splits and Schisms, Nuclear and Social*, PhD thesis, Wisconsin (1946)

Smyth, Henry D., *Atomic Energy for Military Purposes: The Official Report on the Development of the Atomic Bomb under the Auspices of the United States Government, 1940–1945* (Princeton, NJ: Princeton University Press, 1945)

——, 'A general account of the development of methods of using atomic energy for military purposes under the auspices of the United States Government 1940–1945', 1 July 1945, LAC MG30-E533 Vol. 1

Snow, C. P., *The New Men* (London: MacMillan, 1954)

——, *The Two Cultures and the Scientific Revolution: The Rede Lecture, 1959* (Cambridge: Cambridge University Press, 1959)

Soodak, Harry and Edward C. Campbell, *Elementary Pile Theory* (New York: John Wiley, 1950)

Spence, R., 'Twenty-one years at Harwell', *Nature* 314 (1967): 343–8

Spiegel-Rosing, Ina and Derek de Solla Price (eds.), *Science, Technology and Society: A Cross-Disciplinary Perspective* (London: Sage, 1977)

Spiller, James, 'Radiant cuisine: the commercial fate of food irradiation in the United States', *Technology and Culture* 45 (2004): 740–63

Squires, Lombard to B. H. Mackey, letter, 9 August 1950, Hagley 1957 Series II Box 6 folder—Administrative policy correspondence, general, 1950–1963

Stacy, Susan M., 'Proving the Principle', http://www.inl.gov/proving-the-principle/, accessed 12 June 2009

Starr, T. Chauncey, 'The MYTHS of N-Energy', *Nuclear Energy Engineer* 13 (1959): 508–9

Staudenmaier, John M., *Technology's Storytellers: Reweaving the Human Fabric* (Cambridge, Mass.: MIT Press, 1985)

Stephenson, Richard, *Introduction to Nuclear Engineering* (New York: McGraw-Hill, 1954)

Stone & Webster Inc, *Engineering for Atomic Power: The Role of Stone and Webster in Nuclear Development* (New York: Stone & Webster, 1957)

Strickland, Donald A., *Scientists in Politics: The Atomic Scientists Movement, 1945–46* (Lafayette: Purdue University Studies, 1968)

Strinati, Dominic, *An Introduction to Theories of Popular Culture* (London: Routledge, 1995)

Stryker, Sheldon and Peter J. Burke, 'The past, present, and future of an identity theory', *Social Psychological Quarterly* 63 (2000): 284–97

Suite, C. G. to L. R. Boulware, letter, 5 May 1948, Schenectady, DOE DDRS D3800167

Sutton, O. G. to W. B. Lewis, letter, 26 April 1949, QU Lewis fonds Box 1 File 13

Szasz, Ferenc M. and Issei Takechi, 'Atomic heroes and atomic monsters: American and Japanese cartoonists confront the onset of the nuclear age, 1945–80', *The Historian* 69 (2007): 728–52

Teller, Edward and Judith L. Shoolery, *Memoirs: A Twentieth-Century Journey in Science and Politics* (Oxford: Perseus Press, 2001)

Thayer, Harry, *Management of the Hanford Engineer Works in World War II: How the Corps, DuPont and the Metallurgical Laboratory fast tracked the original plutonium works* (Reston, VA: American Society of Civil Engineers Press, 1996)
Thomas, S., 'Young nuclear engineers—an endangered species', *Nuclear Engineering International* 46 (2001): 36
Thomson, L. R. to C. J. Mackenzie, letter, 1 October 1946, LAC RG77 Vol. 283
Thorpe, Charles, *Robert Oppenheimer: The Tragic Intellect* (Chicago: University of Chicago Press, 2006)
Turchetti, Simone, 'Atomic secrets and government lies: nuclear science, politics and security in the Pontecorvo case', *British Journal for the History of Science* 36 (2002): 389–415
Turner, Stephen, 'What are disciplines? And how is interdisciplinarity different?' in: P. Weingart and N. Stehr (eds.), *Practising Interdisciplinarity* (Toronto, 2000), pp. 46–65
——, 'What is the problem with experts?' *Social Studies of Science* 31 (2001): 123–49
Vollmer, Howard M. and Donald L. Mills, 'Nuclear technology and the professionalization of labor', *American Journal of Sociology* 67 (1962): 690–6
Von der Osten, Robert, 'Four generations of Tom Swift: ideology in juvenile science fiction', *The Lion and the Unicorn* 28 (2004): 268–83
Von Halban, Hans, Frederic Joliot and Lew Kowarski, 'Number of neutrons liberated in the nuclear fission of uranium', *Nature* 143 (1939): 680
von Kármán, Theodore, 'Atomic engineering?' *Journal of Applied Physics* 17 (1946): 2–3
VonderLage, F. C. to H. J. Kamack, letter, 21 May 1953, Hagley 1957 Series III Box 12 folder 4 ORSORT 1953–54
Wainwright, Loudon S., 'The heroic disarming of Diablo—Atomic engineers make a suspense-laden climb', *Life*, 16 September 1957: 133–6, 41–44
Walker, J. Samuel, *Permissible Dose: A History of Radiation Protection in the Twentieth Century* (Berkeley: University of California Press, 2000)
——, *Three Mile Island: A Nuclear Crisis in Historical Perspective* (Berkeley: University of California Press, 2005)
Walker, John, 'Training nuclear engineers', *New Scientist*, 18 July 1957, 23–5
Wallace, Philip R., 'Atomic energy in Canada: personal recollections of the wartime years', *Physics in Canada* 56 (2000): 123–31
—— to M. M. R. Williams, letter, 21 July 2000, SFJ collection
Wang, Jessica, *American Science in an Age of Anxiety: Scientists, Anticommunism, and the Cold War* (Chapel Hill, NC: University of North Carolina Press, 1999)
Warneke, Thorsten, *High-precision Isotope Ratio Measurements of Uranium and Plutonium in the Environment*, PhD thesis, School of Ocean and Earth Science, University of Southampton (2002)
Weart, Spencer R., *Scientists in Power* (Cambridge, Mass.: Harvard University Press, 1979)
——, *Nuclear Fear: A History of Images* (Cambridge, Mass.: Harvard University Press, 1988)
Weaver, Lynn to S. F. Johnston, telephone interview, 3 March 2007, SFJ collection
Weinberg, Alvin M., *Reflections on Big Science* (Cambridge, Mass.: MIT Press, 1967)
——, 'Social institutions and nuclear energy', *Science* 177 (1972): 27–34
——, 'Salvaging the atomic age', *Wilson Quarterly* (Summer) (1979): 88–112
——, *The First Nuclear Era: the Life and Times of a Technological Fixer* (New York: AIP Press, 1994)
——, 'Walter Henry Zinn', *Biographical Memoirs of Fellows of the National Academy of Sciences* 85 (2004): 365–74
Wellock, Thomas R., *Critical Masses: Opposition to Nuclear Power in California, 1958–1978* (Madison: University of Wisconsin Press, 1998)
Wende, C. W. J., 'Meetings at Argonne, September 13–15 1950', memo to file, 26 September 1950, Hagley 1957 Series IV Box 44 folder 1
——, 'Visit to Argonne, September 28–29, 1950', memo to file, 20 October 1950, Hagley 1957 Series IV Box 44 folder 1
——, 'Visit to Schenectady November 7 and 8, 1950', memo to file, 27 November 1950, Hagley 1957 Series IV Box 44 folder 2
——, 'Visit to Argonne, January 3 and 4, 1951', memo to file, 12 January 1951, Hagley 1957 Series IV Box 44 folder 2
Wenger, Etienne, *Communities of Practice: Learning, Meaning and Identity* (Cambridge: Cambridge University Press, 1998)
Westwick, Peter J., *The National Labs: Science in an American System, 1947–1974* (Cambridge, Mass.: Harvard University Press, 2003)
White, A. S., 'Chemical engineering research at A.E.R.E.' *The Chemical Engineer* (1962): A66–A73

Whitman, Ann to M. McCrum, letter, 27 January 1956, Eisenhower Presidential Library, http://www.eisenhower.utexas.edu/dl/Atoms_For_Peace/Binder19.pdf

Whitney, Vincent Heath, 'Resistance to innovation: the case of atomic power', *American Journal of Sociology* 56 (1950): 247–54

Whyte Jr., William H., *The Organization Man* (New York: Simon & Schuster, 1957)

Wigner, Eugene P. and Andrew Szanton, *The Recollections of Eugene P. Wigner* (London: Plenum, 1992)

Williams, Michael M. R., 'The development of nuclear reactor theory in the Montreal Laboratory of the National Research Council of Canada (Division of Atomic Energy) 1943–1946', *Progress in Nuclear Energy* 36 (2000): 239–322

—— to S. F. Johnston, email, 31 July 07, SFJ collection

—— to S. F. Johnston, interview, 2 July 2008, London, England, SFJ collection

Williams, Robert Chadwell, *Klaus Fuchs, Atom Spy* (Cambridge, Mass.: Harvard University Press, 1987)

Wilson, A. S., 'Agriculture in an atomic age', report, 8 January 1960, DOE DDRS DA03300049

Wilson, C. G. and D. K. Thomas, 'Teaching nuclear science in the army: The Royal Military College of Science', *Nuclear Energy Engineer* 13 (1959): 443–4

Wilson, Henry W., 'The Scottish Research Reactor Centre', *Nature* 205 (1965): 10–4

Wilson, John F., *British Business History 1720–1994* (Manchester: Manchester University Press, 1995)

Winkler, Allan M., *Life Under a Cloud: American Anxiety about the Atom* (Oxford: Oxford University Press, 1993)

Winner, Langdon, 'Do artifacts have politics?' in: L. Winner (ed.), *The Whale and the Reactor: A Search for Limits in an Age of High Technology* (Chicago: University of Chicago Press, 1986), pp. 19–39

Wisnioski, Matt 'Inside "the system": engineers, scientists, and the boundaries of social protest in the long 1960s', *History and Technology* 19 (2003): 313–33

Wogman, N. A., L. A. Bond, A. E. Waltar and R. E. Leber, 'The nuclear education and staffing challenge: rebuilding critical skills in nuclear science and technology', *Journal of Radioanalytical and Nuclear Chemistry* 263 (2005): 137–43

Woodbury, David O., *Atoms for Peace* (New York: Dodd, Mead & Co., 1955)

Woodhouse, John C., 'Brookhaven fuel elements', memo to file, 21 September 1950, Hagley 1957 Series IV Box 44 folder 1

Worthington, Hood, 'Pile technology—effect of operation on graphite moderator (Wigner and Szilárd effects)—experience to Aug 1, 1945', memo to file, 13 September 1945, Hagley 1957 Series III Box 58 folder 4

—— to W. H. Zinn, letter, 21 December 1950, Hagley 1957 Series IV Box 54 folder 2

Yeo, Frances E. M., *Nuclear Engineering Education in Britain*, MSc thesis, University of Manchester (1997)

Zinn, Walter, 'memo to file', memo, 26 July 1946, UI Box 134

——, 'Minutes of Board Meeting, ANL', memo, 2 May 1948, UI Box 19

—— to W. B. Harrell, letter, 4 February 1948, UI Box 19

—— to H. Worthington, telex, 21 December 1950, Hagley 1957 Series IV Box 54 folder 2

INDEX

Abbott, Andrew 11, 33, 57, 75, 109–10, 144, 161, 170, 224, 264–8
abstraction 109, 267–9
accelerator, particle 120–1, 128, 218, 245, 268
accident 5–6, 15, 125, 210–16, 214, 273
   Chernobyl 129, 255–6, 258–9
   Fukushima 256–9, 262, 267
   NRX 5, 212–3, 238, 253
   Three Mile Island 5, 210, 222, 252–4, 256–9
   Windscale 123–4, 214–5, 220
accidents and professional identity 222, 238, 252–255
AEC *see* United States Atomic Energy Commission
AECL *see* Atomic Energy of Canada Limited
AERE *see* Atomic Energy Research Establishment
AF of L *see* American Federation of Labor
Agar, Jon 242–3
agriculture 162, 165, 240
Akers, Wallace Alan 37–8, 41, 104, 134
Aldermaston 93, 110, 122, 133, 171, 271
Alexander, Robert 130
American Federation of Labor (AF of L) 202, 206, 209
American Federation of Technical Engineers 209
American Institute of Chemical Engineers 47, 53, 163, 173
American Nuclear Society 47, 160, 167, 170, 172, 175, 186, 191–2, 242, 250, 272
American Society for Engineering Education (ASEE)
Anglo-Canadian project 12, 28, 32–43, 54, 58, 71, 77, 90, 98–9, 113, 118, 121, 144, 271
Anglo-Saxons 12–13, 40, 243, 264
ANS *see* American Nuclear Society
AORES *see* Association of Oak Ridge Engineers and Scientists
Arco, ID 86, 113–4
Argonne National Laboratory (ANL) 48, 56, 75–6, 78–9, 81–90, 94–5, 170, 173, 198, 204–6, 245–7

education 137–8, 141–2, 167, 185–91, 220
expertise 85–6, 108, 113–4, 119, 121, 126–8, 149, 155–7, 164, 215, 271
security 99–100, 102, 106, 248–9
staff 40, 67, 71, 77, 86–7, 265–6, 268–9
Argonne Universities Association (AUA) 76, 246–7, 274
Arnold, Lorna 123
Associated Midwest Universities 189, 191, 247
Association of Atomic Energy Technicians and Draftsmen 209
Association of Oak Ridge Engineers and Scientists (AORES) 103, 276
Association of Scientific Workers (AScW) 104, 171, 208
atomic bomb 1, 7, 10, 92–3, 186, 227, 229, 267
   concept 26–30, 82,
   deployment 68, 70, 88, 105, 108, 117, 147, 163, 165, 231
   development 102, 194, 197
   production 71, 75, 78, 157, 268
atomic age 19, 115, 169, 171, 236, 242
AEA *see* UK Atomic Energy Authority
Atomic Energy Act (1946) 77–8, 97, 100–1, 106, 123, 133, 210, 212, 265
Atomic Energy Act (1954) 97, 163
Atomic Energy Commission (AEC) 78–82, 84–7, 117, 163, 212, 229, 245–6, 271–2
   role in defining expertise 79–80, 82, 103, 170, 198, 237, 247
   education 136, 139–43, 167, 244
   interactions with national laboratories 79, 84, 120–1
   interactions with industry 138–44, 147–51, 154, 156–7
   interactions with universities 76, 81, 186, 188–90, 193
   safety 219–20
   security 99–100, 102, 106–7, 205–6, 248–9

Atomic Energy Division (Du Pont SRP) 146–50, 156–7
Atomic Energy of Canada Limited (AECL) 91–2, 125, 165, 187, 193–6, 208–10, 247, 272
Atomic Energy Research Establishment (AERE) 92, 94, 128
  see also Harwell
Atomic Energy Technical Committee 206
Atomic Energy Workers' Union 209
Atomic Engineer and Scientist 166, 276
Atomic Engineers of Oak Ridge 103, 274
Atomic Industrial Forum 198, 246, 274
Atomic Research Workers' Union 209, 274
atomic scientists 1, 2, 96, 103–7, 157, 166, 207, 228–32, 237, 265, 268
Atomic Scientists Association (ASA) 103–4, 106–7, 166, 207, 229
Atomic Scientists' Journal 166, 276
Atomic Scientists' News 104, 166, 228, 276
Atomnaya Energiya 168
Atoms for Peace 97, 142, 161, 163, 165, 170, 174, 186, 210, 236, 243
Atomic Weapons Research Establishment (AWRE) 93, 274
  see also Aldermaston
AUA see Argonne Universities Association
Auger, Pierre Victor 39
Authority see UK Atomic Energy Authority
AWRE see Atomic Weapons Research Establishment

baby boom 242
Battersea College of Technology 178
BCPMA see British Chemical Plant Manufacturers'; Association
Beck, Clifford K. 186
Beloiarsk 236
Bennett, W. J. 91, 103, 195
BEPO (British Experimental Pile zero) 93, 105, 113–4
Berkeley, CA 21, 27, 53, 88, 186, 193, 203, 214
Berkeley Power Station 267
beta radiation 21, 49, 127
Big Science 20, 88, 270
biologists 7, 76, 87, 110, 116, 151, 176, 198, 200, 210, 217–8, 230
bionucleonics 128

bismuth 50, 52
BNEC see British Nuclear Engineering Conference
BNES see British Nuclear Engineering Society
BNFL see British Nuclear Fuels Ltd
BNFL Ltd 261
Bohr, Neils 21–2, 31
boiler, heavy-water 31–2, 40
breach, reactor 6, 212, 257
breeder reactor 6, 80, 86, 112–14, 120, 124, 179, 215–6, 242, 246, 262
Briggs, Lyman 32
Briggs Advisory Committee 32, 97
British Association for the Advancement of Science 207
British Chemical Plant Manufacturers' Association (BCPMA) 173, 274
British Nuclear Energy Conference (BNEC) 168, 174–6, 178, 183, 274
British Nuclear Energy Society (BNES) 176, 250, 274
British Nuclear Fuels Ltd (BNFL) 248, 274, 278
British Thomson-Houston 35, 42
brittleness 4, 70–1, 200
Brookhaven National Laboratory 79, 84, 110, 137, 146, 197
  education 142, 167, 190
  expertise 149, 245, 261, 268
  security 99, 249
  staff 86–8, 184
Bulletin of the Atomic Scientists 104, 166, 227, 276
Bush, Vannevar 27, 33, 41, 42, 76, 229, 269

Calder Hall, Cumbria 125, 143, 163, 164, 166, 172, 174, 182, 197, 199, 215, 234, 236, 253
Cambridge, UK 181–3, 233, 279
Cambridge, MA 215
Canada Labour Relations Board 209
Canadian Association of Nuclear Energy Technicians and Technologists 209–10
Canadian General Electric (CGE) 194, 248, 274
Canadian Nuclear Association 196, 274
Canadian Nuclear Society 196, 250, 274, 277
Canada's Atom Goes to Work 229
CANDU 156, 194, 246, 274
Capenhurst 93, 130, 135, 208

Carter administration 246, 249, 253
cartoons 223–4, 236, 240, 256
cathode rays 20, 21, 22
CEGB *see* Central Electricity Generating Board
censorship 98–9, 101, 135, 166
Central Electricity Generating Board (CEGB) 251, 274
centrifuge 27, 45, 50
Chadwick, James 1, 21–2, 25, 31, 41–3, 113, 207
chain reaction 3–5, 19, 23–5, 27, 30–2, 47–8, 61, 67, 69, 118–19, 124, 128, 182, 217, 252, 272
chain reactor *see* reactor
Chalk River 6, 32, 83, 84, 90–1, 94, 110, 124–6, 128–9, 154, 197, 205, 212, 253, 268–9, 271–2, 278
  education 152, 194–5, 271–2
  expertise 71, 78, 112–5, 118, 121–2, 134–5, 193–5, 238, 265
  staff 41–4, 91, 129, 200, 208–10, 233–4
  security 99, 101, 107–8
Chalk River Nuclear Process Operators' Union 209
Chalk River Reactor School 195, 271
Chambers, F. S. 149
Chapelcross, Dumfriesshire 6, 125, 179, 197, 215
chemical engineers *see* engineer, chemical
Chemical Workers' Union 205, 206, 208
chemistry
  discipline 19, 21–2, 27–8, 30, 53, 57, 126, 138, 184, 194–6
  industry 19, 40, 46–7, 146–50
chemists 7, 22, 27, 46–9, 65, 67, 71, 129, 194, 218–19, 230
  expertise 43, 59, 79, 80, 95, 122, 124, 140–1, 147, 150, 177–81, 266
  nuclear applications 2, 30, 43, 76, 87, 110, 116, 124, 132, 143, 146, 153, 156, 171–2, 198–201, 217–9, 271
  occupation 33–4, 39, 82
Chernobyl, Ukraine 5, 15, 129, 214, 238, 252, 255–9, 262, 267, 273
Cherwell, Lord *see* Lindemann, Frederick Alexander
China 13, 105, 147, 231, 243
cinema 23, 165, 229, 232–3, 254–5
citizenship 44, 90, 102, 140

civil engineers *see* engineer, civil
civil service 36, 90, 94, 104, 106, 113, 129, 131, 133, 143–4, 199, 207, 210, 271
classified knowledge 14, 96, 100–8, 126, 135–43, 150–1, 154, 162–3, 166–9, 183, 186, 223, 227, 230, 250, 269
Clinch College of Nuclear Knowledge 138
Clinton Engineer District 28
Clinton Laboratories 28, 56, 59, 64–5, 67–9, 77–9, 80–1, 85, 103, 114, 135–6, 148–50, 153, 203–4, 260, 268–9
  see also Oak Ridge National Laboratory
Clinton (X-10) Pile 49, 56, 59–61, 67, 118–19, 154–5
Clinton Training School 137–40, 274
CNS *see* Canadian Nuclear Society
cobalt bomb 165, 229
Cockcroft, John Douglas 25, 31, 37, 42–3, 90–2, 94, 99, 104, 107, 111–13, 121, 133–4, 143, 164, 166–8, 172–4, 207, 268, 269
Cold War 6, 8, 14, 95, 150, 162, 262
Cole, Donovan L. 134
college, 159, 177–80, 185, 194, 195, 196, 241, 251, 260
  Churchill 279
  Clinch 138
  community 178
  Imperial 24–5, 29, 159, 181–4, 244, 260, 279
  King's 183
  North Carolina State 186
  Queen Mary 183–5
  Royal Naval 180
  University (London) 31, 207, 251
Collège de France 23
College of Science, Royal Military 137
College of Technology, Leeds 178
College of Technology, Manchester 181
College of Technology, North West Kent 178
Columbia University 23, 26, 46, 48, 53, 58, 75, 79, 83, 106
Commercial Products Division, AECL 165, 209
Commissariat à l'Énergie Atomique 43, 77
Committee on Radiation Hazards 207
Committee on the Scientific Survey of Air Defence 25

Commonwealth-Edison 197
communists 43, 99, 100, 106–7, 204, 207–8, 231
compartmentalization 34, 40, 62, 75, 98–103
Compton, Arthur Holly 8, 22, 26, 28, 33–4, 44, 46–9, 52–60, 64, 67, 78, 126–7, 145, 217, 229, 247
Conant, James Bryant 27, 33, 38, 41–2, 53–4, 229, 237
conductivity, thermal 70
Congress of Industrial Organizations (CIO) 206
containment vessel 6, 210, 215–6, 252, 257
control rods 24, 48, 56, 62, 67, 82, 119, 123, 134, 152, 177, 252, 255
Convair Corporation 85, 187
cooling 4, 55, 70, 71, 113, 119, 123–4, 152, 218, 236, 252, 255, 257, 261
corporatism 174, 232, 270
corrosion 32, 55, 67, 70, 82, 133, 195
CP-1 (Chicago Pile 1) 48, 54, 60, 114, 118
craft workers 124, 143, 172, 177, 179–80, 194–5, 197–8, 200, 202, 205–6, 208–9, 200, 241
crash suit 213
criticality 6, 54, 90, 113, 182, 218, 222
critical mass 27, 109, 129, 218
culture, organizational 1, 14, 33–71
culture, popular 11, 23, 165, 227
Curie, Marie and Pierre 21–3, 43, 238
curricula 170, 180–5, 188–90, 193, 196–7, 244–5, 251, 266–7, 271
Czechoslovakia 6, 40

Davey, Gethin 130, 221
Davison, Boris 196
Department of Scientific and Industrial Research (DSIR) 29, 184, 275
determinism, technological 8, 270–1
Detroit, Michigan 6, 189
deuterium 24, 114, 194
diffusion, gaseous 27, 30, 39, 46, 53, 75, 93, 112, 131–3, 186, 203, 223
discipline 2, 10, 11, 251, 264, 266–9
 emerging 14, 33–4, 46, 53, 60, 64, 99, 109–126, 137, 139, 157, 177–8, 200, 210, 241, 251, 263
 established 33, 53, 67, 144, 174, 178, 180–3, 195, 198, 244–5

 recognition 128–135, 187–93
 subdiscipline 46, 168, 219–20
Disney, Harold V. 131–2
Disney, Walt 236
disputes, jurisdictional *see* jurisdiction
Division of Atomic Energy (UK) 92–4, 100, 104, 116, 129, 134, 143, 174, 221, 265
Dollezhal, Nikolai 113, 129
dosage, radiation *see* radiation exposure
Douglas Point, Ontario 194
Dounreay, Caithness 6, 119, 124, 135, 215, 221–2, 236
Dounreay Fast Reactor (DFR) 119, 179–80, 197, 215–216, 261
Dounreay Prototype Reactor (DPR) 119, 261
Dresden power station, Illinois 198
DSIR *see* Department of Scientific and Industrial Research
Du Pont de Nemours Company 17, 34, 35, 38–9, 41, 44–71, 76, 79–80, 89, 103, 122, 144–57, 197, 202–5, 216–7, 229, 266, 268–9, 272, 274, 279
Dyson, Freeman 187

EBR-1 (Experimental Breeder Reactor 1) 6, 86, 113–4, 119
education, higher 162, 178, 185, 193, 260, 265
 see also universities
education, further 178–9
Einstein, Albert 22–3, 25, 229
Eisenhower administration 4, 88, 97, 161–3, 166, 170, 174, 210, 243
Eldorado Mining and Refining 91–2, 101
electrical craftsman 146, 198
Électricité de France (EDF) 260
emergency 62, 65, 75, 152, 222, 237, 252–5, 257–9
Energy Research and Development Administration (ERDA) 246, 275
engineers versus scientists 53–60, 132, 144–157
engineer
 'atomic' 2, 103, 111, 149, 166, 236, 238–40, 267
 chemical 6, 11, 27, 40, 47, 67, 80, 129, 144, 151, 163, 168, 173–4, 177, 182, 194, 200, 266, 271, 274–5, 279
 civil 6, 35, 45, 122, 128, 133, 165, 175–6

electrical 33, 36, 42, 45, 47, 53, 106, 128, 134, 136, 143, 146, 151, 172, 174, 176, 178, 181–3, 187–8, 191, 194–5, 200, 240, 268–9
mechanical 4, 35, 116, 130, 133–4, 151, 153, 174, 181, 183–5, 195, 200, 241–2, 275
nuclear 1–2, 30, 94, 122, 125, 144, 176, 197, 210, 227, 234, 242–4, 249, 256, 258–9, 262–3, 267–8, 271–2
nuclear (Canada) 44, 195, 232
nuclear (UK) 200, 201, 207, 208, 211, 215, 241, 250–2, 252–3, 260
nuclear (USA) 47, 77, 136, 142, 146, 148, 154, 156, 197–200, 237, 265, 266
process 178
engineering
nuclear 4, 6, 12–13, 19, 20, 28, 34, 99, 110, 116, 126, 155, 159, 161, 165–6, 178, 264, 267, 269, 270
nuclear (Canada) 194, 195, 196
nuclear (UK) 134, 168, 172–5, 178–9, 181–4, 222, 251
nuclear (USA) 47, 53, 56, 60, 70–1, 85, 102, 109, 133, 136, 141, 143–4, 156–7, 163–4, 185–91, 193, 245–8, 254, 266
English Electric Company 94, 171
espionage 13, 77, 97–8, 100, 102, 104–8, 177, 204, 231, 270
European Nuclear Society 222, 275
export 9, 94, 99, 173, 197, 216, 243, 246, 266, 270–1

Fairchild 85
fallout, radioactive 207, 210, 214, 256
fast neutrons 25, 86, 114, 119
fast reactor 93, 112, 119, 122, 124, 193, 215–6
Federal Bureau of Investigation (FBI) 101–2, 106, 151, 204
Federation of Architects, Engineers, Chemists and Technicians (FAECT) 203–4
Federation of Atomic Scientists (FAS) 96, 103–5, 275
female workers 61, 67, 221–2
Fermi 1 reactor 6
Fermi, Enrico 3, 8, 21–6, 30, 34, 44, 47, 54–6, 58–60, 64, 67, 69, 83, 126–7, 145, 164, 200, 229, 268
filling factory 131, 215

filtering of personnel 105–8
Fishenden, R. Martin 105
fissile materials 49, 125, 182
fission 23–4, 26, 75, 182
Ford administration 246
Fox, Terence Robert Corelli 181
France 2, 6, 12–13, 21, 23, 30, 37, 40, 43–4, 77, 79, 90, 99, 107, 113, 124, 193, 202, 210, 243, 260
Frisch, Otto Robert 8, 23, 25, 41, 268
Fuchs, Klaus Emil Julius 97, 101, 105–7
fuel channels 70, 118, 122, 192, 255
fuel cycle 192, 262
fuel elements 5, 62, 70, 82, 93, 149, 178, 180, 238, 242, 257
Fukushima Dai-ichi 5, 15, 214, 238, 252, 256–9, 262, 267, 273

Galison, Peter 11, 100–1, 120
gamma radiation 21, 49, 54, 128, 139, 165, 218
GEC 35
gender 221–2
General Atomics 187
General Electric (GE) 26, 33, 35, 45, 47, 68, 71, 78, 84–5, 80, 94, 116, 124, 126–7, 141, 146–7, 150–1, 156, 194, 203–6, 229, 248, 257, 272
General Nuclear Engineering Corporation 121
general worker 208–9, 213
Geneva International Conference on the Peaceful Uses of Atomic Energy 97, 163–4, 166–7, 174, 193, 210, 234, 237, 265, 271
Germany 13, 32, 23, 25, 30, 35, 40, 44, 193, 207, 259, 267, 270
GE School of Nuclear Engineering 141
Ginns, Dennis W. 41, 134
GLEEP (Graphite Low Energy Experimental Pile) 4, 93, 113–4
Gouzenko, Igor Sergeyevich 101
Government Owned, Company Operated (GOCO) 138, 202
Gowing, Margaret 94, 129
grandchildren 242–59
graphite 23–5, 48–9, 55–6, 60, 67, 69–71, 81, 107, 112–4, 118–9, 121–4, 130, 148, 152, 154–6, 215, 255
great-grandchildren 259–63
Greenewalt, Crawford 39–41, 46, 48–60, 64–5, 68, 70, 145–9, 216, 269

Groves, General Leslie 28–30, 34, 41–3, 45–6, 53–4, 56, 58, 65, 76, 99, 100, 136, 202–3, 217, 228–9
Guéron, Jules 43

Hahn, Otto 23
Halban, Hans H. 23, 25, 30–2, 38, 40, 42–3, 52, 59
Hall, Geoffrey R. 184
Hanford Engineer Works (HEW) 4–6, 98, 103, 113, 197, 214, 228, 237, 266, 268–9, 275
  production 29, 50, 59, 71, 110, 117, 124, 127, 155–6, 217–9
  security 106–8, 204–6
  staff 61, 63–70, 123, 144–53, 150–2, 203–4, 230, 271
  training 137–8, 141
Hanford piles 5, 41, 60, 64, 67–9, 71, 77, 80, 85, 89, 109, 114, 116, 118–9, 122–4, 146, 148–9, 154, 202, 212, 230, 256
Harwell (AERE) 90, 92, 94, 108, 110, 164, 197, 215, 261
  expertise 112–4, 120, 122, 212, 265
  security 102, 105–6
  staff 117, 124, 128–9, 133–4, 172, 174, 195–6, 200, 234–6
  training 143, 167, 177, 179, 181, 242
Harwell Isotope School 13, 128, 143
Harwell Reactor School 167, 177, 179, 242, 271
hazard 49–50, 53, 59, 62, 71, 105, 131, 152, 207, 211–14, 217–19, 249
health and safety 82, 127, 206, 219–21
health physicists *see* physicists, health
heat
  from chain reaction 3–5, 24–5, 27, 136, 139–40, 177, 180, 183, 192, 236
  as danger 5, 52, 123–4, 212, 214, 238, 252, 255, 257
  for electrical power generation 3, 111–12, 120, 169, 230
  as waste by-product 5, 48, 50, 70, 117–8, 129
heavy water 4, 24, 30, 37–41, 52, 82, 118–9, 149, 151, 154, 236, 238
heavy-water boiler 31–2, 40
heavy-water reactor 25, 28, 37–8, 41, 52–3, 59, 78, 82, 85, 107, 113–4, 152, 154–6, 212, 246, 265, 272
Hecht, Gabrielle 2, 12, 13
Heinlein, Robert 240

Heisenberg, Werner 21, 22
helium 4, 48, 59, 55, 62, 66, 119, 124
heroism 228, 238–9, 241, 256, 258, 262
Hilborn, John 187
Hill, John 215
Hinton, Christopher 93–4, 100–1, 128–35, 143–4, 157, 168, 171, 173–4, 178, 181–3, 185, 200, 211–13, 215–16, 221–2, 242, 253, 268–9, 278
Hiroshima 42, 76, 93, 110, 147, 227
historiography 7–10, 19–20, 70, 103
House Un-American Activities Committee (HUAC) 231
Howe, Clarence Decatur 35, 38, 100, 125, 194
Hughes, Thomas Parke 265
Humble Oil 50
humour 223–4, 256

IChemE *see* Institution of Chemical Engineers
ICI *see* Imperial Chemical Industries
Imperial Chemical Industries (ICI) 26–7, 30–1, 33–4, 37, 39, 41, 71, 94, 102–2, 104, 129–30, 133–4, 144, 221, 268
Imperial College London 24–5, 29, 159, 181–4, 244, 260
India 13, 163, 194, 240, 246
Industrial Division 129–30, 133, 157, 174, 181
insects 45, 165
Institution of Chemical Engineers (IChemE) 173–6, 275
Institution of Civil Engineers (ICE) 174–6
Institution of Electrical Engineers (IEE) 174
Institution of Mechanical Engineers (IMechE) 174, 275, 279
Institution of Nuclear Engineers (INucE) 175–7, 208, 222, 250–2, 254, 275–7
instruments 61–2, 66–7, 100, 115, 127, 137–8, 152, 178, 190, 192, 198, 211, 219, 241, 252
Instrument Department, Hanford 66
International Panel on Climate Change (IPCC) 262
interpretive flexibility 3
INucE *see* Institution of Nuclear Engineers
irksome duty 220
irradiation 4, 6, 52, 70, 80, 112, 116–17, 122, 125, 165, 183, 219
isotopes 7, 22, 31, 49, 69, 116, 191, 198, 201, 218, 234

applications 76, 101, 128, 138, 141, 143, 145, 162, 165, 184, 209, 272
   production 79, 80, 86
   separation 28, 45–6, 91, 186–7
   use in reactors 24–5, 27–8, 119, 164–5, 217
isotopics 128
Italy 23, 24, 40, 193, 259

Japan 6, 26, 42, 44, 70, 76, 90, 98, 126, 217, 227–8, 240, 256, 256–60
JASON 184
Jesuits 96
Joliot, Jean Frédéric 21–4, 43, 107
Josephson, Paul 236, 267
Journal of Nuclear Energy 166–8
jurisdiction 11, 30, 33, 45, 109, 128, 161, 171, 173, 264–6

k value 24
Kamack, Harry J. 153–4
Kay, John Menzies 159, 182–3
Kearton, Christopher Frank 133
Kellogg Company 38, 46
Kellex Company 38, 46, 202
Kelvin, Lord 20
Kendall, James W. 132, 215
Keys, David Arnold 31, 107, 114–5, 193–4, 279
King's College, Newcastle 183
Korea 105, 106, 231, 243
Kowarski, Lew 23, 25, 43, 99, 144
Kurchatov, Igor Vasilyevich 21, 25, 82, 113, 164, 271

Laurence, George Craig 24–5, 31, 194
Lawrence, Ernest Orlando 21–2 53, 58, 91, 121, 217, 229
Lawrence Livermore National Laboratory 78, 122, 231
Lawrence Radiation Laboratory 45–6, 91, 244
Leeds College of Technology 178
Leipunskii, Alexandr Il'ich 215
Leverett, Miles C. 47, 50, 55, 60
Lewins, Jeffery 250–2
Lewis, Warren K. 53–4
Lewis, Wilfred Bennett 31, 83, 91, 101, 124, 188, 193
Lindemann, Frederick Alexander (Lord Cherwell) 42, 177

liquid metal 52, 124, 178, 221
Lloyd, Selwyn 207
Long Beach 186
Lonsdale, Kathleen 207
LOPO 114, 119
Los Alamos 10, 28, 41, 58, 64, 67, 77–9, 83, 94, 103, 109, 114, 119–20, 122, 145, 147, 179, 203, 244, 271
lost generation 259
luminous paint 48, 145, 240

Mackenzie, Chalmers Jack 17, 31–2, 35–8, 40–3, 56, 90–1, 99–101, 113, 115, 125, 126, 195, 212, 269, 279
Magnox 215, 248
Manhattan Engineer District 28, 31, 46, 76, 79, 84, 97, 154, 228–30, 202, 271, 275, 278
Manhattan Project 28, 56, 89, 90, 170, 197, 203, 207, 216–7, 227, 237, 259, 265–6, 271–2
   collaborations 30–1, 42–4, 144, 147–9
   management 46, 53, 58–9, 75–8, 84
   security 97–9, 101, 103, 106
   technical staff 34, 37, 44, 68, 70–1, 76, 109, 111, 121, 126, 136–7, 186–7, 230
Marine Engineering Officer (Nuclear) 180
mass spectrometer 27, 45
Massachusetts Institute of Technology (MIT) 27, 53, 136, 186, 215, 275
Massey, Harrie Stewart Wilson 207
mathematicians 93, 143, 177, 200, 201
MAUD committee 17, 25–7, 29–30, 32–3, 37, 41, 53, 94, 97, 129, 268
McCarran Internal Security Act 97, 107
McCarran-Walter Immigration and Nationality Act 97
McCarthy, Joseph 106, 210, 228, 232
McMahon Act *see* Atomic Energy Act (1946)
McNaughton, Andrew George Latta 36
mechanical engineers *see* engineer, mechanical
Meitner, Lise 23
Menius, Arthur C. 186
Metallurgical Laboratory, University of Chicago (Met Lab) 33–4, 64–5, 67–8, 83, 126–7, 170, 247
   expertise 41, 95, 139
   interactions with Du Pont 44–61, 68, 71, 153–5
   staff 76–8, 80, 109, 149, 203, 216, 230, 266, 268–9

metallurgists 4, 6, 47, 67, 116, 143, 171–2, 177, 198–9, 200–1, 230
Metropolitan Vickers 30, 35, 42
Millikan, Robert A. 22–3
Ministry Of Aircraft Production 25, 91
Ministry of Supply 25, 29, 35, 42, 92, 94, 102, 117, 130–1, 134, 143, 174, 177, 208, 229, 265, 271, 279
mistrust 34, 43–4, 101, 121, 227–8, 230–3, 238, 243, 250, 256, 258
MIT *see* Massachusetts Institute of Technology
moderator 24–5, 48, 52, 70–1, 112–13, 118–19, 124, 148, 177, 192, 213
monastic analogy 96, 135, 255
Monsanto Chemical Company 80–1, 126, 138
Montreal Laboratory 37–41, 43–4, 71, 90, 134, 144, 228, 268, 278
Moon, Phillip B. 24–5
Moynihan, Daniel 95, 265
Müller, Hermann Joseph 165
Murray, Raymond L. 186
mutation 165

Nagasaki 42, 68, 76, 110, 122, 147, 228
National Bureau of Standards (NBS) 32, 35
National Defense Research Committee (NRDC) 27, 32, 38, 275
National Laboratories 76, 78–90, 96, 98–9, 106–8, 110, 136–7, 141, 146, 148–9, 151, 155, 161, 170, 185, 187–9, 208, 210, 215, 244–5, 247–8, 265, 268, 272
  see also Argonne National Laboratory; Brookhaven National Laboratory; Oak Ridge National Laboratory
National Physical Laboratory (NPL) 35
National Reactor Testing Station 113
National Research Council of Canada (NRC) 24, 31, 33, 35–7, 39, 41, 56, 71, 91, 97, 113, 125, 208, 266, 268–9, 272, 275, 278
National Union of Atomic Workers 208, 275
National Union of Municipal and General Workers 208
National Union of Scientific Workers 104
NEPA *see* Nuclear Energy for Propulsion of Aircraft
neutron 1, 3–4, 22–5, 28, 31, 48–50, 54, 69–70, 80, 82, 86, 112–14, 116–20, 123–5, 137, 156, 182, 183, 187, 191–3, 196, 200, 217, 218, 272
neutron economy 124
New Zealand 21, 37
Newell, Ronald E. 39–41, 134, 212
Nixon administration 206, 249
Nobel Prize 12, 22–3, 26, 31, 95, 165, 207
North Carolina State College 186
Novovorenezh 236
NPD *see* Nuclear Power Demonstration reactor
NRC *see* National Research Council of Canada
NRDC *see* National Defense Research Committee
NRU (National Research Universal) reactor 114, 125, 193–4, 210
NRX (National Research Experimental) reactor 5, 80, 91, 97, 112–13, 114–15, 118, 125, 154, 193, 210, 212, 238, 253
Nuclear Energy Agency (NEA) 260, 275
Nuclear Energy for Propulsion of Aircraft (NEPA) 47, 85, 275
Nuclear Engineering (journal) 167–9, 171
Nuclear Engineering Society, Windscale 143, 174
Nuclear Institute (NI) 250, 275
Nuclear Nonproliferation Treaty 243
Nuclear Power—the Journal of British Nuclear Engineering 168
Nuclear Power Demonstration reactor (NPD) 193–4
Nuclear Regulatory Commission (NRC) 252, 258, 275
nuclear weapons 8–9, 15, 78, 126, 136, 161–2, 166, 201, 231–2, 243, 245–6, 248, 252, 256, 258, 271
  see also atomic bomb
nucleonics 19, 126–8, 182, 268
Nucleonics (magazine) 107, 166–7, 277
Nunn May, Alan 97, 101, 104–6

Oak Ridge Associated Universities (ORAU) 137, 275
Oak Ridge Engineers and Scientists 103, 166, 274
Oak Ridge Institute of Nuclear Studies (ORINS) 138, 140, 142, 162, 230, 275
Oak Ridge National Laboratory (ORNL) 81, 85–6, 141, 151, 170, 197, 260–1, 269, 275

Oak Ridge School of Reactor Technology (ORSORT) 138–43, 153–4, 156, 185–8, 275
oath, loyalty 99, 106
Obninsk 113, 129, 164
OCAW see Oil, Chemical and Atomic Workers'; Union
OECD see Organization for Economic Co-Operation and Development
Office of Scientific Research and Development (OSRD) 27, 275
Oil, Chemical and Atomic Workers' Union (OCAW) 206
Oliphant, Mark L. 25, 31, 41
Ontario 6, 32, 42, 196, 248, 249, 267
Ontario Hydro 193–5
Oppenheimer, Julius Robert 27, 58, 78, 229, 231–2
ORAU see Oak Ridge Associated Universities
ORINS see Oak Ridge Institute of Nuclear Studies
Organization for Economic Co-Operation and Development (OECD) 260, 275
ORNL see Oak Ridge National Laboratory
ORSORT see Oak Ridge School of Reactor Technology
OSRD see Office of Scientific Research and Development
Ottawa 24–5, 31, 36, 39, 101, 126, 163, 196, 209
Ottawa Atomic Energy Workers 209
Owen, Leonard 133–4, 172, 215
Oxford 29, 31, 39, 92

Pakistan 243
Paneth, F. A. 39
Parr, Joy 241
patriotism 106, 202–3, 206, 238, 270
Peierls, Rudolf Ernst 8, 25, 41, 268
Penney, William George 93, 133–4, 268
physicists 1, 27, 29, 33, 42, 46, 60, 71, 76, 79–80, 91, 103–4, 107, 109–11, 126, 128, 136, 138–9, 148, 150, 194, 219, 230, 237
  expertise 7, 57, 64–5, 67–9, 116, 124, 143–6, 153, 198–201, 220–1
  health 2, 49, 68, 79, 86, 100, 136–7, 139, 144, 151, 168, 183–4, 209, 215, 219–20, 260
  relationship to engineers 28, 38, 48–50, 52–3, 55–7, 65, 85–8, 111, 115–17, 120–1, 133–4, 156, 171–4, 176–7, 188, 232–3, 268–9

physics 4, 11–12, 19, 20–3, 25–6, 31, 39–41, 46–7, 84, 140, 150, 181
  institutions 26, 47, 181, 184–6, 196, 209–10
Physics Today 88
Physikalisch Technische Reichsanstalt (PTR) 35
piles 34, 77, 109, 125, 138, 230
  concept 23–5, 30, 40–2, 80–2, 194
  design and construction 3–4, 28–30, 45–57, 59–60, 82–6, 89–90, 98, 111–16, 118–19, 171, 196, 268
  operation 68–71, 122–3, 164, 193–4, 200, 212–15, 217, 255, 261
  personnel 60–4, 66–8, 131–5, 145-6, 148–56, 202–3, 220–2, 237
  pilot plant 45, 47–9, 53, 55, 56–7, 59, 61, 64, 77, 79–80, 85, 129, 139
  see also Clinton (X-10) pile
Pinawa, Manitoba 209
Placzek, George 39, 41
plutonium 93–5, 116–22, 154, 178, 197, 205, 222, 237, 271
  chain reaction 28, 47, 118
  dangers 164, 218
  extraction 34, 41, 43, 46, 118, 129–30, 150, 166, 202–3, 211, 217–8
  price 125–6, 156
  production 4–5, 29–30, 44–7, 49–50, 52–7, 59–60, 67–8, 70–1, 77, 79, 85, 93–4, 98, 109, 111–14, 134, 145–7, 215, 246, 254, 267
  use in reactors 86, 119–20
polonium 21–2, 71, 165
Pontecorvo, Bruno 97, 105, 106, 107
Portugal 193, 270
Pratt-Whitney 85
priesthood 97
profession 2, 4, 6, 65, 210, 260
  emergence 44, 47, 115, 122, 128–9, 136, 161, 168, 185, 204, 240, 247, 252–6, 269
  expertise 36, 39, 105, 116, 125, 132–3
  jurisdiction 75, 129, 143–4, 156–7, 170–8, 180, 199–200, 245, 268
  representation 10–11, 13–15, 95, 189–92, 194–6, 206–8, 222, 224, 227–8, 232, 235, 238, 258–60, 264–7, 272–3
  responsibility 249–52
professionalization 11, 197, 198, 250–1

Project Candor 162
Project Plowshare 165
protons 24, 127
publishing 23, 77, 88, 101, 135, 166–70, 174–5, 179, 189–90, 207, 224, 228, 234–5, 240, 242, 254

QSEs *see* Qualified Scientists and Engineers
Quakers 207
Qualified Scientists and Engineers (QSEs) 200
quantum mechanics 21–2
Quebec 6, 41, 97, 196, 249
Queen Mary College, London 181, 183–4, 244

Rad Lab *see* Lawrence Radiation Laboratory
radiation 3–4, 6, 12, 22, 49, 50, 54, 61–2, 77, 110, 116–18, 120, 122, 124, 131, 133, 136–40, 165, 180, 181, 183, 191, 193, 198, 200, 209–11, 215
radiation exposure 6, 49, 105, 165, 207, 210, 216–22
 see also irradiation
Radiation Laboratory *see* Lawrence Radiation Laboratory
radioactivity 3, 19, 21–2, 31, 48–50, 56, 61–2, 67, 100, 111, 116, 127, 164, 200, 210–24, 241, 249, 252, 257, 260, 267
radiobiology 164, 217–8
radiochemistry 2, 95, 116, 140, 180, 184, 266, 271
radioisotope 46, 80, 116, 128, 138, 162, 165, 184, 191, 198, 209, 272
radiology 61, 137, 167, 211, 214, 217, 219–20, 261
radium 21–3, 48–9, 110, 116, 165, 217, 238, 240
reactor
 boiling water (BWR) 114, 274
 breach 6, 212, 257
 breeder 6, 80, 86, 112–14, 120, 124, 179, 215–6, 242, 246, 262
 fast 93, 112, 119, 122, 124, 193, 215–6
 heavy-water 25, 28, 37–8, 41, 52–3, 59, 78, 82, 85, 107, 113–4, 152, 154–6, 212, 246, 265, 272
 homogeneous 82, 86, 114, 119
 light-water 265, 270
 NRU 114, 125, 193–4, 210

NRX reactor 5, 80, 91, 97, 112–15, 118, 125, 154, 193, 210, 212, 238, 253
 power 4, 111, 113–14, 116–17, 121, 125, 156, 164, 166, 162–3, 190, 195, 215, 236, 246, 253
 pressurized water (PWR) 180, 248, 275
 production 4, 29, 50, 52, 85, 113–14, 116, 117, 129, 131, 154, 212, 256, 261
 research 87, 113, 114, 116, 155, 164, 181–2, 184, 193, 212
 thermal 114, 119, 122
 see also pile
rem *see* roentgen equivalent (in) man
reprocessing 6, 37, 79, 164, 211, 242
Research Sites Restoration Limited (RSRL) 261, 274
Rickover, Hyman George 138, 185, 265
Risley, Lancashire, UK 93, 116, 122, 130–2, 134–5, 182, 184, 215, 235, 265, 271
 see also Industrial Division
Roberts, Arthur 87–8, 99
Roentgen 49
roentgen equivalent (in) man (rem) 49
Roentgen ray 20
ROF *see* Royal Ordnance Factories
Rolphton, Ontario 193, 267
Roosevelt, Franklin Delano 25, 27, 41, 185
Rosenberg, Julius and Ethel 97, 105–6
Rotblat, Sir Joseph 207
Rowse, David 174
Royal Ordnance Factories 130–1
Russian Federation 262
Rutherford, Ernest 21–2, 26, 31, 42

S-1 32, 97
Santa Susana Field 6, 275
Sargent, Bernice Weldon 31, 194–5
Saunders, Owen A. 181
Savannah River Plant (SRP) 114, 116, 144–57
science fiction 54, 240
scientists 1–2, 6–8, 12, 88, 111, 136, 142, 177, 182, 190, 193, 209–10, 215, 222, 243, 245, 256, 260, 264, 266, 271–2
 American 26–9, 76–82, 126–8, 156–7, 185, 188
 British 25–6, 30–1, 90, 122
 Canadian 37, 39–40, 56, 101, 194–5
 expertise 85, 115–6, 121–2, 138–9, 183, 197–204, 219

French 12, 30, 37, 44, 113
  interactions with engineers 19–20, 33–40, 44–8, 53–4, 57–61, 64–5, 67–8, 70–1, 134, 145, 154, 247–9, 268–9
  movement 65, 95–6, 103–5, 157, 166–7, 207
  representation 169–70, 172–3, 175, 227–34, 236–8, 240–1
  see also biologists, chemists *and* physicists
scientists, atomic *see* atomic scientists
scram 152, 252, 255, 257
secrecy 1, 3–4, 6, 12, 32, 40, 43, 62, 161, 203–4, 210, 224, 250, 264–6
  and identity 10, 14, 88, 95–102, 135–6, 170, 228, 232, 234, 242–3, 250
  and inefficiencies 107–8, 117
  and information transfer 157, 163–4, 166, 168
  commercial 98, 224
  symbolic and functional 95
security 1, 9, 40, 44–5, 61–2, 68, 71, 77–9, 82, 95, 97–108
Seven Days To Noon 231
Seitz, Frederick 139
sievert 49, 259
Sellafield, Cumbria, UK 114, 130, 230
semi-works *see* pilot plant
separation plants 1, 6–7, 27–8, 30, 40–1, 45–6, 49–50, 52–3, 55–6, 6-, 64, 75, 80, 91, 93, 103, 129–31, 133–4, 147, 149–51, 153, 157, 181, 186, 202, 218, 222, 234, 271
Shaw, Martin 185–6
Shils, Edward 95, 135, 270
Shippingport Atomic Power Station 113, 119, 164, 185, 197–8, 217, 237
shortages, labour 94, 131, 135, 171, 177, 180, 188, 190, 194, 215, 241
Silkwood 206, 232, 255
Silkwood, Karen 206, 254
skills 2, 47, 109, 131, 173, 183, 203, 267, 272–3
  emerging 14, 40, 71, 115, 117, 122, 124, 148, 157, 171, 189, 191, 211, 234
  recognized 31, 65, 94, 143, 153, 181, 199, 208–9, 240–1, 248, 262
  training 60–2, 269
slow neutrons 22
SLOWPOKE 187
Smyth, Henry deWolf 1, 7
Smyth report 76–7, 97, 110, 135, 228–30
SNAP (System for Nuclear Auxiliary Power) 187
Snow, Charles Percy 233

social construction 3, 19, 120, 218–9
sociology 75, 109, 197
Soddy, Frederick 21–2, 31
sodium, liquid 6, 119
Soviet Journal of Atomic Energy 168
Soviet Union 12–13, 25–6, 77, 79, 82, 88, 95, 100–1, 105, 108, 113, 129, 162, 164, 168, 193, 215, 218, 231, 236, 250, 255, 267, 270–1
Spain 13, 193, 214, 270
Springfields, Lancashire, UK 179
spy *see* espionage
status, national 113, 125, 227
status, occupational 2, 11, 35, 80–1, 125, 128, 130, 144, 154–5, 157, 209–10, 233, 243, 249–50, 254, 266–7, 269
status, organizational 65, 107, 141, 173, 176, 251–2
Stalin, Josef 25, 105–6, 162, 270
Steacie, E. W. R. 36
Strassman, Fritz (Friedrich Wilhelm) 12
Strickland, Donald 96
subcritical assembly 182
submarines 27, 45, 85–6, 89, 113–14, 119, 120, 129, 164, 180, 185, 188, 236, 254, 265, 271
Sweden 193
Switzerland 6, 97, 193, 248, 259
Szilárd, Leo 3, 25, 34, 44, 50, 57–8, 70, 71, 83, 123, 148, 155

technicians 2, 6, 19, 35, 67, 77, 82, 96, 106, 145, 206–7, 259
  expertise 40, 152, 172, 176, 198, 216, 236–8, 241–2, 253–4, 262
  representation 104, 203–4, 209–10, 227–8, 233–4
  training 136, 179
technological momentum 265–6, 269, 272
technological system 216, 254, 271
Teller, Edward 8, 26, 44, 187, 231
TEPCO *see* Tokyo Electric Power Company
terrorism 259, 261–2
The Beginning Or The End? 229
The China Syndrome 232, 254–5
The New Men 233
The Simpsons 256
The Thief 232
theory 42, 47, 52, 57, 139–40, 151, 153, 177, 181–3, 191–5, 200

thermal reactor 114, 119, 122
thermal runaway 187, 214
Thode, Henry George 196
Thomson, George Paget 24, 25, 37
Thomson, Joseph John (J. J.) 21, 22
Thomson, Lesslie R. 32, 35, 38, 279
thorium 119, 271
Three Mile Island (TMI) 15, 222, 238, 252, 254, 256–9, 263, 273
Tizard, Henry 25, 31
Tokyo Electric Power Company (TEPCO) 258–9
Tongue, Harold 135
Toronto 115, 194, 195, 196
Transport and General Workers Union 208
TRIGA (Training, Research, Isotopes, General Atomics) 187
Tube Alloys Project 29, 31–2, 33–5, 37–9, 41, 92, 97, 101, 103, 104, 129
Tuohy, Tom 134
Turner, Charles J. 130, 134
Turner, Stephen B. 2, 139, 177

U-235 25, 27–8, 30, 45–6, 52, 75, 93, 119, 217
U-238 27, 46, 119
UK Atomic Energy Authority (UKAEA) 117, 123, 130, 143–4, 168, 172, 174, 177, 181–2, 185, 200, 207–8, 210, 214–5, 220, 223, 241, 244, 248, 251, 261, 269, 275, 278
Underwood, Newton 186
UNESCO 229
unions, labour 201–10, 246, 266, 272
United Nations 161–3, 167, 229, 262
United States Atomic Energy Commission (USAEC) *see* Atomic Energy Commission
University
 Heriot Watt 181
 Iowa State 193, 244
 Kansas State 142, 193, 244, 245
 Liverpool 105, 184, 207
 Manchester 21, 184
 McGill 21, 91, 195
 McMaster 101, 196
 Queen's 83, 195, 196, 279
 Rice 244
 Texas A&M 244
University of Arizona 186

University of Birmingham 25, 196
University of California at Los Angeles 193
University of Chicago 26, 47, 54, 60–1, 77, 79, 80–1, 95, 100–1, 114, 203–4
 see also Metallurgical Laboratory
University of Glasgow 183
University of Illinois 187, 279
University of London Reactor Centre 184
University of Michigan 184, 189, 193
University of Oklahoma 244
University of Tennessee 137, 244
University of Toronto 115, 195, 196
uranium 32, 60, 97, 222
 bomb 42, 271
 chain reaction 23–30, 128
 dangers 50, 52, 123–4, 214, 217–18, 252
 production 30, 44–6, 64, 91, 93, 101, 103, 129–30, 132–3, 45, 147, 151, 164, 166, 179, 202, 243, 246
 radiation 21, 23
 use in reactors 4–5, 38–40, 48–50, 55–6, 67, 69–71, 82, 86, 111–12, 114, 117–21, 169, 171
Urey, Harold Clayton 46, 58
USAEC *see* United States Atomic Energy Commission
USSR *see* Soviet Union

Waltner, Arthur 186
waste 125, 147, 164, 188, 195, 214, 218, 247, 254, 257, 260–1
water, heavy *see* heavy water
Weaver, Lynn 187
Weinberg, Alvin Martin 59–60, 73, 79, 80–2, 97, 109, 119, 145, 148, 163, 164, 249–52, 256, 260–1, 265, 268–9
Wende, C. W. J. 149–50, 155–6
Westinghouse 26, 33, 35, 45, 84–5, 153, 248
Wheeler, John Archibald 26, 46, 56, 57, 69, 145, 149, 155
White, Arthur S. 133, 174
Whiteshell Nuclear Research Establishment (WNRE) 194, 209, 275
Wigner disease 70, 122
Wigner energy 214
Wigner growth 122–3, 214

Wigner, Eugene Paul 3, 26, 44, 46, 50, 57–60, 68, 71, 77, 80, 85, 109, 122, 129, 138, 139, 144, 155, 269
Windscale, Cumbria 108, 110, 113, 116, 143, 164, 166, 197, 230, 236
  design 119, 130, 132
  development 122–5, 212–13
  operation 6, 70, 93–4, 134–5, 213–25, 217, 220–2, 238, 255, 261
Windscale accident 6, 70, 124, 215, 220, 222
Winner, Langdon 270
WNRE *see* Whiteshell Nuclear Research Establishment
Woodhouse, John 149
Worthington, Hood 70, 148–9, 155

X-10 *see* Clinton (X-10) reactor
x-rays 20, 22–3, 26, 165, 217
xenon poisoning 50, 69, 71, 155

Yankee Atomic Power Station 198
Young, Gale 50

Zedong, Mao 105, 147
ZEEP 91, 113–14, 118, 135, 212
ZETA 120
Zinn, Walter Henry 60, 79, 83–90, 94, 100, 119, 121, 126–8, 148–9, 155–7, 164, 170, 173, 185, 213, 237, 247, 268–9
ZOE 113